ADDENDUM TO

H. Leipholz, *Theory of elasticity*,
ISBN 90 286 0193 7
page IV of prelims

originally published in 1968 by

G. Braun Verlag at Karlsruhe, W.-Germany

entitled *"Einführung in die Elastizitätstheorie"*

AF777597

Theory of elasticity

LIBRARY COPY
LEWIS LIBRARY
CLEVELAND, OHIO

Monographs and textbooks on mechanics of solids and fluids

editor-in-chief: G. Æ Oravas

Mechanics of elastic stability

editor: H. Leipholz

1. H. Leipholz
   Theory of elasticity

2. L. Librescu
   The elasto-statics and kinetics of anisotropic and heterogeneous shell-type structures

3. C. L. Dym
   Stability theory and its applications to structural mechanics

Theory of elasticity

H. Leipholz

University of Waterloo
Faculty of Engineering
Solid Mechanics Division
Waterloo, Ontario

LIBRARY COPY
RETURN TO
LEWIS LIBRARY
CLEVELAND, OHIO

NOORDHOFF INTERNATIONAL PUBLISHING LEYDEN

Q
91
L442
c 1

© 1974 Noordhoff International Publishing
A division of A. W. Sijthoff International Publishing Company B.V.
Leyden, The Netherlands

All rights reserved. No part of this publication may be reproduced, stored in a retrieval system, or transmitted, in any form or by any means, electronic, mechanical, photocopying, recording or otherwise, without the prior permission of the copyright owner.

ISBN 90 286 0193 7
Library of Congress Catalog Card Number 73–84402

Printed in The Netherlands by
Drukkerij Dijkstra Niemeyer bv - Groningen

Contents

Preface

This book is designed for use by students and teachers in the field of applied mechanics and mathematics, and for practitioners in civil and mechanical engineering.

Since tensor calculus is an indispensable prerequisite when dealing with the theory of elasticity in a modern way, the first part of the book consists in an introduction into this subject.

In the second part, the physical foundations of the theory of elasticity are given, including nonlinearities. The excursion into the field of geometric and physical nonlinearities is done in order to prepare the reader for further advances into the most recent developments of the theory. The book itself, in the remainder, is restricted to linear problems only.

The third part of the book deals with the mathematical theory of linear elasticity in full extent. Curvilinear problems, two- and three-dimensional problems are included. Stress has been put on working out a systematic approach to the solutions of all kinds of stress states, not neglecting triaxial problems. Also, energy methods have been dealt with, taking into account the generalization and extension of these methods by Rüdiger and Reissner.

The fourth and last part of the book consists in an application of the general methods, as outlined in part 3, to special structures like plates and shells, thus giving hopefully something of interest to the practising engineer.

Many examples illustrating the theory are contained in the text making the book useful for self-studying. The numerous references may enable the reader either to complete his background of knowledge necessary for an intelligent reading of the book or to find access to farther leading literature.

May the book be well accepted, which due to the valuable cooperation of Noordhoff International Publishing can be presented to the public in such an excellent outfit.

1

Fundamentals of tensor analysis

1.1 Cartesian tensors

1.1.1 *Definitions*

By *tensors* we mean certain physical quantities that are governed by *particular transformation laws* when the co-ordinate system is transformed. We shall discuss these laws at a later stage. If in the transformation of the co-ordinate system, we are specifically dealing with the *rotation of a cartesian system of axes*, then we are also dealing with *cartesian tensors*. Let us work with these more closely. We distinguish between certain orders of tensors. The *tensors of zero order* are called *scalars*, the *tensors of the first order* are called *vectors*, the *tensors of second order* are called *dyadics* and the tensors of higher orders are generally called *tensors of the n-th order*.

Scalars need not be discussed further, and vectors will be discussed at a later stage. Dyadics are physical quantities, occurring for instance in the theory of elasticity as *stress-tensors*, and hence the name tensor. Similarly to vectors, dyadics can be represented mathematically by using a *matrix*. A dyadic, or a second order tensor may be written as a 3,3-matrix. The elements of such an *array of numbers* are called *tensor components*. Whenever it is to our advantage, we shall notate tensors in matrix notation, using them in accordance with the rules of matrix calculation. This method is especially suitable for engineers and is explained in detail in [1].

1.1.2 *Vectors*

We shall in our calculations assume vectors as special tensors, i.e., we shall take them as tensors of the first order. Several different possibilities of representation are possible: in *symbolic notation* they are represented by boldfaced letters (e.g., $\boldsymbol{A}$), in *analytic notation* they are represented by means of indices (e.g., A_i, $i = 1, 2, 3$), the general component A_i of the vector taking on the signification of the vector itself. Furthermore, they may be represented by way of *matrix notation* (e.g., matrix $\boldsymbol{A} = (\mathrm{A}_1, \mathrm{A}_2, \mathrm{A}_3)$) as well as by way of *component notation* (e.g., $\boldsymbol{A} = A_1\boldsymbol{e}_1 + A_2\boldsymbol{e}_2 + A_3''\boldsymbol{e}_3$, when the unit vectors $\boldsymbol{e}_i$, $i = 1, 2, 3$, are used as a *basis*). All these possibilities of representing tensors may also apply to dyadics whenever appropriate. A representation of tensors higher than the second order by matrix notation becomes impossible. All other methods of representation remain valid.

To begin with, vector $\boldsymbol{A}$ occurs as a *well-ordered number triplet*, e.g., as the triplet of its component A_i, $i = 1, 2, 3$. It will be shown that in addition a transformation law characterizes vector $\boldsymbol{A}$, and governs its components. Concerning the transformation we shall limit ourselves to the rotation of a cartesian system of axes and hence to cartesian tensors. As an introduction and for sake of simplicity, let us work with a plane problem (Figure 1).

From Figure 1 we read off the following equations:

$$\begin{aligned} A_1' &= A_1 \cos\varphi + A_2 \sin\varphi = A_1 \cos\varphi + A_2 \cos(90° - \varphi), \\ A_2' &= -A_1 \sin\varphi + A_2 \cos\varphi = A_1 \cos(90° + \varphi) + A_2 \cos\varphi. \end{aligned} \tag{1.1.1}$$

These relate to the connection between component A_i and A'_i of the vectors. The first component is found in connection with the original system of axes x_i, the latter one with the system x'_i rotated through the angle φ.

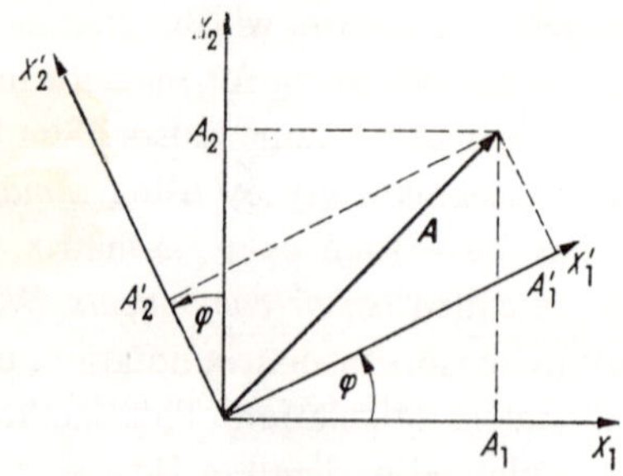

Figure 1

The relation

$$a_{ij} = \cos(x_i', x_j) \tag{1.1.2}$$

represents the cosine of that angle, which is formed by the two intersecting axes x_i and x'_i of the original and the rotated reference frame. Hence, we are also able to write for the co-efficients of the first line in (1.1.1)

$$\cos \varphi = a_{11}, \cos (90° - \varphi) = a_{12}$$

and for the second line

$$\cos (90° + \varphi) = a_{21}, \cos \varphi = a_{22}.$$

Therefore, (1.1.1) changes into

$$\begin{aligned} A'_1 &= a_{11} A_1 + a_{12} A_2, \\ A'_2 &= a_{21} A_1 + a_{22} A_2. \end{aligned}$$

Generalizing these facts with respect to a three-dimensional vector A_i, $i = 1, 2, 3$, and to the three-dimensional cartesian systems of axes x_i, $i = 1, 2, 3$, (original system), and x'_i, $i = 1, 2, 3$, (rotated system), then we obtain,

$$\begin{aligned} A'_1 &= a_{11} A_1 + a_{12} A_2 + a_{13} A_3, \\ A'_2 &= a_{21} A_1 + a_{22} A_2 + a_{23} A_3, \\ A'_3 &= a_{31} A_1 + a_{32} A_2 + a_{33} A_3. \end{aligned} \tag{1.1.3}$$

The system of equations (1.1.3) in its shortened version is

$$A'_i = \sum_{j=1}^{3} a_{ij} A_j, \quad i = 1, 2, 3. \tag{1.1.4}$$

By using *Einstein's summation convention*, (1.1.4) may yet be shortened: omitting the summation sign, we write

$$A'_i = a_{ij} A_j, \quad i = 1, 2, 3, \tag{1.1.5}$$

agreeing upon adding over the *repeated indices* [in this instance the index j on the right side of (1.1.5)]. These repeated or *summation indices* are also called *dummy indices*, which can be renamed arbitrarily in the course of the calculation. If, in an exceptional case, despite the indices occurring twice, we are not required to add, then this will be marked by bracketing the index concerned or by special mention, e.g.,

$$\sigma_{(i)} n_i, \ \sigma_i n_i \text{ (do not add), respectively.}$$

Let us say in addition, those indices over which we are not required to add, are called *free indices*. In (1.1.5), the index i is the free index, and denotes the respective equation number.

At times we are working with two-dimensional quantities, then the indices do not go through the numbers 1, 2, 3 but only through 1, 2. Should this be the case, then it is common to use small Greek letters as indices instead of the usual small Latin letters, e.g.,

$$A_\alpha, \ \alpha = 1, 2.$$

In equation (1.1.5) we have actually found the transformation law valid for cartesian vectors. We shall for the sake of completeness solve the equations in (1.1.5) for the unprimed quantities A_j. We could do this in the usual manner as common for linear, algebraic equations. We shall, however, choose a more elegant method. We substitute primed by unprimed quantities and vice-versa in (1.1.5), then A'_i changes simply into A_i, and A_j changes into A'_j and from

$$a_{ij} = \cos(x'_i, x_j)$$

we obtain

$$\cos(x_i, x'_j) \Rightarrow \cos(x'_j, x_i) = a_{ji},$$

i.e., this change signifies for the co-efficients of the equation system an exchange in the order of the indices. Thus, from (1.1.5) an inversion takes place resulting in

$$A_i = a_{ji} A'_j, \ i = 1, 2, 3. \tag{1.1.6}$$

The relations (1.1.5) and (1.1.6) represent in full the transformation law for cartesian vectors. Hence, a complete definition of vectors can be given as follows: *Vectors are well-ordered number triplets, with the components* A_i,

$i = 1, 2, 3$, *relating to the cartesian system of axis* x_i, $i = 1, 2, 3$, *and the components* A'_i, $i = 1, 2, 3$, *relating to the rotated system* x'_i, $i = 1, 2, 3$. *Connecting the components* A_i *and* A'_i *are the transformation formulae* (1.1.5) *and* (1.1.6).

To find out whether a given quantity having three components is a tensor or not, the validity of the transformation law (1.1.5) and (1.1.6) respectively, has to be tested. As an example, let us look at the *gradient* of a scalar function f. In symbolic notation it is given by

$$\operatorname{grad} f \equiv \nabla f, \tag{1.1.7}$$

and in analytic notation by

$$\frac{\partial f}{\partial x_i}, \quad i = 1, 2, 3. \tag{1.1.8}$$

Relation (1.1.8) is a number triplet. A comparison of (1.1.7) and (1.1.8) shows that for the often-used *Nabla-symbol* the relation

$$\nabla = \frac{\partial}{\partial x_i}, \quad i = 1, 2, 3, \tag{1.1.9}$$

is also valid. Let us prove that (1.1.8) is a vector. We use the rotated system x'_i, for which grad f is given in analytic notation by

$$\frac{\partial f}{\partial x'_i}, \quad i = 1, 2, 3. \tag{1.1.10}$$

According to the chain rule for derivatives, we get

$$\frac{\partial f}{\partial x_i} = \frac{\partial f}{\partial x'_1}\frac{\partial x'_1}{\partial x_i} + \frac{\partial f}{\partial x'_2}\frac{\partial x'_2}{\partial x_i} + \frac{\partial f}{\partial x'_3}\frac{\partial x'_3}{\partial x_i},$$

which can also be written as

$$\frac{\partial f}{\partial x_i} = \frac{\partial f}{\partial x'_j}\frac{\partial x'_j}{\partial x_i} \tag{1.1.11}$$

following the summation convention.

The relation of co-ordinates of the reference systems is determined by

$$x'_i = a_{ij}x_j, \; i = 1, 2, 3. \tag{1.1.12}$$

To examine its validity we can expand (1.1.12) by observing (1.1.2), thus obtaining a well known formula of analytical geometry. From (1.1.12) follows

$$\frac{\partial x'_j}{\partial x_j} = a_{ij},$$

which, when inserted into (1.1.11), yields

$$\frac{\partial f}{\partial x_i} = a_{ji}\frac{\partial f}{\partial x'_j}. \tag{1.1.13}$$

The equation (1.1.13) represents the transformation law (1.1.6) for the triplet $\partial f/\partial x_i$.Therefore, grad f is not only a well-ordered triplet, but even a vector. Hence the Nabla-operator (1.1.9) is formally a vector. This fact may be used in the following.

1.1.3 *Dyadics and higher order tensors*

Dyadics may be encountered when dealing with the *indeterminate product of two vectors,* or when considering the co-efficients of a *linear vector function.*

In the first case, let us relate vectors u_i and v_i in the form of a dyad from which, following the rules of matrix calculation,

$$\begin{pmatrix} u_1 \\ u_2 \\ u_3 \end{pmatrix} (v_1, v_2, v_3) = \begin{pmatrix} u_1 v_1 & u_1 v_2 & u_1 v_3 \\ u_2 v_1 & u_2 v_2 & u_2 v_3 \\ u_3 v_1 & u_3 v_2 & u_3 v_3 \end{pmatrix}, \tag{1.1.14}$$

we obtain the right hand matrix in (1.1.14). The elements of this matrix are notated as

$$t_{ij} = u_i v_j \tag{1.1.15}$$

and yield in symbolic notation

$$T = (t_{ij}).$$

By assumption, u_i and v_j are vectors. Then, the transformation formula (1.1.6) is applicable, and (1.1.15) changes to

$$t_{ij} = a_{ki} a_{lj} u'_k v'_l.$$

In agreement with (1.1.15) it is written

$$u'_k v'_l = t'_{kl}.$$

Thus, the transformation formula

$$t_{ij} = a_{ki} a_{lj} t'_{kl} \tag{1.1.16}$$

is obtained. The inverse formula is obtained by the transition of primed to unprimed quantities, and vice-versa. We obtain

$$t'_{ij} = a_{ik} a_{jl} t_{kl}. \tag{1.1.17}$$

It becomes evident that the array (1.1.14), symbolically notated by T, which has been obtained by the indeterminate product of two vectors, is more than a mere 3,3-matrix, i.e., a well-ordered scheme of numbers. It represents a second order tensor, a dyadic, since the components t_{ij} are governed by the transformation law laid down in (1.1.16) and (1.1.17). Therefore, the following definition is applicable to second order cartesian tensors in a three-dimensional space:

Dyads are well-ordered number schemes having $3^2 = 9$ *elements, representable by* 3,3-*matrices. With respect to the cartesian system of axes* x_i, $i = 1, 2, 3$, *they have components* t_{ij}, $i, j = 1, 2, 3$, *and with respect to the rotated system* x'_i, $i = 1, 2, 3$, *they have components* t'_{ij}, $i, j = 1, 2, 3$. *The relation between* t_{ij} *and* t'_{ij} *is given by the transformation formulae* (1.1.16) *and* (1.1.17).

The dyadic, whose symbolic notation by T and by matrix

$$\begin{pmatrix} t_{11} & t_{12} & t_{13} \\ t_{21} & t_{22} & t_{23} \\ t_{31} & t_{32} & t_{33} \end{pmatrix}$$

has been discussed, is written *analytically* as t_{ij}, according to *indicial notation*.

In the second case, let $\boldsymbol{D} = (D_1, D_2, D_3)$ be the *angular momentum vector* and let $\boldsymbol{\omega} = (\omega_1, \omega_2, \omega_3)$ be the *angular velocity vector*. The components of these vectors are known to be related by the linear vector function

$$\begin{aligned} D_1 &= \quad A\omega_1 - F\omega_2 - E\omega_3, \\ D_2 &= -F\omega_1 + B\omega_2 - D\omega_3, \\ D_3 &= -E\omega_1 - D\omega_2 + C\omega_3. \end{aligned} \tag{1.1.18}$$

Using the co-efficients of (1.1.18) in setting up the matrix

$$\boldsymbol{\Theta} = \begin{pmatrix} A & -F & -E \\ -F & B & -D \\ -E & -D & C \end{pmatrix}, \tag{1.1.19}$$

this matrix may be used to shorten (1.1.18) into

$$\boldsymbol{D} = \boldsymbol{\Theta}\boldsymbol{\omega}. \tag{1.1.20}$$

We actually arrive back at (1.1.18) from (1.1.20), if the matrix multiplication specified on the right-hand side in (1.1.20) is carried out.

Further examination, which will be discussed later, shows that $\boldsymbol{\Theta}$ is more than a matrix. We are, in fact, dealing with the *dyadic of inertia*. This example indicates that a dyadic can also be the co-efficient scheme of a linear vector function, and additionally, the example shows that a dyadic can produce a vector if used a a *linear operator* on a vector. In this case it means that: If dyad $\boldsymbol{\Theta}$ acts upon vector $\boldsymbol{\omega}$, then vector $\boldsymbol{D}$ results. The 'linear operator' concept will be discussed later.

We have stressed in 1.1.2, that the essential feature of a vector is in the transformation formulae (1.1.5) and (1.1.6). For the dyadics too, the transformation formulae (1.1.16) and (1.1.17) have been of special importance. The next step is to generalize the transformation formulae and thus to introduce tensors of a higher order. For example, let us assume a quantity of the n-th order, having the 3^n well-ordered components $t_{ijk\ldots n}$ in the three-dimensional cartesian space. Let us assume this quantity to be a *cartesian tensor of the n-th order*, when for its components the transformation law

$$\begin{aligned} t_{ijk\ldots n} &= a_{oi}\,a_{pj}\,a_{qk}\ldots a_{un}\,t'_{opq\ldots u} \\ t'_{ijk\ldots n} &= a_{io}\,a_{jp}\,a_{kq}\ldots a_{nu}\,t_{opq\ldots u}, \quad \text{respectively} \end{aligned} \tag{1.1.21}$$

has been satisfied.

1.1.4 *The transformation matrix and its properties*

Since the transformation law in connection with tensors is of significance, we shall discuss the transformation matrix (a_{ij}). The latter is substantially related to the transformation law of tensors. Let us look at Figure 2, which represents the cartesian systems x_i, x'_i and the unit vectors $\bar{e}_i$, $\bar{e}'_i$ situated in the axes. For the unit vectors of the x'_i system, the following is obviously valid

$$\begin{aligned} \bar{e}'_i &= (a_{i1}, a_{i2}, a_{i3}), \\ a_{ij} &= \cos(x'_i, x_j), \end{aligned} \tag{1.1.22}$$

and hence, by constructing the scalar product of the $\bar{e}'_i$, we are able to obtain the relations

$$\begin{aligned} \bar{e}'^2_i &= a_{(i)k}\, a_{ik} = 1, \\ \bar{e}'_i \bar{e}'_j &= a_{ik}\, a_{jk} = 0, \; i \neq j. \end{aligned}$$

These can be summarized by using *Kronecker's symbol*

$$\delta_{ij} = \begin{cases} 1 & \text{for } i = j, \\ 0 & \text{for } i \neq j \end{cases}$$

to

$$\bar{e}'_i \bar{e}'_j = a_{ik}\, a_{jk} = \delta_{ij}.$$

Of further interest is the relation

$$a_{ik} a_{jk} = \delta_{ij}, \tag{1.1.23}$$

which, by changing from primed to unprimed quantities and vice-versa, and by $\delta'_{ij} \equiv \delta_{ij}$, may be rewritten as

$$a_{ki} a_{kj} = \delta_{ij}. \tag{1.1.24}$$

Multiplying the transformation matrix (a_{ij}) with its transposed one (a_{ji}), we obtain by observing (1.1.23)

$$(a_{ij})(a_{ji}) = (a_{ik}\, a_{jk}) = (\delta_{ij}), \tag{1.1.25}$$

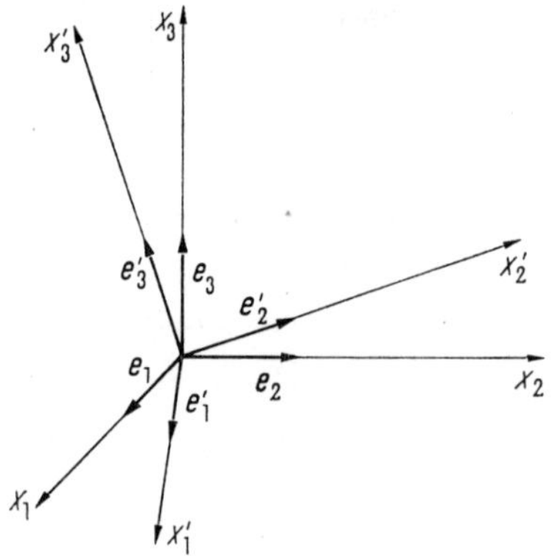

Figure 2

which obviously represents the unit matrix. Establishing the determinant on both sides of (1.1.25), it yields

$$\det(a_{ij}) \cdot (\det(a_{ji}) = \det(\delta_{ij}).$$

But since $\det(a_{ji}) = \det(a_{ij})$ and $\det(\delta_{ij}) = 1$, then

$$[\det(a_{ij})]^2 = 1,$$

from which, following limitation to cartesian right-handed systems, we obtain the relation

$$\det(a_{ij}) = 1. \tag{1.1.26}$$

1.1.5 *The* $\boldsymbol{\delta}$*-tensor and the* $\boldsymbol{\varepsilon}$*-tensor*

In connection with equation (1.1.25) we have been acquainted with the $\boldsymbol{\delta}$-tensor. It is called the *unit tensor* or *substitution tensor*. It has as its elements the Kronecker symbols δ_{ij}, so that

$$\delta = (\delta_{ij}) = \begin{pmatrix} \delta_{11} & \delta_{12} & \delta_{13} \\ \delta_{21} & \delta_{22} & \delta_{23} \\ \delta_{31} & \delta_{32} & \delta_{33} \end{pmatrix} = \begin{pmatrix} 1 & 0 & 0 \\ 0 & 1 & 0 \\ 0 & 0 & 1 \end{pmatrix}$$

is valid. In symbolic notation, this tensor has been written as $\boldsymbol{\delta}$. In analytic notation it is written as δ_{ij} (indicial notation). In the second manner of notation, special attention should be given to whether by δ_{ij} the Kronecker symbol is meant as the component of the tensor or whether the tensor itself is meant. Thus, the change from j to i in δ_{ij} can simply result in $\delta_{ii} = 1$, if for the index i we have in mind a fixed numeral, e.g., $i = 2$. If we have the tensor in mind, then, following the summation convention, δ_{ii} is the sum $\delta_{ii} = \delta_{11}+\delta_{22}+\delta_{33} = 3$, thus representing something entirely different from the Kronecker symbol having two equal indices.

If the $\boldsymbol{\delta}$-tensor acts on a vector $\boldsymbol{A}$, then we obtain

$$\delta_{ij}A_j = A_i, \tag{1.1.27}$$

i.e., the vector itself. For this reason, the tensor was named unit or substitution tensor. Basically, (1.1.27) represents the case of a tensor (the $\boldsymbol{\delta}$-tensor) acting on a vector (the vector $\boldsymbol{A}$) as an operator. Without discussing the significance of operators, we are able to check the accuracy of (1.1.27) by adding over j on the left hand side and by observing the characteristic of Kronecker's symbols. This yields the right-hand side of (1.1.27).

The tensor character of δ_{ij} can he proven by using (1.1.23) and (1.1.24) respectively. Let us write

$$a_{ik}a_{jl}\delta_{kl} = \delta_{ij}$$

which is obviously equal to (1.1.23). Because of $\delta_{ik} \equiv \delta'_{ik}$ this can also be written as

$$a_{ik}a_{jl}\delta_{kl} = \delta'_{ij}$$

which is no less than the transformation formula (1.1.17) for the second order tensor δ_{ij}.

Another tensor, which is important in calculation, is the $\boldsymbol{\varepsilon}$-tensor. It is of the third order, i.e., a so-called *triadic* and is determined by the following:

$$\varepsilon_{ijk} = \begin{cases} 0, & \text{if two indices are identical,} \\ +1, & \text{if the indices are an even permutation of 1, 2, 3,} \\ -1, & \text{if the indices are an uneven permutation of 1, 2, 3.} \end{cases}$$

Let us prove that it is a tensor. We show that ε_{ijk} is governed by the transformation law

$$\begin{aligned} \varepsilon'_{ijk} &= a_{il} a_{jm} a_{kn} \varepsilon_{lmn}, \\ \varepsilon_{ijk} &= a_{li} a_{mj} a_{nk} \varepsilon'_{lmn}. \end{aligned} \tag{1.1.28}$$

We write in full the statement $a_{il}a_{jm}a_{kn}\varepsilon_{lmn}$, observing the summation convention and the characteristics of the ε-tensor, and thus arrive at

$$\begin{aligned} a_{il} a_{jm} a_{kn} \varepsilon_{lmn} &= a_{i1} a_{j2} a_{k3} + a_{i2} a_{j3} a_{k1} + a_{i3} a_{j1} a_{k2} \\ &\quad - a_{i2} a_{j1} a_{k3} - a_{i1} a_{j3} a_{k2} - a_{i3} a_{j2} a_{k1} \\ &= \begin{vmatrix} a_{i1} & a_{i2} & a_{i3} \\ a_{j1} & a_{j2} & a_{j3} \\ a_{k1} & a_{k2} & a_{k3} \end{vmatrix}. \end{aligned}$$

But for the determinant thus obtained, the following is valid: that for two identical rows, i.e., if two of the indices i,j,k are equal, the determinant is equal to nought. Also that the determinant represents for i,j,k, as an even permutation of 1,2,3, the transformation matrix and that therefore, according to (1.1.26), it is equal to 1. Furthermore that the determinant changes the sign, thus being equal to -1, if any two of its rows are interchanged, i.e., if the indices i, j, k form an uneven permutation of 1, 2, 3. All these facts prove that the determinant is identical with ε_{ijk}. Therefore we write

$$a_{il} a_{jm} a_{kn} \varepsilon_{lmn} = \begin{vmatrix} a_{i1} & a_{i2} & a_{i3} \\ a_{j1} & a_{j2} & a_{j3} \\ a_{k1} & a_{k2} & a_{k3} \end{vmatrix} \equiv \varepsilon_{ijk}.$$

But similarly to the δ-tensor, $\varepsilon'_{ijk} \equiv \varepsilon_{ijk}$. This enables us to write finally

$$a_{il} a_{jm} a_{kn} \varepsilon_{lmn} = \varepsilon'_{ijk}.$$

This is in agreement with the transformation law, proving that in ε_{ijk} we are dealing with a tensor. The ε-tensor is used to express the vector product of two vectors A_j and B_k in analytic notation. We set up the relation

$$C_i = \varepsilon_{ijk} A_j B_k. \tag{1.1.29}$$

Its result is equal to

$$\boldsymbol{C} = \boldsymbol{A} \times \boldsymbol{B}, \tag{1.1.29a}$$

which is the symbolic representation of the vector product. In order to prove this, let us show that the quantity on the left-hand side of (1.1.29) is in fact a vector.

The right-hand side of (1.1.29) is rearranged according to (1.1.6). It yields

$$C_i = \varepsilon_{ijk} a_{lj} A'_l a_{mk} B'_m .$$

Then, both sides are multiplied by a_{ni}, and after simple rearrangement and because of (1.1.28),

$$a_{ni} C_i = \varepsilon_{ijk} a_{ni} a_{lj} a_{mk} A'_l B'_m = \varepsilon'_{nlm} A'_l B'_m . \tag{1.1.30}$$

Writing down (1.1.29) for primed quantities yields

$$C'_n = \varepsilon'_{nlm} A'_l B'_m . \tag{1.1.31}$$

Comparing (1.1.31) with (1.1.30) indicates that

$$a_{ni} C_i = C'_n$$

must be true, which because of (1.1.5) denotes that C_i is a vector.

If for example we want to calculate the *mixed triple product* $\boldsymbol{A} \cdot (\boldsymbol{B} \times \boldsymbol{C})$ using the $\boldsymbol{\varepsilon}$-tensor, then the scalar product of the vectors A_l and $D_i = \varepsilon_{ijk} B_j C_k$ has to be formed, i.e., we write $A_i D_i$:

$$A_i D_i = \varepsilon_{ijk} A_i B_j C_k .$$

If the triple-sum on the right-hand side is carried out, then the determinant

$$\varepsilon_{ijk} A_i B_j C_k = \begin{vmatrix} A_1 & A_2 & A_3 \\ B_1 & B_2 & B_3 \\ C_1 & C_2 & C_3 \end{vmatrix}$$

is obtained which is the known result for the mixed triple product.

Finally, let us look at an important connection between the $\boldsymbol{\varepsilon}$- and the $\boldsymbol{\delta}$-tensor, which is the $\boldsymbol{\varepsilon}$, ***δ-identity***. It is as follows:

$$\varepsilon_{ijk}\varepsilon_{ist} = \begin{vmatrix} \delta_{js} & \delta_{jt} \\ \delta_{ks} & \delta_{kt} \end{vmatrix} = \delta_{js}\delta_{kt} - \delta_{jt}\delta_{ks} \tag{1.1.32}$$

which can be verified by writing out in full.

1.1.6 *Examples*

Following are two examples. The first example sets out to prove the statement in 1.1.3, that the quantity $\boldsymbol{\Theta}$ appearing as a co-efficient scheme of the linear vector function, (1.1.18), is a dyadic, i.e., a tensor. The second example will yield an important formula of vector analysis, which is used quite often in connection with the theory of elasticity.

Going back to (1.1.19), we notice that according to definition we may write

$$A = \int_V \rho(r^2 - x_1^2)\,\mathrm{d}V,\ B = \int_V \rho(r^2 - x_2^2)\,\mathrm{d}V,\ C = \int_V \rho(r^2 - x_3^2)\,\mathrm{d}V,$$
$$D = \int_V \rho x_2 x_3\,\mathrm{d}V,\ E = \int_V \rho x_1 x_3\,\mathrm{d}V,\ F = \int_V \rho x_1 x_2\,\mathrm{d}V$$

where V is the volume, ρ is the density of the particular body, x_i are the coordinates of a mass point and $r = (x_i x_i)^{\frac{1}{2}}$ is the radius vector. Therefore (1.1.19) is written in its short version as

$$\boldsymbol{\Theta} = (J_{ij}),\ J_{ij} = \int_V \rho(\delta_{ij} r^2 - x_i x_j)\,\mathrm{d}V, \tag{1.1.33}$$

and (1.1.20) as

$$D_i = J_{ij}\omega_j.$$

Let us change from system x_i to system x'_i. Thus, we obtain

$$x_i = a_{ki} x'_k,$$
$$\mathrm{d}V = |a_{lk}|\,\mathrm{d}V' \equiv \mathrm{d}V',\ \text{because of } |a_{lk}| = 1,$$
$$\delta_{ij} = a_{ki} a_{lj} \delta'_{kl}.$$

Also for scalars r and ρ we are allowed to write

$$r = r', \quad \rho = \rho'.$$

Thus, J_{ij} changes at first into

$$J_{ij} = \int_V \rho'(a_{ki}a_{lj}\delta'_{kl}r'^2 - a_{ki}x'_k a_{lj}x'_l)\,\mathrm{d}V',$$

and after simple rearrangement it results in

$$J_{ij} = a_{ki}a_{lj}\int_V \rho'(\delta'_{kl}r'^2 - x'_k x'_l)\,\mathrm{d}V'.$$

According to definition

$$\int_V \rho'(\delta'_{kl}r'^2 - x'_k x'_l)\,\mathrm{d}V' = J'_{kl},$$

hence we arrive at

$$J_{ij} = a_{ki}a_{lj}J'_{kl}.$$

This is the transformation law for a second-order tensor, and proves that $\boldsymbol{\Theta} = (J_{ij})$ is a tensor.

The second example deals with the formula

$$\nabla^2 \boldsymbol{w} = \text{grad div } \boldsymbol{w} - \text{curl curl } \boldsymbol{w}, \tag{1.1.34}$$

which in symbolic notation is

$$\nabla^2 \boldsymbol{w} = \nabla(\nabla \cdot \boldsymbol{w}) - \nabla \times (\nabla \times \boldsymbol{w}). \tag{1.1.35}$$

By using $\boldsymbol{w} = (w_i)$, (1.1.9), as well as (1.1.29), the latter specifically for representation of the vector product using the ε-tensor, we are able to use indicial notation for the equation (1.1.35) and obtain

$$\frac{\partial^2 w_i}{\partial x_j \partial x_j} = \frac{\partial}{\partial x_i}\frac{\partial w_j}{\partial x_j} - \varepsilon_{ijk}\frac{\partial}{\partial x_j}\left(\varepsilon_{kpq}\frac{\partial}{\partial x_p}w_q\right).$$

Let us show both sides of this relation to be identical. A simple change yields

$$\frac{\partial}{\partial x_j}\frac{\partial}{\partial x_j}w_i=\frac{\partial}{\partial x_i}\frac{\partial}{\partial x_j}w_j-\varepsilon_{ijk}\varepsilon_{kpq}\frac{\partial}{\partial x_j}\frac{\partial}{\partial x_p}w_q.$$

Using the ε,δ-identity on the right-hand side, we obtain

$$\frac{\partial}{\partial x_j}\frac{\partial}{\partial x_j}w_i=\frac{\partial}{\partial x_i}\frac{\partial}{\partial x_j}w_j-(\delta_{ip}\delta_{jq}-\delta_{iq}\delta_{jp})\frac{\partial}{\partial x_k}\frac{\partial}{\partial x_p}w_q.$$

Observing the character of the Kronecker-symbols, we are able to calculate the equation further

$$\frac{\partial}{\partial x_j}\frac{\partial}{\partial x_j}w_i=\frac{\partial}{\partial x_i}\frac{\partial}{\partial x_j}w_j-\frac{\partial}{\partial x_i}\frac{\partial}{\partial x_q}w_q+\frac{\partial}{\partial x_j}\frac{\partial}{\partial x_j}w_i.$$

By renaming the dummy indices q of the second term on the right-hand side into j (this is permissible for dummy indices), then, the first and second term will disappear from the right-hand side, and we obtain an identity. The accuracy of formulae (1.1.34), (1.1.35) has thus been proven.

1.1.7 *Tensors as linear operators*

The combination of a tensor t_{ij} and a vector A_i of the kind

$$C_i=t_{ij}A_j \tag{1.1.36}$$

has already been looked at. We have carried out a *tensor multiplication*, which involves also a *contraction*. A contraction results when in a tensor multiplication, two indices of different factors are made equal. In the above example it is the index j, which occurs in the dyadic, which represents one factor, as well as in the vector, which represents the other factor. The simplest case of a contraction is in the *scalar product* of two vectors. In this case, the product of two first order tensors (vectors) is being given in the form A_iB_i, in which the indices of the two factors have just been made equal. The result is known to be a scalar.

The facts described in (1.1.36) may also be interpreted as a linear vector function, in which tensor t_{ij} acts on vector A_j as an *operator*, thus, producing vector C_i. In symbolic notation it is

$$\boldsymbol{TA} = \boldsymbol{C}.$$

In this connection, the particular example contained in (1.1.27) can be symbolically notated as

$$\boldsymbol{\delta A} = \boldsymbol{A}.$$

In this way the characteristic of the $\boldsymbol{\delta}$-tensor as unit or substitution tensor becomes apparent, since vector $\boldsymbol{A}$ has been unchanged when operator $\boldsymbol{\delta}$ was in effect.

Let us point out that a tensor is a linear operator since it satisfies the two conditions

$$\begin{aligned} \boldsymbol{T}(\boldsymbol{A}+\boldsymbol{B}) &= \boldsymbol{TA}+\boldsymbol{TB}, \\ \boldsymbol{T}(m\boldsymbol{A}) &= m\boldsymbol{TA}, \end{aligned} \tag{1.1.37}$$

characteristic of a linear operator. In (1.1.37), $\boldsymbol{A}$, $\boldsymbol{B}$ are vectors, and m is a scalar. The accuracy of equations (1.1.37) is checked by changing over to indicial notation and verifying the equality of the right sides with the left-hand sides through calculation. As an exercise let us show that a vector $\boldsymbol{C}$ actually results from operation $\boldsymbol{TA}$. Let us use indicial notation, changing on the right-hand side from (1.1.36), using the transformation formulae, to the primed quantities. This yields

$$C_i = a_{ki}a_{lj}t'_{kl}a_{mj}A'_m = a_{ki}a_{lj}a_{mj}t'_{kl}A'_m.$$

But $a_{lj}a_{mj} = \delta_{lm}$, so that we can write

$$C_i = a_{ki}\delta_{lm}t'_{kl}A'_m = a_{ki}t'_{km}A'_m.$$

Because of $t'_{km}A'_m = C'_k$ we finally obtain

$$C_i = a_{ki}C'_k,$$

which represents the transformation formula for a vector. Thus, there is a vector on the left-hand side of (1.1.36), i.e., the effect of operator $\boldsymbol{T}$ on vector $\boldsymbol{A}$ does result in another vector, i.e., $\boldsymbol{C}$.

1.1.8 *Transposed, symmetric, antisymmetric tensors*

Let a tensor $\boldsymbol{T}$ be represented as $\boldsymbol{T} = (t_{ij})$. We are now able to transpose matrix (t_{ij}) i.e., we can change over to matrix (t_{ji}). It can be shown that this transposed matrix again represents a tensor, i.e., $\boldsymbol{T}^{(T)} = (t_{ji})$. We call $\boldsymbol{T}^{(T)}$ the *transposed tensor* or the tensor *conjugated* to T.

To show that $\boldsymbol{T}^{(T)}$ does have tensor character, the reader can prove for himself that for $\boldsymbol{T}^{(T)}$ the corresponding transformation formulae are valid.

If a tensor $\boldsymbol{T}$ is equal to its conjugated tensor, i.e., if $t_{ij} = t_{ji}$ is valid, it is called *symmetric*. If instead it is $t_{ij} = -t_{ji}$, then the tensor is called *antisymmetric* or *skew-symmetric*. The diagonal components for an antisymmetric tensor are obviously equal to nought. Let us write the identity

$$t_{ij} \equiv \tfrac{1}{2}(t_{ij} + t_{ji}) + \tfrac{1}{2}(t_{ij} - t_{ji})$$

for any tensor. It shows that any tensor can be represented as the sum of a symmetric and of an antisymmetric tensor, since $\frac{1}{2}(t_{ij} + t_{ji})$ is a symmetric and $\frac{1}{2}(t_{ij} - t_{ji})$ is an antisymmetric tensor. Let us introduce vector

$$\boldsymbol{\omega} = -(t_{23}\boldsymbol{e}_1 + t_{31}\boldsymbol{e}_2 + t_{12}\boldsymbol{e}_3),$$

whose components are taken from those of an antisymmetric tensor

$$\boldsymbol{T}_{(\mathrm{as})} = \begin{pmatrix} 0 & t_{12} & -t_{31} \\ -t_{12} & 0 & t_{23} \\ t_{31} & -t_{23} & 0 \end{pmatrix}.$$

$\boldsymbol{e}_i$ are the unit-vectors located in the co-ordinate axes. It is easily found that

$$t_{ij(\mathrm{as})} = -\varepsilon_{ijk}\,\omega_k. \tag{1.1.38}$$

Using this relation an interesting fact becomes apparent: Let us assume an antisymmetric tensor $\boldsymbol{T}_{(as)}$ to act as operator on a vector $\boldsymbol{A}$, thus producing vector $\boldsymbol{C}$. In symbolic notation this is

$$\boldsymbol{T}_{(\mathrm{as})}\boldsymbol{A} = \boldsymbol{C},$$

and in analytic notation it is

$$t_{ij(\mathrm{as})} A_j = C_i . \tag{1.1.39}$$

Because of (1.1.38), we are able to change (1.1.39) into

$$-\varepsilon_{ijk}\, \omega_k A_j = C_i ,$$

which again because of $\varepsilon_{ijk} = -\varepsilon_{ikj}$ is changed into

$$\varepsilon_{ikj} \omega_k A_j = C_i . \tag{1.1.40}$$

Comparing (1.1.40) with (1.1.29) shows that equation (1.1.40) in agreement with (1.1.29a), can be notated as

$$\boldsymbol{\omega} \times \boldsymbol{A} = \boldsymbol{C}. \tag{1.1.40a}$$

When vector $\boldsymbol{\omega}$ is termed as *vector of the antisymmetric tensor* $\boldsymbol{T}_{(\mathrm{as})}$, then because of (1.1.40a) we are able to state that the effect of tensor $\boldsymbol{T}_{(\mathrm{as})}$ as operator on vector $\boldsymbol{A}$ can be substituted by the vector product 'of the tensor's vector' $\boldsymbol{\omega}$ with vector $\boldsymbol{A}$.

1.1.9 *Characteristic value, characteristic vector, principal axes, invariants of a symmetric tensor*

For simplicity's sake, let us assume a second order tensor. Then we require that the effect of this tensor $\boldsymbol{T}$ as operator on a vector $\boldsymbol{A}$ yields especially

$$\boldsymbol{T}\boldsymbol{A} = \lambda \boldsymbol{A} \tag{1.1.41}$$

In (1.1.41), λ is a temporarily undetermined scalar. Let us multiply on the right-hand side of (1.1.41) with the unit-tensor $\boldsymbol{\delta}$, without making any changes, which yields

$$\boldsymbol{T}\boldsymbol{A} = \lambda \boldsymbol{\delta} \boldsymbol{A}.$$

This equation can be changed into the so-called *eigenvalue equation* of tensor $\boldsymbol{T}$, i.e., into

$$(\boldsymbol{T}-\lambda\boldsymbol{\delta})\boldsymbol{A} = 0.$$

Changing from the symbolic to the analytic representation, then we obtain

$$(t_{ij}-\lambda\delta_{ij})A_j = 0. \tag{1.1.42}$$

Let us introduce unit vector

$$\boldsymbol{e} = \frac{\boldsymbol{A}}{|\boldsymbol{A}|},$$

having components $l_i = A_i/|\boldsymbol{A}|$. Since $\boldsymbol{e}$ is a unit-vector, then the l_i are direction cosines, and the following is valid for them,

$$l_i l_i = 1. \tag{1.1.43}$$

By dividing all equations of the system (1.1.42) by $|\boldsymbol{A}|$ we arrive at

$$(t_{ij}-\lambda\delta_{ij})l_j = 0. \tag{1.1.44}$$

This is a linear, algebraic equation system for the calculation of the l_j. Since it is homogenous, then for the existence of non-trivial solutions we require

$$\det(t_{ij}-\lambda\delta_{ij}) = 0. \tag{1.1.45}$$

Equation (1.1.45) is called the *characteristic equation* of the tensor. Expanding the determinant (1.1.45) yields the cubic equation

$$\lambda^3-J_1\lambda^2+J_2\lambda-J_3 = 0. \tag{1.1.46}$$

Roots $\lambda^{(k)}$, $k = 1, 2, 3$, of this equation are the *characteristic values* of the tensor. Since we have assumed a symmetric tensor, they are always real. The coefficients J_k, $k = 1, 2, 3$, of (1.1.46) are the *invariants of the tensor* because their values remain the same regardless of the co-ordinate system used to represent the tensor. They are related to the components t_{ij} of the tensor as follows:

$$\begin{aligned} J_1 &= t_{ii}, \\ J_2 &= \tfrac{1}{2}(t_{ii}t_{jj}-t_{ij}t_{ij}), \\ J_3 &= \varepsilon_{ijk}t_{i1}t_{j2}t_{k3}. \end{aligned} \tag{1.1.47}$$

It is noteworthy that J_1 results from contraction of the tensor. By contraction is meant an operation by which two indices of a tensor are equalled, and a summation must be carried out. Therefore, the tensor is lowered by two orders. In our special case, the dyadic becomes a scalar by contraction, e.g., $t_{11}+t_{22}+t_{33}$, i.e., the sum of the diagonal components. This sum is also called 'trace' of tensor (tr. $\boldsymbol{T}$).

Having solved (1.1.46), we find the characteristic values $\lambda^{(k)}$ which are inserted successively in (1.1.44) and the l_j are calculated from

$$(t_{ij}-\lambda^{(k)}\delta_{ij}) = 0, \quad l_j^{(k)}l_j^{(k)} = 1. \tag{1.1.48}$$

We have added (1.1.43) so that enough independent equations are available.

For each specific $\lambda^{(k)}$ we obtain a triplet $l_j^{(k)}, j = 1, 2, 3$. If the triplet is assumed to be a set of direction cosine, then it determines a *principal direction* of the tensor. If, on the other hand, we assume the $l_j^{(k)}$ to be components, then they determine a *characteristic vector* whose line of application corresponds to the particular principal direction. Since we have three $\lambda^{(k)}$, then we obtain three principal directions, characteristic vectors, respectively. They are orthogonal to each other forming the so-called *principle axes system* (PAS) of the tensor. The orthogonality can easily be proved under the condition of three distinct characteristic values:

Let

$$\boldsymbol{A}\cdot\boldsymbol{TB}-\boldsymbol{B}\cdot\boldsymbol{TA} = t_{ij}B_jA_i-t_{ij}A_jB_i = (t_{ij}-t_{ji})B_jA_i = 0$$

be valid for the symmetric tensor $\boldsymbol{T}$ and two vectors $\boldsymbol{A}$ and $\boldsymbol{B}$, since we were allowed to rename dummy indices where suitable, and to use the relation $t_{ij} = t_{ji}$ (symmetry).

In the same manner

$$\boldsymbol{e}_1\cdot\boldsymbol{Te}_2-\boldsymbol{e}_2\cdot\boldsymbol{Te}_1 = 0$$

is valid for two characteristic vectors $\boldsymbol{e}_1$ and $\boldsymbol{e}_2$. By definition, $\boldsymbol{Te}_1 = \lambda^{(1)}\boldsymbol{e}_1$ and $\boldsymbol{Te}_2 = \lambda^{(2)}\boldsymbol{e}_2$ is valid for characteristic vectors. When substituting this in the above relation, we obtain

$$(\lambda^{(2)}-\lambda^{(1)})\boldsymbol{e}_1\cdot\boldsymbol{e}_2 = 0.$$

Because of $\lambda^{(2)} \neq \lambda^{(1)}$, it can be deduced that the two characteristic vectors are orthogonal to each other.

If the PAS is used as the reference system, then the characteristic vectors $\boldsymbol{e}_k$ have the components δ_{kj}. Then $\boldsymbol{Te}_k$ changes to

$$\boldsymbol{Te}_k = t_{ij}\delta_{kj} = t_{ik}. \tag{1.1.49}$$

Since we are dealing with characteristic vectors, then also

$$\boldsymbol{Te}_k = \lambda^{(k)}\boldsymbol{e}_k = \lambda^{(k)}\delta_{ki}. \tag{1.1.50}$$

Equating the right-hand side of (1.1.49) and (1.1.50), it yields

$$t_{ik} = \lambda^{(k)}\delta_{ki}, \tag{1.1.51}$$

which will be written in full. Thus, owing to the characteristic of the Kronecker-symbol, we arrive from (1.1.51) at

$$\begin{aligned} &t_{11} = \lambda^{(1)}, \quad t_{12} = 0, \quad t_{13} = 0,\\ &t_{21} = 0, \quad t_{22} = \lambda^{(2)}, \quad t_{23} = 0,\\ &t_{31} = 0, \quad t_{32} = 0, \quad t_{33} = \lambda^{(3)}, \end{aligned}$$

and deduce that the tensor in PAS is simply represented by the *diagonal matrix*

$$\boldsymbol{T} = \begin{pmatrix} \lambda^{(1)} & 0 & 0 \\ 0 & \lambda^{(2)} & 0 \\ 0 & 0 & \lambda^{(3)} \end{pmatrix}.$$

On this fact is based the great significance of the PAS. Those non vanishing components of $\boldsymbol{T}$ relative to PAS are the characteristic values of the tensor, and they result as roots of the characteristic equation (1.1.44).

We are dealing with a special case, when not all of the characteristic values $\lambda^{(k)}$ are different from each other. If two of the characteristic values are equal, but are different from a third, then the third characteristic value determines a first principal direction, a characteristic vector, respectively. A second principal direction, a characteristic vector, respectively, corresponds to the double-valued characteristic value. We find the missing characteristic direction, the missing characteristic vector, respectively, by supplementing the already existing system of axes by a third axis, such that an orthogonal right handed system is obtained. Both those directions, which are vertical to the characteristic direction of the initial, single-valued characteristic value, span a plane. It can be shown that any other orthogonal system of two axes in this plane which forms a right-handed system together with the third characteristic direction, is also applicable to the construction of a PAS. If $\lambda^{(1)} = \lambda^{(2)}$, then

$$\boldsymbol{T}\boldsymbol{e}_1 = \lambda^{(1)}\boldsymbol{e}_1, \; \boldsymbol{T}\boldsymbol{e}_2 = \lambda^{(1)}\boldsymbol{e}_2$$

is valid, and let $a\boldsymbol{e}_1 + b\boldsymbol{e}_2$ be any vector in the plane spanned by $\boldsymbol{e}_1, \boldsymbol{e}_2$. Then, relation

$$\boldsymbol{T}(a\boldsymbol{e}_1 + b\boldsymbol{e}_2) = \lambda^{(1)}(a\boldsymbol{e}_1 + b\boldsymbol{e}_2)$$

is easily set up and we deduce that, in fact, any vector in the plane set up by $\boldsymbol{e}_1, \boldsymbol{e}_2$ is usable as a characteristic vector and can also indicate a principle direction in its line of application.

If especially all three characteristic values are equal, then any orthogonal right handed system can be taken as the PAS.

1.2 General tensors

1.2.1 *Definitions*

We are dealing with tensors of a more general kind than the cartesian tensors, if we do not limit ourselves to the three-dimensional space, but allow *affine* and *metric* n-dimensional spaces, whereby the latter may be *euclidean* or *riemannian*, and if we allow, in addition, any curvilinear co-ordinates and general co-ordinate transformations. During the discussion of the general tensors, we shall closely adhere to I. S. Sokolnikoff [2] as far as the presentation and the notations are concerned.

Let x^i, $i = 1, 2, \ldots, n$, be one co-ordinate system and let y^i, $i = 1, 2, \ldots n$, be another one. Before we start, let us point out that starting from now and differing from 1.1, the variables, like x^i and y^i, are to be marked by a raised index (or superscript).

Let there be a specification which allows the change from one system to the other, e.g.,

$$y^i = y^i(x^1, x^2, \ldots, x^n) \quad i = 1, 2, \ldots, n, \tag{1.2.1}$$

which we shall call a *transformation*. Henceforth, we shall discuss only so-called *admissible* transformations. These are reversible, so that inverse

$$x^i = x^i(y^1, y^2, \ldots, y^n), \quad i = 1, 2, \ldots, n, \tag{1.2.2}$$

exists, and these are one-to-one. Furthermore, it is assumed for these that the functions $y^i(x)$ together with their first derivations are continuous, (i.e., they should at least belong to class C^1), that its *Jacobi's determinant*

$$J = \left| \frac{\partial y^i}{\partial x^j} \right| \tag{1.2.3}$$

exists and that J differs everywhere from nought.

Let $y^i(x)$ be an admissible transformation having a Jacobian J and let $x^i(y)$ be its inverse having the Jacobian J^{-1}. Then, from the identity

$$y^i = y^i[x^1(y^1, y^2, \ldots, y^n), \ldots, x^n(y^1, y^2, \ldots, y^n)],$$

we obtain through differentiation[1]

$$\frac{\partial y^i}{\partial y^j} = \delta_{ij} = \frac{\partial y^i}{\partial x^\alpha}\frac{\partial x^\alpha}{\partial y^j}, \; (\alpha = 1, 2, \ldots, n). \tag{1.2.4}$$

Consider

$$\det\left(\frac{\partial y^i}{\partial x^\alpha}\frac{\partial x^\alpha}{\partial y^j}\right) \equiv \det\left(\frac{\partial y^i}{\partial x^\alpha}\right) \cdot \det\left(\frac{\partial x^i}{\partial y^\alpha}\right). \tag{1.2.5}$$

From (1.2.4) is yielded

[1] In this instance and in the following, Greek letters are used in this section as indices, when they do not only go over 1, 2 but even over 1, 2, . . ., n. But the representation makes mistakes impossible.

$$\det\left(\frac{\partial y^i}{\partial x^\alpha}\frac{\partial x^\alpha}{\partial y^j}\right) = \det(\delta_{ij}) = 1, \tag{1.2.6}$$

and by definition

$$\det\left(\frac{\partial y^i}{\partial x^\alpha}\right) = J,\ \det\left(\frac{\partial x}{\partial y^\alpha}\right) = J^{-1} \tag{1.2.7}$$

is obtained. The summary of (1.2.5), (1.2.6) and (1.2.7) yields

$$J \cdot J^{-1} = 1. \tag{1.2.8}$$

In this way an important relation between Jacobi's determinant of an admissible transformation and its inverse has been set up.

Let us look at the transformations $y^i = y^i(x)$ and $z^i = z^i(y)$, having the Jacobians J_1 and J_2. Let us carry out both transformations one after the other, i.e., we set up

$$z^i = z^i[y^1(x^1, x^2, \ldots, x^n), \ldots, y^n(x^1, x^2, \ldots, x^n)],$$

which is termed a *product-transformation*. The Jacobian

$$J_3 = \left|\frac{\partial z^i}{\partial y^\alpha}\frac{\partial y^\alpha}{\partial x^j}\right| = \left|\frac{\partial z^i}{\partial y^j}\right| \cdot \left|\frac{\partial y^i}{\partial x^j}\right| = J_1 J_2 \tag{1.2.9}$$

corresponds to this product-transformation. From (1.2.9), we read off that for admissible transformations, the Jacobian of a product-transformation is a product, the Jacobians of the two transformations being the factors of the product. Since the concept 'product of transformations' has been introduced, let us add that the set of all admissible transformations form a *group*.

Let us discuss two kinds of transformations: one being the transformation by *covariance* and the other the transformation by *contravariance*.

The transformation of co-ordinates is specified by $x^i = x^i(y)$. Let us look at how the gradient

$$\frac{\partial f}{\partial x^k} \equiv f_{,x^k}$$

of a function $f(x^k)$ may be transformed. The chain rule of differentiation yields the transformation formula

$$\frac{\partial f}{\partial y^i} = \frac{\partial x^\alpha}{\partial y^i}\frac{\partial f}{\partial x^\alpha}. \tag{1.2.10}$$

This result is generalized by changing over to vectors. Then we write

$$B_i(y) = \frac{\partial x^\alpha}{\partial y^i} A_\alpha(x), \tag{1.2.11}$$

where (1.2.11) corresponds formally to law (1.2.10). Let us call (1.2.11) a *covariant transformation*, $B_i(y)$ the *covariant components* of a vector in the y-system and $A_i(x)$ the covariant components of the *same* vector in the x-system. In short: *By a covariant vector we understand the well-ordered set of components, which are transformed in accordance with the covariant law* (1.2.11). Comparison of (1.2.11) with (1.2.10) shows that gradient $f,_{x^k}$ represents a covariant vector following this definition.

Let the transformation of co-ordinates be given by $x^i = x^i(y)$. Let us examine how the differential $\mathrm{d}x^k$ is transformed. The answer is

$$\mathrm{d}y^i = \frac{\partial y^i}{\partial x^\alpha}\mathrm{d}x^\alpha. \tag{1.2.12}$$

This relation represents a transformation which differs from that of (1.2.11) and which we, therefore, call *contra-variant*. When it is used on the components of vectors, and is thus generalized, it leads to

$$B^i(y) = \frac{\partial y^i}{\partial x^\alpha} A^\alpha(x). \tag{1.2.13}$$

Similarly to the preceding, let us call $B^i(y)$ the contra-variant components of a vector in the y-system and $A^i(x)$ the contra-variant components of the same vector in the x-system, and *by a contra-variant vector, we understand the well-ordered set of components which are transformed in accordance with the contra-variant law* (1.2.13).

Although we have talked about 'covariant' and 'contra-variant' vectors for the sake of shortness, which we shall continue to do, it does not mean that we are dealing with two fundamentally different kinds of vectors. It becomes a question of selecting the manner of representation, whether the very same vector is represented using its covariant components at one time, or its contra-variant components at the next. A so-called covariant vector can, in certain circumstances, be equal to a contra-variant vector insofar as both are different representations of the same vector. This will become apparent later on when the relation between the covariant and contra-variant components of a vector (or tensor) are presented.

We change over especially from the general co-ordinate transformations to the linear transformations

$$y^i = a_{ij}x^j, \quad x^j = a_{ij}y^i, \quad i, j = 1, 2, 3, \tag{1.2.14}$$

of cartesian co-ordinates. If for the coefficients a_{ij} of the transformation, $a_{ij}a_{ik} = \delta_{jk}$ is valid, then we call it orthogonal, and if in addition $|a_{ij}| = +1$ is valid, then it represents a rotation. This is the kind of transformation we have assumed in 1.1.

From (1.2.14) we calculate

$$\frac{\partial y^i}{\partial x^j} = a_{ij}, \quad \frac{\partial x^j}{\partial y^i} = a_{ij}. \tag{1.2.15}$$

Using this, we write the covariant transformation for a vector

$$B_i = \frac{\partial x^j}{\partial y^i} A_j = a_{ij} A_j. \tag{1.2.16}$$

Correspondingly, the contra-variant transformation is

$$B^i = \frac{\partial y^i}{\partial x^j} A^j = a_{ij} A^j. \tag{1.2.17}$$

The comparison of (1.2.16) with (1.2.17) shows that for linear, orthogonal transformation of cartesian co-ordinates, the difference between co- and contra-variant transformation of vectors (tensors, respectively) vanishes completely. For this reason we did not distinguish in 1.1. between co- and contra-variant quantities, and we were able to calculate, using cartesian vectors and tensors, in such a simple way.

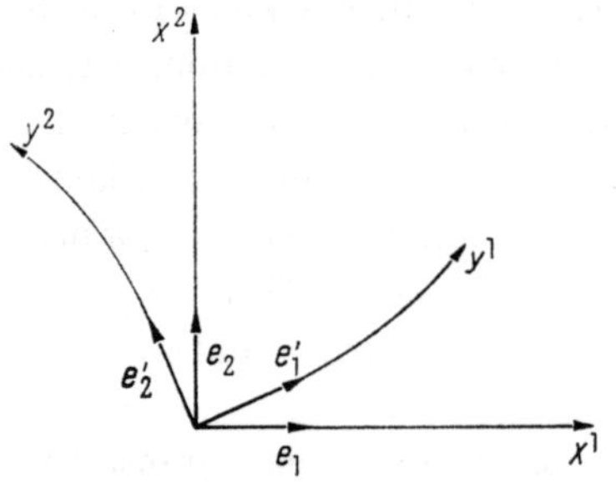

Figure 3

Let us look at the two reference systems x^i and y^i in Figure 3, which are related to each other by the transformation formula
$y^i = y^i(x)$. Hence

$$\mathrm{d}x^\alpha = \frac{\partial x^\alpha}{\partial y^i}\,\mathrm{d}y^i \tag{1.2.18}$$

is valid. Also

$$\mathrm{d}y^i \boldsymbol{e}_i' = \mathrm{d}x^\alpha \boldsymbol{e}_\alpha . \tag{1.2.19}$$

Through (1.2.18), (1.2.19) can be changed into

$$\mathrm{d}y^i \boldsymbol{e}_i' = \frac{\partial x^\alpha}{\partial y^i}\,\mathrm{d}y^i \boldsymbol{e}_\alpha ,$$

from which, because $\partial y^i \neq 0$,

$$\boldsymbol{e}_i' = \frac{\partial x^\alpha}{\partial y^i}\,\boldsymbol{e}_\alpha . \tag{1.2.20}$$

This is the (co-variant) transformation law for the unit-vectors of the reference system; using this transformation law we have largely orientated ourselves in the choice of notations. We have called those vectors, tensors, respectively, *co-variant* which have behaved in the *same manner* as unit-vectors of an n-dimensional coordinate system. Those which have changed in an *opposite manner* we have called *contra-variant.*

The definitions obtained so far are generalized as follows: Let us assume an *n-dimensional reference system* x^i called for short the x-system, which could be changed into an n-dimensional y-system through the transformation formula $y^i = y^i(x)$. Let us say that:

A well-ordered set of n^r quantities $A_{i}, \ldots, {}_{ir}(x)$, given relative to the x-system, represents the set of components of a covariant tensor of the order r, if the corresponding set $B_{i}, \ldots, {}_{ir}(y)$, relative to the y-system, is related with the first set by the transformation formula

$$B_{i_1 \ldots i_r}(y) = \frac{\partial x^{\alpha_1}}{\partial y^{i_1}} \frac{\partial x^{\alpha_2}}{\partial y^{i_2}} \cdots \frac{\partial x^{\alpha_r}}{\partial y^{i_r}} A_{\alpha_1 \ldots \alpha_r}(x). \tag{1.2.21}$$

Quite similarly, a *contra-variant tensor of the order r* is defined by the transformation formula

$$B^{i_1 \ldots i_r}(y) = \frac{\partial y^{i_1}}{\partial x^{\alpha_1}} \frac{\partial y^{i_2}}{\partial x^{\alpha_2}} \cdots \frac{\partial y^{i_r}}{\partial x^{\alpha_r}} A^{\alpha_1 \ldots \alpha_r}(x), \tag{1.2.22}$$

and a *mixed tensor, which is a covariant of the order r, and a contra-variant of the order s,* is defined by the transformation formula

$$B_{i_1 \ldots i_r}^{j_1 \ldots j_s}(y) = \frac{\partial x^{\alpha_1}}{\partial y^{i_1}} \cdots \frac{\partial x^{\alpha_r}}{\partial y^{i_r}} \frac{\partial y^{j_s}}{\partial x^{\beta_1}} \cdots \frac{\partial y^{j_s}}{\partial x^{\beta_s}} A_{\alpha_1 \ldots \alpha_r}^{\beta_1 \ldots \beta_s}(x). \tag{1.2.23}$$

Using the latter transformation formula, let us show, in what manner an inversion of formula is made:

$$A_{\alpha_1 \ldots \alpha_r}^{\beta_1 \ldots \beta_s}(x) = \frac{\partial y^{i_1}}{\partial x^{\alpha_1}} \cdots \frac{\partial y^{i_r}}{\partial x^{\alpha_r}} \frac{\partial x^{\beta_1}}{\partial y^{j_1}} \cdots \frac{\partial x^{\beta_s}}{\partial y^{j_s}} B_{i_1 \ldots i_r}^{j_1 \ldots j_s}(y). \tag{1.2.24}$$

The equation (1.2.24) is such an inversion of (1.2.23). From this we are able to see how formula (1.2.21) and (1.2.22) are similarly reversible, since covariant and contra-variant tensors are definable as special cases of mixed tensors.

1.2.2 *Tensor algebra*

Addition and *subtraction* of tensors of the same type and the same order are defined in the following way: Let the tensors

$$B^{j_1\ldots j_s}_{i_1\ldots i_r} = \frac{\partial x^{\alpha_1}}{\partial x^{i_1}} \cdots \frac{\partial x^{\alpha_r}}{\partial y^{i_r}} \frac{\partial y^{j_1}}{\partial x^{\beta_1}} \cdots \frac{\partial y^{j_s}}{\partial x^{\beta_s}} A^{\beta_1\ldots\beta_s}_{\alpha_1\ldots\alpha_r}$$

and

$$\tilde{B}^{j_1\ldots j_s}_{i_1\ldots i_r} = \frac{\partial x^{\alpha_1}}{\partial y^{i_1}} \cdots \frac{\partial x^{\alpha_r}}{\partial y^{i_r}} \frac{\partial y^{j_1}}{\partial x^{\beta_1}} \cdots \frac{\partial y^{j_s}}{\partial x^{\beta_s}} \tilde{A}^{\beta_1\ldots\beta_s}_{\alpha_1\ldots\alpha_r}$$

be given.
Their sum and their difference are defined by

$$B^{j_1\ldots j_s}_{i_1\ldots i_r} \pm \tilde{B}^{j_1\ldots j_s}_{i_1\ldots i_1} = \frac{\partial x^{\alpha_1}}{\partial y^{i_1}} \cdots \frac{\partial x^{\alpha_r}}{\partial y^{i_r}} \frac{\partial y^{j_1}}{\partial x^{\beta_1}} \cdots \frac{\partial y^{j_s}}{\partial x^{\beta_s}} \left(A^{\beta_1\ldots\beta_s}_{\alpha_1\ldots\alpha_r} \pm \tilde{A}^{\beta_1\ldots\beta_s}_{\alpha_1\ldots\alpha_r}\right) \tag{1.2.25}$$

which represent a tensor of the same type and of the same order as those previously given. Addition and subtraction are commutative.

The representation has been given for mixed tensors, since they correspond to the most general of cases. Those relations valid in regard to addition and subtraction of covariant, contra-variant tensors, respectively, follow as special cases from (2.1.25), which the reader may easily read off for himself. Therefore, we shall deal with mixed tensors also in the following.

Let us change over to the *tensor product* or *outer product*. We have already discussed its special case in 1.1., i.e., the dyad. Let the tensors

$B^{j_1\ldots j_q}_{i_1\ldots i_p}$, covariant of order p, contra-variant of order q

and

$\tilde{B}^{l_1\ldots l_s}_{k_1\ldots k_r}$, covariant of order r, contra-variant of order s.

Its tensor product is defined as being

$$\mathfrak{B}^{j_1\ldots j_q l_1\ldots l_s}_{i_1\ldots i_p k_1\ldots k_r} = B^{j_1\ldots j_q}_{i_1\ldots i_p} \tilde{B}^{l_1\ldots l_s}_{k_1\ldots k_r}. \tag{1.2.26}$$

This is a tensor, which is covariant of order $p+r$, and contra-variant of order $q+s$. The product in indicial notation (which has been used in this context) is commutative; it is also associative and distributive.

Another operation of calculation is *contraction.* By this we mean the equating of two indices in a tensor. Thus, a summation has been forced, so that the tensor is 'contracted' by two orders. If, for example, the tensor

$$B^i_{jkl} = \frac{\partial y^i}{\partial x^\alpha}\frac{\partial x^\beta}{\partial y^j}\frac{\partial x^\gamma}{\partial y^k}\frac{\partial x^\delta}{\partial y^l}A^\alpha_{\beta\gamma\delta}$$

has been given, then the contraction may consist of setting down $k = i$. The result of this operation

$$B^i_{jil} = \frac{\partial y^i}{\partial x^\alpha}\frac{\partial x^\beta}{\partial y^j}\frac{\partial x^\gamma}{\partial y^i}\frac{\partial x^\delta}{\partial y^l}A^\alpha_{\beta\gamma\delta}$$

is

$$B^i_{jil} = \frac{\partial x^\beta}{\partial y^j}\frac{\partial x^\delta}{\partial y^l}\delta^\gamma_\alpha A^\alpha_{\beta\gamma\delta} = \frac{\partial x^\beta}{\partial y^j}\frac{\partial x^\delta}{\partial y^l}A^\alpha_{\beta\alpha\delta} \equiv \frac{\partial x^\beta}{\partial y^j}\frac{\partial x^\delta}{\partial y^l}\tilde{A}_{\beta\delta},$$

i.e.,

$$B^i_{jil} = \tilde{B}_{jl}. \tag{1.2.27}$$

Hence it can be stated:

If in a tensor, which is s-times covariant and r-times contra-variant, a covariant index is made equal to a contra-variant index, and the addition is carried out indicated by this 'repeated' index, then we obtain a tensor, which is $(s-1)$-times covariant and $(r-1)$-times contra-variant.

We are able to deduce that a corresponding occurrence will take place if two covariant or two contra-variant indices are equalled. The result would be a tensor, which would be lower covariant, contra-variant, by two orders, respectively. Worth noting is that for the deduction of (1.2.27), we have used δ^i_j. This is nothing more than the Kronecker-symbol in its original meaning, now written as a mixed tensor.

A generalization of contraction is the *contracting product.* This follows when the equating of two indices in two different factors of a tensorial product takes place. Let the tensors

$$B_{ij} = \frac{\partial x^\alpha}{\partial y^i}\frac{\partial x^\beta}{\partial y^j}A_{\alpha\beta}, \quad B^k = \frac{\partial y^k}{\partial x^\gamma}A^\gamma,$$

with which the tensor product

$$B_{ij} B^k = \frac{\partial x^\alpha}{\partial y^i} \frac{\partial x^\beta}{\partial y^j} \frac{\partial y^k}{\partial x^\gamma} A_{\alpha\beta} A^\gamma$$

is made up. The contraction may consist of putting $k = i$. This leads to

$$B_{ij} B^i = \frac{\partial x^\alpha}{\partial y^i} \frac{\partial x^\beta}{\partial y^j} \frac{\partial y^i}{\partial x^\gamma} A_{\alpha\beta} A^\gamma = \frac{\partial x^\beta}{\partial y^j} \delta^\alpha_\gamma A_{\alpha\beta} A^\gamma = \frac{\partial x^\beta}{\partial y^j} \tilde{A}_\beta$$

and to

$$B_{ij} B^i = \tilde{B}_j \tag{1.2.28}$$

respectively, i.e., to a tensor of the first order. The tensor product was of the third order, thus the order has been lowered by two orders.

1.2.3 *The quotient law*

This law is used to prove the tensor character of a given quantity. Let there be quantity $A(i, j, k)$ and any vector v^α. The only fact known of quantity $A(i, j, k)$ is that it is dependent from three indices; let us prove its tensor character. We also know that the product of $A(i, j, k)$ with any vector v^α yields a second order tensor, e.g., A^j_k. For $A^j_k = A(\alpha, j, k)v^\alpha$, the transformation law

$$B^j_k = B(\alpha, j, k)w^\alpha = \frac{\partial y^j}{\partial x^\beta} \frac{\partial x^\gamma}{\partial y^k} A^\beta_\gamma = \frac{\partial y^j}{\partial x^\beta} \frac{\partial x^\gamma}{\partial y^k} [A(\lambda, \beta, \gamma)v^\lambda] \tag{1.2.29}$$

holds true. Since v^α is a vector, a transformation law applies likewise,

$$v^\lambda = \frac{\partial x^\lambda}{\partial y^\alpha} w^\alpha \tag{1.2.30}$$

If this is inserted in the right-hand side of (1.2.29), then after leaving vector w^α outside the bracket, it yields

$$\left[B(\alpha, j, k) - \frac{\partial x^\lambda}{\partial y^\alpha} \frac{\partial x^\gamma}{\partial y^k} \frac{\partial y^j}{\partial x^\beta} A(\lambda, \beta, \gamma)\right] w^\alpha = 0. \tag{1.2.31}$$

Since v^α and hence w^α by definition are any vectors, then (1.2.31) is only valid if the square bracket in (1.2.31) is equal to nought. Thus we have obtained

$$B(\alpha, j, k) = \frac{\partial x^\lambda}{\partial y^\alpha} \frac{\partial x^\gamma}{\partial y^k} \frac{\partial y^j}{\partial x^\beta} A(\lambda, \beta, \gamma),$$

i.e., the transformation law for a third order tensor. Hence, the tensor character of $A(i, j, k)$ has been proven, since

$$A(i, j, k) = A^j_{ik};$$

thus the given quantity is a mixed tensor.

These detailed calculations may be shortened by the so-called *quotient law*. Let us carry out no real, but a 'fictitious division' (formally), by giving

$$A(\alpha, j, k)v^\alpha = A^j_k$$

as

$$A(\alpha, j, k) = \frac{A^j_k}{v^\alpha} = A^j_k \overset{*}{v}_\alpha$$

then $\overset{*}{v}_\alpha$ is supposed to be 'fictitious reciprocal' (formally) to v^α, thus being represented as

$$\frac{1}{v^\alpha} = \overset{*}{v}_\alpha .$$

Correspondingly, we write:

$$\frac{1}{v_\alpha} = \overset{*}{v}{}^\alpha$$

From $A^j_k \overset{*}{v}_\alpha$ is deduced (in the sense of the tensor product) on $A(\alpha, j, k) = A^j_k \overset{*}{v}_\alpha = A^j_{\alpha k}$, thus on the tensor character of $A(\alpha, j, k)$.

Similarly, we obtain for example the further following relations:

a) from $A(i, j, k, \alpha)v_\alpha = A^{ijk}$ we arrive at

$$A(i, j, k, \alpha) = \frac{A^{ijk}}{v_\alpha} = A^{ijk}\overset{*}{v}{}^\alpha = A^{ijk\alpha},$$

b) from $A(i, j, k, \alpha)v_\alpha = A^i_{jk}$ we can deduce

$$A(i, j, k, \alpha) = \frac{A^i_{jk}}{v_\alpha} = A^i_{jk}\overset{*}{v}{}^\alpha = A^{i\alpha}_{jk},$$

and

c) $A(i, \alpha)v_{\alpha j} = A_{ij}$ leads to

$$A(i, \alpha) = \frac{A_{ij}}{v_{\alpha j}} = A_{ij}\overset{*}{v}{}^{\alpha j} = A^{\alpha j}_{ij} = A^{\alpha}_{i\cdot}.$$

We have already mentioned that in all formulae, the division is only to be taken as formal. If, in particular, an invariant

$$J = A(\alpha_1, \ldots, \alpha_r)\xi_{\alpha_1} \ldots \xi_{\alpha_r}$$

is given, in which the ξ_{α_n} are vector components, then according to the quotient rule,

$$A(\alpha_1, \ldots, \alpha_r) = A^{\alpha_1 \ldots \alpha_r}$$

must be a contra-variant tensor of the r-th order. Correspondingly, it follows from

$$J = A(\alpha_1, \ldots, \alpha_r)\xi^{\alpha_1} \ldots \xi^{\alpha_r},$$

that

$$A(\alpha_1, \ldots, \alpha_r) = A_{\alpha_1 \ldots \alpha_r}$$

is a covariant tensor of the r-th order.

1.2.4 *Relative tensors*

Let the co-ordinate transformation $x^i = x^i(y)$ be prescribed. Accordingly, a general scalar is governed by the transformation law

$$g(y^1, \ldots, y^n) = f[x^1(y), \ldots, x^n(y)]. \tag{1.2.32}$$

We are also able to introduce

$$g(y) = f[x(y)] \cdot \left|\frac{\partial x^i}{\partial y^j}\right|^W \tag{1.2.33}$$

as the obvious modification of (1.2.32). Let us term those quantities which transform according to the preceding law (1.2.33) as *relative scalars of weight W*. Then W can be positive or negative.

Let us look at, for example, the volume integral

$$M = \int_V \rho(x^1, x^2, x^3)\mathrm{d}x^1\mathrm{d}x^2\mathrm{d}x^3, \tag{1.2.34}$$

which gives the mass M of a body of volume V if $\rho(x^1, x^2, x^3)$ is its density. When the co-ordinate transformation $x^i = x^i(y^1, y^2, y^3)$ is carried out, then the volume element transforms according to

$$\mathrm{d}x^1\mathrm{d}x^2\mathrm{d}x^3 = \left|\frac{\partial x^i}{\partial y^j}\right| \mathrm{d}y^1\mathrm{d}y^2\mathrm{d}y^3. \tag{1.2.35}$$

We set up

$$\mathrm{d}y^1\mathrm{d}y^2\mathrm{d}y^3 = \mathrm{d}x^1\mathrm{d}x^2\mathrm{d}x^3 \, . \left|\frac{\partial x^i}{\partial y^j}\right|^{-1}, \tag{1.2.36}$$

and comparison of (1.2.36) with (1.2.33) indicates that the *volume is a relative scalar of weight* $W = -1$. When (1.2.35) is inserted in (1.2.34) then it yields

$$\begin{aligned} M &= \int_V \rho(x^1, x^2, x^3)\mathrm{d}x^1\mathrm{d}x^2\mathrm{d}x^3 = \int_V \rho[x(y)] \left|\frac{\partial x^i}{\partial y^j}\right| \mathrm{d}y^1\mathrm{d}y^2\mathrm{d}y^3 \\ &= \int_V \tilde{\rho}(y^1, y^2, y^3)\mathrm{d}y^1\mathrm{d}y^2\mathrm{d}y^3, \end{aligned}$$

from which follows

$$\tilde{\rho}(y^1, y^2, y^3) = \rho(x^1, x^2, x^3) \cdot \left| \frac{\partial x^i}{\partial y^j} \right|^{+1}. \tag{1.2.37}$$

The comparison of (1.2.37) with (1.2.33) indicates that *the density is a relative scalar of weight* $W = +1$.

These facts can be generalized for tensors and

$$B^{j_1 \ldots j_s}_{i_1 \ldots i_r}(y) = \frac{\partial y^{j_1}}{\partial x^{\alpha_1}} \cdots \frac{\partial y^{j_s}}{\partial x^{\alpha_s}} \frac{\partial x^{\beta}}{\partial y^{i_1}} \cdots \frac{\partial x^{\beta_1}}{\partial y^{i_r}} \left| \frac{\partial x^i}{\partial y^j} \right|^{w} A^{\alpha_1 \ldots \alpha_s}_{\beta_1 \ldots \beta_r}(x) \tag{1.2.38}$$

can be interpreted as a *relative tensor of weight W*.

Equivalent to what is valid for density and volume, we refer to a relative tensor of weight $+1$ as *tensor density* and to a tensor of weight -1 as *tensor volume*. By the same token a relative tensor of weight nought is a true tensor. In order to differentiate it from a relative tensor we call it an *absolute tensor*. When we talk about tensors in the following, we always mean absolute tensors.

Several calculation rules are valid for relative tensors. They are so simple that they may be produced without actual proof:

a) The product of two relative tensors of weights W and V is a relative tensor of weight $W + V$.

b) The product of two relative tensors of weights $+W$ and $-W$ is an absolute tensor.

c) Each absolute tensor of k-th order can be represented as the product of a relative tensor of the same order of weight W, together with a relative scalar of weight $-W$.

d) The sum (difference) of relative tensors of the same type, same order and same weight leads to a relative tensor of the same structure and the same weight.

e) Contraction does not alter the weights in question in relative tensors.

1.2.5 *The metric tensor*

The components of the metric tensor are obtained by considering the line element in an Euclidean space. This space is *metric* having the *metric*

$$ds^2 = dx^i dx^i. \tag{1.2.39}$$

If a co-ordinate transformation $x^i = x^i(y^1, \ldots, y^n)$ is carried out, then (1.2.39) changes into (1.2.40),

$$ds^2 = \frac{\partial x^i}{\partial y^\alpha} \frac{\partial x^i}{\partial y^\beta} dy^\alpha dy^\beta. \tag{1.2.40}$$

The co-efficients

$$g_{\alpha\beta} = \frac{\partial x^i}{\partial y^\alpha} \frac{\partial x^i}{\partial y^\beta} \tag{1.2.41}$$

of the quadratic form (1.2.40) are called *metric co-efficients*, since they determine the metric of the space concerned, and they may be interpreted as the components of a tensor, i.e., the *metric tensor*. This tensor is symmetric. This fact can be read off (1.2.41).

Conversely, if the $g_{\alpha\beta}$ were prescribed, then through (1.2.41) we would obtain a set of $\frac{1}{2}n(n+1)$ partial differential equations for the unknown functions $x^i(y^1, \ldots, y^n)$, $i = 1, 2, \ldots, n$. If a solution exists for this system of differential equations, then the quadratic form $ds^2 = g_{\alpha\beta} dy^\alpha dy^\beta$ can be changed into $ds^2 = dx^i dx^i$, indicating that the space is not only metric, but even specially Euclidean. If, however, no solution exists, the space remains metric, but *is not Euclidean any more.*

Let us set up conditions for the following: Let the spaces in question be metric, so that $ds^2 = g_{ij} dx^i dx^j$ is the square of their line elements; but they are not necessarily Euclidean. Let, for the metric co-efficients g_{ij}, the following be valid, let them be differentiable and together with their differential quotients let them be continuous ($g_{ij} \in C^1$), and in each point of the space concerned let $\det(g_{ij}) \equiv |g_{ij}| \equiv g \neq 0$ be valid.

If a determinant

$$\det(a^i_j) = |a^i_j| = \begin{vmatrix} a^1_1 & \cdots & a^1_n \\ \vdots & a^i_j & \vdots \\ a^n_1 & \cdots & a^n_n \end{vmatrix} = a$$

be given, then the following is valid,

$$\left.\begin{aligned} a^i_j \cdot \mathrm{co}\,(a^k_j) &= a\delta^i_{k}, \\ a^i_j \,\mathrm{co}\,(a^i_k) &= a\delta^k_{j}. \end{aligned}\right\}. \tag{1.2.42}$$

$\mathrm{co}(a^i_k)$ is the *co-factor* of the element a^i_k. It is the minor of a^i_k multiplied by $(-1)^{i+k}$. The accuracy of the first formula (1.2.42) is proven by writing in full:

$$a^i_1 \,\mathrm{co}\,(a^k_1) + \ldots + a^i_n \,\mathrm{co}\,(a^k_n) = \begin{cases} a & \text{for } k = 1, \\ 0 & \text{for } k \neq i; \end{cases}$$

we have followed the expansion of the determinant by the i-th row for $k = i$, while for $k \neq i$, according to a known principle of the determinant theory, we have obtained nought.

The second formula (1.2.42) is written as

$$a^1_j \,\mathrm{co}\,(a^1_k) + \ldots a^n_j \,\mathrm{co}\,(a^n_k) = \begin{cases} a & \text{for } k = j \\ 0 & \text{for } k \neq j, \end{cases}$$

which indicates that the determinant has been expanded by the j-th column for $k = j$. This, of course, leads to a ,while for $k \neq j$, in accordance with a theorem of the determinant theory, nought is obtained.

Let us assign the contra-variant, symmetric metric tensor g^{ij} to the covariant symmetric metric tensor g_{ij}. They are known as the *fundamental tensors*. Basically, the g_{ij} and g^{ij} are the components of one and the same tensor, relative to different bases. But it is common to treat the g_{ij} and g^{ij} as being different tensors.

In order to obtain the g^{ij}, we set up the quantity

$$g(i, j) = \frac{\mathrm{co}\,(g_{ij})}{\det\,(g_{ij})} = \frac{\mathrm{co}\,(g_{ij})}{g}. \tag{1.2.43}$$

We shall show that the $g(i, j)$ defined in this way represent the components g^{ij} of a symmetric contra-variant tensor.

The symmetry becomes evident since it follows via (1.2.43) immediately from the symmetry of g_{ij} and g. The tensor character is proved by using the quotient law. Starting from identity

$$v_i = g_{\alpha i} v^\alpha, \tag{1.2.44}$$

both sides of (1.2.44) are multipled by $g(\beta, i) = \mathrm{co}(g_{\beta i})/g$. Thus we obtain

$$g(\beta, i)v_i = \frac{\mathrm{co}\,(g_{\beta i})}{g} g_{\alpha i} v^\alpha. \tag{1.2.45}$$

But because of (1.2.42), $\mathrm{co}(g_{\beta i})g_{\alpha i} = g\delta^\beta_\alpha$, so that (1.2.45) changes into

$$g(\beta, i)v_i = v^\beta.$$

In accordance with the quotient law this equation changes to

$$g(\beta, i) = \frac{v^\beta}{v_i} = v^\beta \overset{*}{v}{}^i = g^{\beta i}, \tag{1.2.46}$$

thus indicating that those quantities established by (1.2.43) are actually components of a contra-variant tensor.

If the previously used relation

$$\mathrm{co}\,(g_{\beta i})g_{\alpha i} = g\delta^\beta_\alpha \tag{1.2.47}$$

is divided by g on both sides, then

$$\frac{\mathrm{co}\,(g_{\beta i})}{g} g_{\alpha i} = \delta^\beta_\alpha \tag{1.2.48}$$

is obtained. Comparison of (1.2.46) with (1.2.43) indicates that

$$g^{ij} = \frac{\mathrm{co}\,(g_{ij})}{g} \tag{1.2.49}$$

is valid. Inserting (1.2.49) in (1.2.48), the essential formula

$$g^{\beta i} g_{\alpha i} = \delta^\beta_\alpha$$

is obtained, which leads to

$$g^{ij} g_{kj} = \delta^i_k = \begin{cases} 1 \text{ for } i = k, \\ 0 \text{ for } i \neq k \end{cases} \tag{1.2.50}$$

by renaming the indices.

Using the fundamental tensors, we are able to carry out an important calculation, which is known as the *raising and lowering of indices*. Let us set up the tensor product of a tensor with one of the fundamental tensors, thus obtaining the so-called *associated tensor*. We carry out a contraction in this tensor product. This operation causes the formal effect of raising and lowering the indices, e.g., if in

$$g^{\alpha i} A_{ijk} = A^{\alpha}_{\cdot jk}, \; g_{j\beta} A^{ij\alpha}_{lm} = A^{i\cdot\alpha}_{lm\beta}, \text{ respectively,}$$

the position of the indices of factors A_{ijk}, $A^{ij\alpha}_{lm}$, respectively, on the left-hand side is compared to the position of the indices of the associated tensors on the right-hand side, it becomes plain that in the first case the index i has been raised and changed to α and that in the second case the index j has been lowered and changed to β. The actual justification for this formal operation lies in contraction, and the simultaneous consideration of the summation convention. The tensors obtained by the raising and lowering of indices, which are tensors associated with a given tensor, may be taken as the representations of one and the same tensor relative to different reference systems.

Let us mention further, that in general, the original spot where a raised or lowered index used to be, should be indicated by a dot, i.e., see above. Since

$$g^{i\alpha} A_{j\alpha} = A^{i}_{j\cdot} \neq g^{i\alpha} A_{\alpha j} = A^{i}_{\cdot j}$$

is valid, if not particularly $A_{ji} = A_{ij}$, i.e., symmetry of the original tensor is given. If the dot were left out, then in both cases simply A^i_j would have been obtained and hence, the difference between $A^{i}_{j\cdot}$ and $A^{i}_{\cdot j}$ were not perceptible.

To give an example for the calculation of the metric tensor g_{ij}, let the transformation

$$x = r \sin \Theta \cos \varphi, \qquad y = r \sin \Theta \sin \varphi, \qquad z = r \cos \Theta \tag{1.2.51}$$

be used. This leads from the cartesian co-ordinates $x^1 = x$, $x^2 = y$, $x^3 = z$, to the spherical co-ordinates $y^1 = r$, $y^2 = \Theta$, $y^3 = \varphi$. Instead of (1.2.51) we can also write

$$\begin{aligned} x^1 &= y^1 \sin y^2 \cos y^3, \\ x^2 &= y^1 \sin y^2 \sin y^3, \\ x^3 &= y^1 \cos y^2, \end{aligned} \tag{1.2.52}$$

and calculate the coefficients g_{ij} by using (1.2.52) and (1.2.41). Thus we have

$$g_{\alpha\beta} = \frac{\partial x^1}{\partial y^\alpha}\frac{\partial x^1}{\partial y^\beta} + \frac{\partial x^2}{\partial y^\alpha}\frac{\partial x^2}{\partial y^\beta} + \frac{\partial x^3}{\partial y^\alpha}\frac{\partial x^3}{\partial y^\beta}, \tag{1.2.53}$$

indicating that by using (1.2.53) in (1.2.52), $g_{\alpha\beta} = 0$ always for $\alpha \neq \beta$, i.e., in our case the metric tensor g_{ij} is diagonal. It is always diagonal if, as in our example, we use orthogonal co-ordinate systems.

Finally we find for the diagonal components of the metric tensor

$$\begin{aligned} g_{11} &= \left(\frac{\partial x^1}{\partial y^1}\right)^2 + \left(\frac{\partial x^2}{\partial y^1}\right)^2 + \left(\frac{\partial x^3}{\partial y^1}\right)^2 = (\sin y^2 \cos y^3)^2 + (\sin y^2 \sin y^3)^2 + \\ &\quad + (\cos y^2)^2, \\ g_{11} &= (\sin y^2)^2 + (\cos y^2)^2 = 1, \\ g_{22} &= \left(\frac{\partial x^1}{\partial y^2}\right)^2 + \left(\frac{\partial x^2}{\partial y^2}\right)^2 + \left(\frac{\partial x^3}{\partial y^2}\right)^2 = (y^1 \cos y^3 \cos y^2)^2 + \\ &\quad + (y^1 \sin y^3 \cos y^2)^2 + (-y^1 \sin y^2)^2, \\ g_{22} &= (y^1 \cos y^2)^2 + (-y^1 \sin y^2)^2 = (y^1)^2 \equiv r^2, \\ g_{33} &= \left(\frac{\partial x^1}{\partial y^3}\right)^2 + \left(\frac{\partial x^2}{\partial y^3}\right)^2 + \left(\frac{\partial x^3}{\partial y^3}\right)^2 = (-y^1 \sin y^2 \sin y^3)^2 + \\ &\quad + (y^1 \sin y^2 \cos y^3)^2, \\ g_{33} &= (y^1 \sin y^2)^2 \equiv r^2 \sin^2 \Theta. \end{aligned}$$

The metric tensor itself is thus represented by the matrix

$$(g_{ij}) = \begin{pmatrix} g_{11} & 0 & 0 \\ 0 & g_{22} & 0 \\ 0 & 0 & g_{33} \end{pmatrix} = \begin{pmatrix} 1 & 0 & 0 \\ 0 & r^2 & 0 \\ 0 & 0 & r^2 \sin^2 \Theta \end{pmatrix}$$

1.2.6 *Christoffel symbols*

These symbols, also known as *Christoffel's triple indicial symbols* or *Chris-*

toffel brackets, were introduced into mathematics in the year 1869 in an essay printed in Crelles Journal, by E. B. Christoffel. There are two kinds of Christoffel symbols. The symbol of the *first kind* is defined by

$$[ij, k] = \frac{1}{2}\left(\frac{\partial g_{ik}}{\partial x^j} + \frac{\partial g_{jk}}{\partial x^i} - \frac{\partial g_{ij}}{\partial x^k}\right), \tag{1.2.54}$$

for which the following notation is also commonly used,

$$[ij, k] \equiv \begin{bmatrix} ij \\ k \end{bmatrix} \equiv \Gamma_{kij} \equiv \Gamma_{k,ij}.$$

The symbol of the *second kind* is derived from the symbol of the first kind through the relation

$$\begin{Bmatrix} k \\ ij \end{Bmatrix} = g^{k\alpha}[ij, \alpha]. \tag{1.2.55}$$

Other commonly used notations are also applicable to this case, for example,

$$\begin{Bmatrix} k \\ ij \end{Bmatrix} \equiv \begin{Bmatrix} ij \\ k \end{Bmatrix} \equiv \Gamma^k_{.ij} \equiv \Gamma^k_{ij}.$$

In order to demonstrate the occurrence of Christoffel symbols, let us recalculate Lagrange's equations of the second kind

$$\frac{\mathrm{d}}{\mathrm{d}t}\left(\frac{\partial T}{\partial \dot{q}_r}\right) - \frac{\partial T}{\partial q_r} = Q_r, \quad r = 1, 2, \ldots, n, \tag{1.2.56}$$

which describe the behaviour of a mechanical system, as they contain Christoffel symbols of the first kind. In the kinetic energy

$$T = \tfrac{1}{2} m_{ij}\dot{q}_i\dot{q}_j, \tag{1.2.57}$$

the coefficients m_{ij} are symmetric, i.e., $m_{ij} = m_{ji}$, and in addition, they are only functions of co-ordinates q_i, but not of time t. If, therefore, $\partial T/\partial \dot{q}_r = m_{sr}\dot{q}_s$ and (1.2.57) are inserted in (1.2.56), then

$$m_{sr}\ddot{q}_s+\frac{\partial m_{sr}}{\partial q_m}\dot{q}_m\dot{q}_s-\frac{1}{2}\frac{\partial m_{ij}}{\partial q_r}\dot{q}_i\dot{q}_j=Q_r,$$

and

$$m_{sr}\ddot{q}_s+\frac{1}{2}\left(\frac{\partial m_{sr}}{\partial q_m}+\frac{\partial m_{mr}}{\partial q_s}\right)\dot{q}_m\dot{q}_s-\frac{1}{2}\frac{\partial m_{ij}}{\partial q_r}\dot{q}_i\dot{q}_j=Q_r.$$

Those terms which are in the round bracket have their dummy indices renamed (s in i, m in j) and hence

$$m_{sr}\ddot{q}_s+\frac{1}{2}\left(\frac{\partial m_{ir}}{\partial q_j}+\frac{\partial m_{jr}}{\partial q_i}-\frac{\partial m_{ij}}{\partial q_r}\right)\dot{q}_i\dot{q}_j=Q_r$$

is obtained. Using (1.2.54) we write finally

$$m_{sr}\ddot{q}_s+[ij,r]\dot{q}_i\dot{q}_j=Q_r.$$

Thus (1.2.56) has been changed and the Christoffel symbols have been introduced.

In preparation of calculations to come, let us indicate the characteristics and calculation rules of the Christoffel symbols, as far as we shall make use of them later on.

a) From definitions (1.2.54) and (1.2.55) we read off

$$[ij,k]=[ji,k],\quad \begin{Bmatrix}k\\ij\end{Bmatrix}=\begin{Bmatrix}k\\ji\end{Bmatrix}, \tag{1.2.58}$$

i.e., the symmetry with respect to indices, i, j.

b) The change from a symbol of one kind to a symbol of the second kind is determined by the definition equation (1.2.55) on one hand, and on the other hand by:

$$[ij,\beta]=g_{k\beta}\begin{Bmatrix}k\\ij\end{Bmatrix}. \tag{1.2.59}$$

To prove (1.2.59), both sides of (1.2.55) are multiplied by $g_{k\beta}$. It yields

$$g_{k\beta}\begin{Bmatrix}k\\ij\end{Bmatrix} = g_{k\beta}\,g^{k\alpha}[ij, \alpha]. \tag{1.2.60}$$

Because of (1.2.50), we have $g_{k\beta}g^{k\alpha} = \delta^{\alpha}_{\beta}$, by which (1.2.60) changes into (1.2.59). Thus proof for the validity of (1.2.59) has been given.

From (1.2.55) and (1.2.59) we read off that the fundamental tensors influence the Christoffel symbols in such a way that the indices are either raised or lowered. In spite of this fact, we should not hasten to conclude that Christoffel symbols are tensors. In general they are not tensors which we shall see later on.

c) Because of the symmetry of g_{ij} and because of (1.2.59), the *identities*

$$\frac{\partial g_{ij}}{\partial x^k} = [ik, j]+[jk, i] = g_{\alpha j}\begin{Bmatrix}\alpha\\ik\end{Bmatrix}+g_{\alpha i}\begin{Bmatrix}\alpha\\jk\end{Bmatrix} \tag{1.2.61}$$

are valid which are easily verified by the reader. Another identity is obtained from (1.2.50), i.e., from $g_{i\alpha}g^{\alpha j} = \delta^j_i$. When this relation is differentiated, we obtain

$$\frac{\partial g_{i\alpha}}{\partial x^k}g^{\alpha j}+g_{i\alpha}\frac{\partial g^{\alpha j}}{\partial x^k} = 0,\ g_{i\alpha}\frac{\partial g^{\alpha j}}{\partial x^k} = -g^{\alpha j}\frac{\partial g_{i\alpha}}{\partial x^k},\ \text{respectively.} \tag{1.2.62}$$

Multiplying both sides of (1.2.62) with $g^{i\beta}$, we get

$$g^{i\beta}g_{i\alpha}\frac{\partial g^{\alpha j}}{\partial x^k} = -g^{i\beta}g^{\alpha j}\frac{\partial g_{i\alpha}}{\partial x^k},$$

and from this because of (1.2.50) and (1.2.61)

$$\frac{\partial g^{\beta j}}{\partial x^k} = -g^{i\beta}g^{\alpha j}([ik, \alpha]+[\alpha k, i]).$$

Using (1.2.55), this equation is changed into

$$\frac{\partial g^{\beta j}}{\partial x^k} = -g^{i\beta}\begin{Bmatrix}j\\ik\end{Bmatrix}-g^{\alpha j}\begin{Bmatrix}\beta\\\alpha k\end{Bmatrix},$$

and by change of indices β into i, i into α, into

$$\frac{\partial g^{ij}}{\partial x^k} = -g^{\alpha i}\begin{Bmatrix} j \\ \alpha k \end{Bmatrix} - g^{\alpha j}\begin{Bmatrix} i \\ \alpha k \end{Bmatrix}. \tag{1.2.63}$$

d) Let us deduce the *transformation formulae*, which govern the Christoffel symbols in a co-ordinate transformation. Let $g_{ij}(x)$ for the metric tensor in an x-system be given, and relative to a y-system let it have the components $h_{ij}(y)$. Let the co-ordinate transformation be carried out according to $y^i = y^i(x^1, \ldots, x^n)$, and the transformation of the metric coefficients is therefore determined by

$$h_{ij}(y) = \frac{\partial x^\alpha}{\partial y^i}\frac{\partial x^\beta}{\partial y^j} g_{\alpha\beta}(x). \tag{1.2.64}$$

The Christoffel symbol of the first kind is given based on definition (1.2.54) by

$$[ij, k]_y = \frac{1}{2}\left(\frac{\partial h_{ik}}{\partial y^j} + \frac{h_{jk}}{\partial y^i} - \frac{\partial h_{ij}}{\partial y^k}\right). \tag{1.2.65}$$

When (1.2.64) is inserted in (1.2.65), and we observe the symmetry of g_{ij}, then after lengthy calculation the sought transformation formula

$$[ij, k]_y = \frac{\partial x^\alpha}{\partial y^i}\frac{\partial x^\beta}{\partial y^j}\frac{\partial x^\gamma}{\partial y^k}[\alpha, \beta, \gamma]_x + \frac{\partial^2 x^\alpha}{\partial y^i \partial y^j}\frac{\partial x^\beta}{\partial y^k} g_{\alpha\beta} \tag{1.2.66}$$

is obtained. We deduce that the $[ij, k]$ do not in general transform like covariant tensors. It is prevented by the second term on the right-hand side of (1.2.66).

Quite similarly this occurs in the Christoffel symbol of the second kind. In

$$\begin{Bmatrix} k \\ ij \end{Bmatrix}_y = h^{k\mu}[ij, \mu]^y$$

we insert

$$h^{k\mu} = \frac{\partial y^k}{\partial x^\rho}\frac{\partial y^\mu}{\partial x^\sigma} g^{\rho\sigma}$$

and (1.2.66). After several changes we obtain the transformation formula

$$\begin{Bmatrix} k \\ ij \end{Bmatrix}_y = \frac{\partial y^k}{\partial x^\rho} \frac{\partial x^\alpha}{\partial y^i} \frac{\partial x^\beta}{\partial y^j} \begin{Bmatrix} \rho \\ \alpha\beta \end{Bmatrix}_x + \frac{\partial^2 x^\alpha}{\partial y^i \partial y^j} \frac{\partial y^k}{\partial x^\alpha}, \tag{1.2.67}$$

from which we read off that the symbol of the second kind does *not* transform as a mixed, double covariant, single contra-variant tensor, because again the second term on the right-hand side of (1.2.67) interferes. Only if the co-ordinate transformation is linear, so that $x^\alpha = c^\alpha_j y^j$, $c^\alpha_j = \text{const.}$ and hence

$$\frac{\partial^2 x^\alpha}{\partial y^i \partial y^j} = 0$$

is valid, then the interfering terms in (1.2.66) and (1.2.67) disappear. Then the Christoffel symbols transform in the same way as tensors would.

e) Let us deduce the so-called *Christoffel formulae*, which are valid for the second derivatives of the co-ordinates. We obtained the first formula by multiplying both sides of (1.2.67) by $\partial x^m/\partial y^k$. Because of

$$\frac{\partial x^m}{\partial y^k} \frac{\partial y^k}{\partial x^\alpha} = \delta^m_\alpha,$$

and by simple rearrangement, we obtain

$$\frac{\partial^2 x^m}{\partial y^i \partial y^j} = \frac{\partial x^m}{\partial y^k} \begin{Bmatrix} k \\ ij \end{Bmatrix}_y - \frac{\partial x^\alpha}{\partial y^i} \frac{\partial x^\beta}{\partial y^j} \begin{Bmatrix} m \\ \alpha\beta \end{Bmatrix}_x \tag{1.2.68}$$

which represents the sought formula. By exchange of x with y and vice-versa, (1.2.68) becomes the second formula

$$\frac{\partial^2 y^m}{\partial x^i \partial x^j} = \frac{\partial y^m}{\partial x^k} \begin{Bmatrix} k \\ ij \end{Bmatrix}_x - \frac{\partial y^\alpha}{\partial x^i} \frac{\partial y^\beta}{\partial x^j} \begin{Bmatrix} m \\ \alpha\beta \end{Bmatrix}_y. \tag{1.2.69}$$

f) Finally, let us deal with a formula important for section 1.2.10, which is obtained by differentiation of $g = \det(g_{ij})$. It is $g = g_{(i)\alpha} \cdot \text{co}(g_{i\alpha})$ because of (1.2.48). Following differentiation

$$\frac{\partial g}{\partial g_{ij}} = \frac{\partial[g_{(i)\alpha} \cdot \text{co}(g_{i\alpha})]}{\partial g_{ij}} = \text{co}(g_{i\alpha}) \frac{\partial g_{(i)\alpha}}{\partial g_{ij}},$$

(do not add with respect to i),

and because of

$$\partial g_{(i)\alpha}/\partial g_{ij} = \delta^j_\alpha,$$

also

$$\frac{\partial g}{\partial g_{ij}} = \text{co}(g_{ij}).$$

Therefore,

$$\frac{\partial g}{\partial x^i} = \frac{\partial g}{\partial g_{\alpha\beta}} \frac{\partial g_{\alpha\beta}}{\partial x^i} = \text{co}(g_{\alpha\beta}) \frac{\partial g_{\alpha\beta}}{\partial x^i},$$

and because of (1.2.49),

$$\frac{\partial g}{\partial x^i} = g g^{\alpha\beta} \frac{\partial g_{\alpha\beta}}{\partial x^i}.$$

Using (1.2.61) and (1.2.50), then we obtain

$$\frac{\partial g}{\partial x^i} = g g^{\alpha\beta} \left[g_{\gamma\beta} \begin{Bmatrix} \gamma \\ \alpha i \end{Bmatrix} + g_{\gamma\alpha} \begin{Bmatrix} \gamma \\ \beta i \end{Bmatrix} \right] = g \left[\begin{Bmatrix} \alpha \\ \alpha i \end{Bmatrix} + \begin{Bmatrix} \beta \\ \beta i \end{Bmatrix} \right],$$

which, after renaming indices, changes to

$$\frac{\partial g}{\partial x^i} = 2g \begin{Bmatrix} \alpha \\ \alpha i \end{Bmatrix}, \quad \frac{1}{2g} \frac{\partial g}{\partial x^i} = \begin{Bmatrix} \alpha \\ \alpha i \end{Bmatrix}, \text{ respectively.}$$

It can be rewritten as

$$\frac{\partial}{\partial x^i}(ln\sqrt{g}) = \frac{1}{\sqrt{g}} \frac{\partial \sqrt{g}}{\partial x^i} = \begin{Bmatrix} \alpha \\ \alpha i \end{Bmatrix}.$$

Hence,

$$\frac{\partial\sqrt{g}}{\partial x^i} = \sqrt{g}\begin{Bmatrix}\alpha\\ \alpha i\end{Bmatrix}. \tag{1.2.70}$$

1.2.7 *Covariant differentiation*

Let us deduce it for the simplest of cases, i.e., for vectors, and deal with the covariant vector

$$B_i(y) = \frac{\partial x^\alpha}{\partial y^i} A_\alpha(x). \tag{1.2.71}$$

Partially differentiated it is

$$\frac{\partial B_i}{\partial y^j} = \frac{\partial x^\alpha}{\partial y^i}\frac{\partial x^\beta}{\partial y^j}\frac{\mathrm{d}A_\alpha}{\partial x^\beta} + \frac{\partial^2 x^\alpha}{\partial y^j \partial y^i} A_\alpha. \tag{1.2.72}$$

We read off from (1.2.72) that differential quotient $\partial B_i/\partial y^j$ is not a tensor, since it is prevented by the second term on the right-hand side of (1.2.72) that this formula represents the transformation law for a covariant tensor of second order. For this reason, and by using the first Christoffel formula (1.2.68), we change (1.2.72). It yields

$$\frac{\partial B_i}{\partial y^j} = \frac{\partial x^\alpha}{\partial y^i}\frac{\partial x^\beta}{\partial y^j}\frac{\partial A_\alpha}{\partial x^\beta} + \frac{\partial x^\alpha}{\partial y^k}\begin{Bmatrix}k\\ ij\end{Bmatrix}_y A_\alpha - \frac{\partial x^k}{\partial y^i}\frac{\partial x^\beta}{\partial y^j}\begin{Bmatrix}\alpha\\ k\beta\end{Bmatrix}_x A_\alpha. \tag{1.2.73}$$

Because of (1.2.71),

$$\frac{\partial x^\alpha}{\partial y^k} A_\alpha = B_k,$$

so that (1.2.73) is thus changed into

$$\frac{\partial B_i}{\partial y^j} - \begin{Bmatrix}k\\ ij\end{Bmatrix}_y B_k = \frac{\partial x^\alpha}{\partial y^i}\frac{\mathrm{d}x^\beta}{\partial y^j}\left(\frac{\partial A_\alpha}{\partial x^\beta} - \begin{Bmatrix}k\\ \alpha\beta\end{Bmatrix}_x A_k\right). \tag{1.2.74}$$

This equation represents the transformation law for a second order, covariant tensor, i.e., for the tensor

$$B_{i|j} \equiv \frac{\partial B_i}{\partial y^j} - \begin{Bmatrix} k \\ ij \end{Bmatrix}_y B_k \tag{1.2.75}$$

which we call the *covariant derivative of the covariant vector* B_i. Correspondingly, the covariant derivative of the contra-variant vector

$$B^i(y) = \frac{\partial y^i}{\partial x^\alpha} A^\alpha(x) \tag{1.2.76}$$

can be set up. By partial differentiation we obtain from (1.2.76)

$$\frac{\partial B^i}{\partial y^j} = \frac{\partial y^i}{\partial x^\alpha} \frac{\partial x^\beta}{\partial y^j} \frac{\partial A^\alpha}{\partial x^\beta} + \frac{\partial^2 y^i}{\partial x^\alpha \partial x^\beta} \frac{\partial x^\beta}{\partial y^j} A^\alpha. \tag{1.2.77}$$

This relation is changed by using the second Christoffel formula (1.2.69) into

$$\frac{\partial B^i}{\partial y^j} = \frac{\partial y^i}{\partial x^\alpha} \frac{\partial x^\beta}{\partial y^j} \frac{\partial A^\alpha}{\partial x^\beta} + \frac{\partial y^i}{\partial x^k} \frac{\partial x^\beta}{\partial y^j} \begin{Bmatrix} k \\ \alpha\beta \end{Bmatrix} A^\alpha - \frac{\partial y^\gamma}{\partial x^\alpha} \frac{\partial y^\varepsilon}{\partial x^\beta} \frac{\partial x^\beta}{\partial y^j} \begin{Bmatrix} i \\ \gamma\varepsilon \end{Bmatrix}_y A^\alpha. \tag{1.2.78}$$

The last term on the right-hand side of (1.2.78), because of

$$\frac{\partial y^\varepsilon}{\partial x^\beta} \frac{\partial x^\beta}{\partial y^j} = \delta_j^\varepsilon$$

and (1.2.76), is further changed into

$$\frac{\partial y^\gamma}{\partial x^\alpha} \frac{\partial y^\varepsilon}{\partial x^\beta} \frac{\partial x^\beta}{\partial y^j} \begin{Bmatrix} i \\ \gamma\varepsilon \end{Bmatrix}_y A^\alpha = \frac{\partial y^\gamma}{\partial x^\alpha} \begin{Bmatrix} i \\ \gamma j \end{Bmatrix}_y A^\alpha = \begin{Bmatrix} i \\ \gamma j \end{Bmatrix}_y B^\gamma. \tag{1.2.79}$$

Inserting (1.2.79) into (1.2.78), and following rearrangement, we obtain

$$\frac{\partial B^i}{\partial y^j} + \begin{Bmatrix} i \\ \gamma j \end{Bmatrix}_y B^\gamma = \frac{\partial x^\beta}{\partial y^j} \frac{\partial y^i}{\partial x^\alpha} \frac{\partial A^\alpha}{\partial x^\beta} + \frac{\partial y^i}{\partial x^k} \frac{\partial x^\beta}{\partial y^j} \begin{Bmatrix} k \\ \alpha\beta \end{Bmatrix}_x A^\alpha. \tag{1.2.80}$$

For the second term on the right-hand side of (1.2.80), the dummy indices are renamed, e.g., k into α and α into γ so that finally (1.2.80) is changed into

$$\frac{\partial B^i}{\partial y^j} + \begin{Bmatrix} i \\ \gamma j \end{Bmatrix}_y B^\gamma = \frac{\partial x^\beta}{\partial y^j} \frac{\partial y^i}{\partial x^\alpha} \left(\frac{\partial A^\alpha}{\partial x^\beta} + \begin{Bmatrix} \alpha \\ \gamma\beta \end{Bmatrix}_x A^\gamma \right). \tag{1.2.81}$$

This represents the transformation formula for the simple covariant and simple contra-variant tensor

$$B^i|_j \equiv \frac{\partial B^i}{\partial y^j} + \begin{Bmatrix} i \\ \gamma j \end{Bmatrix}_y B^\gamma, \tag{1.2.82}$$

which we term the *covariant derivative of the contra-variant vector* B^i.

The formulae set up for the covariant derivative of vectors is further generalized, so that the covariant derivative for tensors of any order can be indicated. Let us write down the *covariant derivative for a mixed tensor of higher order*, then as a special case, the derivative for co- or contra-variant tensors of higher order is obtained. By

$$\begin{aligned} A^{j_1\ldots j_s}_{i_1\ldots i_r}|_l = {} & \frac{\partial A^{j_1\ldots j_s}_{i_1\ldots i_r}}{\partial x^l} - \begin{Bmatrix} \alpha \\ i_1\, l \end{Bmatrix} A^{j_1\ldots j_s}_{\alpha\ldots i_r} - \begin{Bmatrix} \alpha \\ i_2\, l \end{Bmatrix} A^{j_1\ldots j_s}_{i_1\alpha\ldots i_r} \\ & - \begin{Bmatrix} \alpha \\ i_r\, l \end{Bmatrix} A^{j_1\ldots j_s}_{i_1\ldots\alpha} + \begin{Bmatrix} j_1 \\ \alpha l \end{Bmatrix} A^{\alpha\ldots j_s}_{i_1\ldots i_r} + \begin{Bmatrix} j_2 \\ \alpha l \end{Bmatrix} A^{j_1\alpha\ldots j_s}_{i_1\ldots i_r} + \ldots \\ & + \begin{Bmatrix} j_s \\ \alpha l \end{Bmatrix} A^{j_1\ldots\alpha}_{i_1\ldots i_r}, \end{aligned} \tag{1.2.83}$$

the covariant derivative for a mixed tensor, covariant of order r and contra-variant of order s, is determined.

Let the covariant derivative of sums and products be given without proof, by the following formula:

$$\begin{aligned} &(A_i^j \pm B_i^j)|_l = A_i^j|_l \pm B_i^j|_l, \\ &(A_i^j B_r^k)|_l = A_i^j|_l B_r^k + A_i^j B_r^k|_l, \\ &(A^{j_1j_2}B_{i_1i_2})|_l = A^{j_1j_2}|_l B_{i_1i_2} + A^{j_1j_2}B_{i_1i_2}|_l. \end{aligned} \tag{1.2.84}$$

We are able to read off all the essential facts from these formulae, which are easily generalized.

For a *relative scalar* of weight W

$$g(y) = f(x)\left|\frac{\partial x^i}{\partial y^j}\right|^W,$$

is the covariant derivative

$$f|_l = \frac{\partial f}{\partial x^l} - Wf\begin{Bmatrix} \alpha \\ l\alpha \end{Bmatrix},$$

which indicates that for an *absolute scalar* of weight $W = 0$,

$$f|_l \equiv \frac{\partial f}{\partial x^l} \tag{1.2.85}$$

is valid. In this case, the covariant and common partial derivative are identical.

Finally, let us discuss the covariant derivative of unit vectors and fundamental tensors. Then, according to (1.2.83)

$$\delta^i_j|_l = \frac{\partial \delta^i_j}{\partial x^l} - \begin{Bmatrix} \alpha \\ jl \end{Bmatrix} \delta^i_\alpha + \begin{Bmatrix} i \\ \alpha l \end{Bmatrix} \delta^\alpha_j .$$

Because of $\partial\delta^i_j/\partial x^i \equiv 0$, and the special characteristics of the Kronecker-symbols, it changes to

$$\delta^i_j|_l = -\begin{Bmatrix} i \\ jl \end{Bmatrix} + \begin{Bmatrix} i \\ jl \end{Bmatrix} \equiv 0. \tag{1.2.86}$$

Thus, for δ^i_j, the covariant derivative, as well as the common partial derivative, is equal to nought.

The same is valid for g_{ij}. Following (1.2.61) is

$$\frac{\partial g_{ij}}{\partial x^k} = g_{\alpha j} \begin{Bmatrix} \alpha \\ ik \end{Bmatrix} + g_{\alpha i} \begin{Bmatrix} \alpha \\ jk \end{Bmatrix},$$

and following (1.2.83) is

$$g_{ij}|_k = \frac{\partial g_{ij}}{\partial x^k} - \begin{Bmatrix} \alpha \\ ik \end{Bmatrix} g_{\alpha j} - \begin{Bmatrix} \alpha \\ jk \end{Bmatrix} g_{i\alpha} . \tag{1.2.87}$$

Inserting (1.2.61) in (1.2.87), and considering $g_{i\alpha} = g_{\alpha i}$, then we obtain

$$g_{ij}|_k = 0. \tag{1.2.88}$$

Finally, from $g^{i\alpha} g_{\alpha j} = \delta^i_j$, and because of (1.2.84) and (1.2.86), we obtain

$$(g^{i\alpha} g_{\alpha j})|_l = g^{i\alpha}|_l g_{\alpha j} + g^{i\alpha} g_{\alpha j}|_l = \delta^i_j|_l \equiv 0. \tag{1.2.89}$$

Through (1.2.88) we arrive from (1.2.89) to

$$g^{i\alpha}|_l g_{\alpha j} = 0.$$

This represents a homogeneous, algebraic system of equations, whose coefficient determinant, by definition, differs from nought, since det $(g_{\alpha j}) \neq 0$ is valid. Thus, the system can only assume the trivial solution

$$g^{i\alpha}|_l \equiv 0. \tag{1.2.90}$$

Using (1.2.88) and (1.2.90), it has been shown that the covariant derivative of the fundamental tensors is equal to nought.

1.2.8 *Derivation of the base vectors. The different components of a tensor*

By way of transformation $x^i = x^i(y^1, y^2, y^3)$, $i = 1, 2, 3$, we change over from the coordinate system x^i to system y^i (Figure 4), introducing the *base vectors*

$$\boldsymbol{b}_i = \frac{\partial \boldsymbol{r}}{\partial y^i} \tag{1.2.91}$$

for this system. Then, $r = (x^1, x^2, x^3)$ represents the radius vector assumed relative to the x^i-system. Let us set up formally

$$\mathrm{d}\boldsymbol{r} = \frac{\partial \boldsymbol{r}}{\partial y^i}\,\mathrm{d}y^i, \tag{1.2.92}$$

which may be rewritten as $\mathrm{d}\boldsymbol{r} = \mathrm{d}y^i \boldsymbol{b}_i$, after using (1.2.91). This indicates that vector $\mathrm{d}\boldsymbol{r}$ has the components $\mathrm{d}y^i$ relative to the base $\boldsymbol{b}_i$. Using (1.2.92), we calculate

$$\mathrm{d}s^2 = \mathrm{d}\boldsymbol{r}^2 = \frac{\partial \boldsymbol{r}}{\partial y^i}\frac{\partial \boldsymbol{r}}{\partial y^j}\,\mathrm{d}y^i \mathrm{d}y^j,$$

which because of (1.2.40) and (1.2.41) is

$$\mathrm{d}s^2 = g_{ij}\,\mathrm{d}y^i \mathrm{d}y^j,$$

and because of (1.2.91) yields

$$ds^2 = \boldsymbol{b}_i \boldsymbol{b}_j \mathrm{d}y^i \mathrm{d}y^j.$$

Through comparison, the important relation

$$g_{ij} = \frac{\partial \boldsymbol{r}}{\partial y^i} \frac{\partial \boldsymbol{r}}{\partial y^j} = \boldsymbol{b}_i \boldsymbol{b}_j \tag{1.2.93}$$

is obtained.

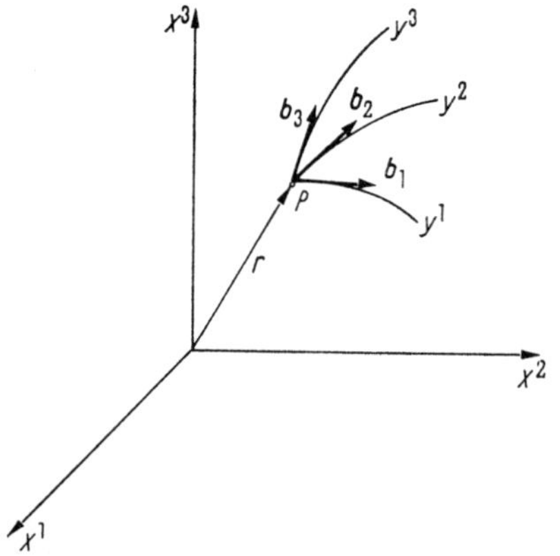

Figure 4

In the same way, vector $\mathrm{d}\boldsymbol{r}$ has been represented relative to base $\boldsymbol{b}_i$, any vector $\boldsymbol{A}$ can be represented by

$$\boldsymbol{A} = A^i \boldsymbol{b}_i. \tag{1.2.94}$$

Since the base vectors $\boldsymbol{b}_i$ shall be used for the *contra-variant representation of vectors*, (1.2.94) is the 'component representation' of $\boldsymbol{A}$.

Starting from base $\boldsymbol{b}_i$, let us set up the *reciprocal system of base vectors*

$$\boldsymbol{b}^1 = \frac{\boldsymbol{b}_2 \times \boldsymbol{b}_3}{\boldsymbol{b}_1 \boldsymbol{b}_2 \boldsymbol{b}_3}, \quad \boldsymbol{b}^2 = \frac{\boldsymbol{b}_3 \times \boldsymbol{b}_1}{\boldsymbol{b}_1 \boldsymbol{b}_2 \boldsymbol{b}_3}, \quad \boldsymbol{b}^3 = \frac{\boldsymbol{b}_1 \times \boldsymbol{b}_2}{\boldsymbol{b}_1 \boldsymbol{b}_2 \boldsymbol{b}_3} \tag{1.2.95}$$

with which

$$\mathrm{d}\boldsymbol{r} = \mathrm{d}y_i \boldsymbol{b}^i.$$

In this formula we have used the covariant components $\mathrm{d}y_i$ of $\mathrm{d}\boldsymbol{r}$. It follows

$$\mathrm{d}s^2 = \boldsymbol{b}^i\boldsymbol{b}^j\mathrm{d}y_i\mathrm{d}y_j,$$

and because of

$$\mathrm{d}s^2 = g^{ij}\mathrm{d}y_i\mathrm{d}y_j$$

is thus

$$g^{ij} = \boldsymbol{b}^i\boldsymbol{b}^j. \tag{1.2.96}$$

For any vector $\boldsymbol{A}$ the component notation

$$\boldsymbol{A} = A_i\boldsymbol{b}^i \tag{1.2.97}$$

is valid, if the base vectors $\boldsymbol{b}^i$ are used. The base $\boldsymbol{b}^i$ serves for the *covariant representation of vectors.* Since the formula (1.2.94) and (1.2.97) may be equated, we realize that both possibilities of a shortened indicial notation, e.g., A_i for the 'covariant vectors' and $\boldsymbol{A}^i$ for the 'contra-variant vector', do actually represent one and the same vector, $\boldsymbol{A}$. The covariant components A_i and the contra-variant components A^i are components of one vector, but taken relative to different bases.

What has been said about vectors may be generalized for tensors of a higher order.

If equating of (1.2.94) and (1.2.97) is actually carried out, it yields

$$A^i\boldsymbol{b}_i = A_i\boldsymbol{b}^i. \tag{1.2.98}$$

Let us multiply both sides of (1.2.98) by $\boldsymbol{b}_j$ considering (1.2.93); thus we obtain:

$$\boldsymbol{b}_i\boldsymbol{b}_jA^i = g_{ij}A^i = \boldsymbol{b}^i\boldsymbol{b}_jA_i. \tag{1.2.99}$$

From 1.2.5, which explained the lowering of indices, we learn that

$$g_{ij}A^i = A_j. \tag{1.2.100}$$

Hence, the agreement of (1.2.100) and (1.2.99) is obtained only if

$$\boldsymbol{b}^i \boldsymbol{b}_j = \delta^i_j. \tag{1.2.101}$$

Thus, we have obtained another remarkable formula in the use of base vectors.

Let us differentiate (1.2.94). We obtain

$$\frac{\partial A}{\partial y^j} = \boldsymbol{b}_i \frac{\partial A^i}{\partial y^j} + A^i \frac{\partial \boldsymbol{b}_i}{\partial y^j}. \tag{1.2.102}$$

To recalculate further, let us discuss the significance of $\partial \boldsymbol{b}_i / \partial y^j$. Starting from (1.2.93), we find

$$\frac{\partial g_{ij}}{\partial y^k} = \frac{\partial \boldsymbol{b}_i}{\partial y^k} \boldsymbol{b}_j + \frac{\partial \boldsymbol{b}_j}{\partial y^k} \boldsymbol{b}_i,$$

and through permutation of indices

$$\frac{\partial g_{ik}}{\partial y^j} = \frac{\partial \boldsymbol{b}_i}{\partial y^j} \boldsymbol{b}_k + \frac{\partial \boldsymbol{b}_k}{\partial y^j} \boldsymbol{b}_i,$$

$$\frac{\partial g_{jk}}{\partial y^i} = \frac{\partial \boldsymbol{b}_j}{\partial y^i} \boldsymbol{b}_k + \frac{\partial \boldsymbol{b}_k}{\partial y^i} \boldsymbol{b}_j.$$

Inserting this into (1.2.54), after rewriting x into y, it shows that when interchanging the order of differentiation, i.e., using

$$\frac{\partial \boldsymbol{b}_i}{\partial y^j} = \frac{\partial}{\partial y^j}\left(\frac{\partial \boldsymbol{r}}{\partial y^i}\right) = \frac{\partial}{\partial y^i}\left(\frac{\partial \boldsymbol{r}}{\partial y^j}\right) = \frac{\partial \boldsymbol{b}_j}{\partial y^i}, \text{ etc.,}$$

the relation

$$[ij, k] = \frac{\partial \boldsymbol{b}_i}{\partial y^j} \boldsymbol{b}_k \tag{1.2.103}$$

can be set up.

Let us multiply (1.2.103) on both sides by $\boldsymbol{b}^k$ considering (1.2.101), thus arriving at

$$\frac{\partial \boldsymbol{b}_i}{\partial y^j} = [ij, k]\boldsymbol{b}^k. \tag{1.2.104}$$

At this point, both sides are multiplied by $\boldsymbol{b}^\alpha$ which, because of (1.2.96), yields

$$\frac{\partial \boldsymbol{b}_i}{\partial y^j}\boldsymbol{b}^\alpha = g^{\alpha k}[ij, k],$$

and because of (1.2.55)

$$\frac{\partial \boldsymbol{b}_i}{\partial y^j}\boldsymbol{b}^\alpha = \begin{Bmatrix} \alpha \\ i\,j \end{Bmatrix}.$$

This relation is finally multiplied by $\boldsymbol{b}_\alpha$ which because of (1.2.101) yields

$$\frac{\partial \boldsymbol{b}_i}{\partial y^j} = \begin{Bmatrix} \alpha \\ i\,j \end{Bmatrix} \boldsymbol{b}_\alpha. \tag{1.2.105}$$

The formulae (1.2.104) and (1.2.105) serve to change (1.2.102). Using (1.2.105), we obtain from (1.2.102)

$$\frac{\partial \boldsymbol{A}}{\partial y^j} = \frac{\partial A^i}{\partial y^j}\boldsymbol{b}_i + \begin{Bmatrix} \alpha \\ i\,j \end{Bmatrix} A^i \boldsymbol{b}_\alpha,$$

moreover, by renaming dummy indices and because of (1.2.82),

$$\frac{\partial \boldsymbol{A}}{\partial y^j} = \left[\frac{\partial A^\alpha}{\partial y^j} + \begin{Bmatrix} \alpha \\ i\,j \end{Bmatrix} A^i\right] \boldsymbol{b}_\alpha = A^\alpha|_j\, \boldsymbol{b}_\alpha. \tag{1.2.106}$$

Similarly

$$\frac{\partial \boldsymbol{A}}{\partial y^j} = A_\alpha|_j\, \boldsymbol{b}^\alpha \tag{1.2.107}$$

is calculated; we notice from (1.2.106) and (1.2.107), that the covariant derivations $A^\alpha|_j$, i.e., $A_\alpha|_j$, are the components of tensor $\partial \boldsymbol{A}/\partial y^j$ relative to bases $\boldsymbol{b}_\alpha$ and $\boldsymbol{b}^\alpha$. Thus, we have obtained a clear explanation for the covariant derivative of vectors.

Let us set up the relation existing between base vectors of different reference systems. Let k_i be the base vectors in the x^i-system and $\boldsymbol{b}_i$ the base vectors in the y_i-system. Both,

$$\mathrm{d}\boldsymbol{r} = \frac{\partial \boldsymbol{r}}{\partial x^i}\,\mathrm{d}x^i = \boldsymbol{k}_i\,\mathrm{d}x^i,$$

and

$$\mathrm{d}\boldsymbol{r} = \frac{\partial \boldsymbol{r}}{\partial y^i}\,\mathrm{d}y^i = \boldsymbol{b}_i\,\mathrm{d}y^i,$$

are valid for vector $\mathrm{d}\boldsymbol{r}$.

But then,

$$\boldsymbol{k}_i\,\mathrm{d}x^i = \boldsymbol{b}_i\,\mathrm{d}y^i = \boldsymbol{b}_i\,\frac{\partial y^i}{\partial x^\alpha}\,\mathrm{d}x^\alpha$$

must be valid. After renaming the dummy indices on the left-hand side, the equation

$$\boldsymbol{k}_\alpha = \frac{\partial y^i}{\partial x^\alpha}\,\boldsymbol{b}_i \tag{1.2.108}$$

is obtained. Multiplying this equation on both sides by $\boldsymbol{b}^i\boldsymbol{k}^\alpha$, then because of (1.2.101), we obtain

$$\boldsymbol{b}^i = \frac{\partial y^i}{\partial x^\alpha}\,\boldsymbol{k}^\alpha,$$

which yields

$$\boldsymbol{k}^i = \frac{\partial x^i}{\partial y^j}\,\boldsymbol{b}^j, \tag{1.2.109}$$

following the renaming of indices and solving for $\boldsymbol{k}^\alpha$.

Let us relate the base vectors $\boldsymbol{b}_i$ and $\boldsymbol{b}^i$ which apply to the curvilinear coordinates y^i, to unit vectors $\boldsymbol{e}_i$ and $\boldsymbol{e}^i$. These have the same line of application as the $\boldsymbol{b}_i$ and $\boldsymbol{b}^i$. Let us take, for example, from (1.2.93)

$$\boldsymbol{b}_i \boldsymbol{b}_{(i)} = g_{i(i)}, \; |\boldsymbol{b}_i| = \sqrt{g_{i(i)}}, \text{ respectively.}$$

Based on this relation we are permitted to write

$$\boldsymbol{b}_i = \sqrt{g_{(ii)}}\, \boldsymbol{e}_i. \tag{1.2.110}$$

To abbreviate, let us introduce

$$h_i = \sqrt{g_{i(i)}}, \tag{1.2.111}$$

which yields

$$\boldsymbol{b}_i = h_{(i)} \boldsymbol{e}_i. \tag{1.2.112}$$

Similarly, from (1.2.96) follows

$$\boldsymbol{b}^i \boldsymbol{b}^{(i)} = g^{(i)i}, \; |\boldsymbol{b}^i| = \sqrt{g^{i(i)}}, \text{ respectively,}$$

which because of (1.2.50), i.e.

$$g^{i(i)} = \frac{1}{g_{i(i)}},$$

and because of (1.2.111), also yields

$$|\boldsymbol{b}^i| = \frac{1}{\sqrt{g_{i(i)}}} = \frac{1}{h_i}$$

for orthogonal coordinates. Thus

$$\boldsymbol{b}^i = \sqrt{g^{(ii)}}\boldsymbol{e}^i = \frac{\boldsymbol{e}^i}{\sqrt{g_{(ii)}}} = \frac{\boldsymbol{e}^i}{h_{(i)}} \tag{1.2.113}$$

is true. Inserting (1.2.110) in (1.2.94), we obtain

$$\boldsymbol{A} = A^i \sqrt{g_{(ii)}}\, \boldsymbol{e}_i.$$

We call the components

$$A^{*i} = A^i \sqrt{g_{(ii)}} \tag{1.2.114}$$

of vector $\boldsymbol{A}$, which were taken relative to system $\boldsymbol{e}_i$, the *physical contra-variant components* of this vector. Because of (1.2.112) we may write

$$A^{*i} = h_{(i)} A^i \tag{1.2.115}$$

for (1.2.114). If on the other hand, we insert (1.2.113) in (1.2.97) it follows that

$$\boldsymbol{A} = A_i \sqrt{g^{(ii)}} \boldsymbol{e}^i = \frac{A_i}{\sqrt{g_{(ii)}}} \boldsymbol{e}^i = \frac{A_i}{h_{(i)}} \boldsymbol{e}^i, \tag{1.2.116}$$

from which the quantities

$$A_i^* = A_i \sqrt{g^{(ii)}} = \frac{A_i}{\sqrt{g_{(ii)}}} = \frac{A_i}{h_{(i)}} \tag{1.2.117}$$

are obtained. These are the *physical covariant components* of vector $\boldsymbol{A}$.

The definition as just given for physical components of vectors can be generalized for tensors if these are represented in component notation, e.g., let tensor $\boldsymbol{T}$ be obtained from the dyad of two vectors $\boldsymbol{u} = u^i \boldsymbol{b}_i$ and $\boldsymbol{v} = v^j \boldsymbol{b}_j$, i.e., from

$$\boldsymbol{T} = u^i v^j \boldsymbol{b}_i \boldsymbol{b}_j$$

so that

$$\boldsymbol{T} = t^{ij} \boldsymbol{b}_i \boldsymbol{b}_j \tag{1.2.118}$$

be valid with $t^{ij} = u^i v^j$. This example of component-notation of a second-order tensor is easily generalized to tensors of a higher order, for example, to $T = t^{ijk} \boldsymbol{b}_i \boldsymbol{b}_j \boldsymbol{b}_k$ for a contra-variant third order tensor or to $\boldsymbol{T} = t^{ij\cdot\cdot\cdot}_{\cdot\cdot\cdot rs} \boldsymbol{b}_i \boldsymbol{b}_j \ldots \boldsymbol{b}^r \boldsymbol{b}^s$ for a mixed tensor of any orders.

Using as an example (1.2.118), let us change over to unit-vectors and because of (1.2.110) let us find

$$\boldsymbol{T} = t^{ij} \sqrt{g_{(ii)}} \sqrt{g_{(jj)}} \, \boldsymbol{e}_i \boldsymbol{e}_j. \tag{1.2.119}$$

Let us term

$$t^{*ij} = t^{ij}\sqrt{g_{(ii)}g_{(jj)}} = t^{ij}h_{(i)}h_{(j)} \tag{1.2.120}$$

the physical contra-variant components of a second order tensor $\boldsymbol{T}$. In the same way, we obtain more of the physical mixed and covariant components, i.e.,

$$t_j^{*i} = t_j^{\,i}\frac{h_{(i)}}{h_{(j)}},\; t_i^{*j} = t_i^{\,j}\frac{h_{(j)}}{h_{(i)}},\; t_{ij}^{*} = \frac{t_{ij}}{h_{(i)}h_{(j)}} \tag{1.2.121}$$

for the second order tensors by referring to unit-vectors.

1.2.9 *The generalized* $\boldsymbol{\delta}$*- and* $\boldsymbol{\varepsilon}$*-tensors*

Both these tensors had been introduced in their simplest form in 1.1, as being δ_{ij} and ε_{ijk}. So far, the *mixed Kronecker-tensor* δ_j^i has been used in 1.2, but without any detailed explanation. Let us show, therefore, that for the $\boldsymbol{\delta}$- as well as for the $\boldsymbol{\varepsilon}$-tensor, more general definitions than those given in 1.1, are possible.

By adding to the definition in 1.1.5, let us say that for a system of quantities which represents the components of the generalized $\boldsymbol{\varepsilon}$-tensor, the following conditions apply:

$$\varepsilon_{i_1\ldots i_n}(\text{or } \varepsilon^{i_1\ldots i_n}) = \begin{cases} +1, \text{ if } i_1,\ldots, i_n \text{ is an even permutation of the numbers } 1, 2, \ldots, n, \\ -1, \text{ if } i_1,\ldots, i_n \text{ is an uneven permutation of the numbers } 1, 2, \ldots, n, \\ 0 \text{ for all other cases.} \end{cases}$$

In the same way, should the quantities $\delta_{j_1\ldots j_k}^{i_1\ldots i_k}$ be the components of the generalized $\boldsymbol{\delta}$-tensor (Kronecker-tensor); the following applies to these quantities:

$$\delta^{i_1 \ldots i_k}_{j_1 \ldots j_k} = \begin{cases} \text{completely antisymmetric relative to the upper and lower indices,} \\ +1, \text{ if the upper indices are different from each other and if the lower indices represent an even permutation of them,} \\ -1, \text{ if the upper indices are different from each other and the lower indices represent an uneven permutation of them,} \\ 0 \text{ in all other cases.} \end{cases}$$

To illustrate this, let us look at δ^{ij}_{lk}. This quantity is equal to nought, if $i = j$, $k = l$ or if i, j are different from k, l. The quantity equals to plus one or to minus one, according to whether k, l are an even or uneven permutation of i, j:

$$\begin{aligned} \delta^{11}_{kl} &= \delta^{22}_{kl} = \delta^{12}_{13} = \ldots = 0, \\ \delta^{12}_{12} &= \delta^{13}_{13} = \delta^{21}_{21} = \ldots = +1, \\ \delta^{12}_{21} &= \delta^{13}_{31} = \delta^{21}_{12} = \ldots = -1. \end{aligned}$$

It is easily shown that

$$\varepsilon^{\alpha\beta\gamma}\varepsilon_{ijk} = \delta^{\alpha\beta\gamma}_{ijk} \tag{1.2.122}$$

is valid. It is followed, at simultaneous generalization, that

$$\begin{aligned} \varepsilon^{i_1 \ldots i_n} &= \delta^{i_1 \ldots i_n}_{1 \ldots n}, \\ \varepsilon_{i_1 \ldots i_n} &= \delta^{1 \ldots n}_{i_1 \ldots i_n}, \end{aligned}$$

thus establishing the relation between generalized **ε**- and **δ**-tensor.

Let

$$\det \left(\frac{\partial y^i}{\partial x^j}\right) \equiv \left|\frac{\partial y}{\partial x}\right| = J, \; J^{-1} = \left|\frac{\partial x}{\partial y}\right| \tag{1.2.123}$$

be valid for the co-ordinate transformation $y^i = y^i(x^1 \ldots x^n)$. From the definition of the ε-tensor, the relations

$$\varepsilon^{\alpha\beta \ldots \gamma} \det \left(\frac{\partial y^i}{\partial x^j}\right) = \varepsilon^{ij \ldots k} \frac{\partial y^\alpha}{\partial x^i} \frac{\partial y^\beta}{\partial x^j} \cdots \frac{\partial y^\gamma}{\partial x^k},$$

$$\varepsilon_{ij\ldots k}\det\left(\frac{\partial y^i}{\partial x^j}\right)=\varepsilon_{\alpha\beta\ldots\gamma}\frac{\partial y^\alpha}{\partial x^i}\frac{\partial y^\beta}{\partial x^j}\cdots\frac{\partial y^\gamma}{\partial x^k}$$

are obtained, which because of relation (1.2.123) can be rewritten as

$$\varepsilon^{\alpha\beta\ldots\gamma}=\frac{\partial y^\alpha}{\partial x^i}\cdots\frac{\partial y^\gamma}{\partial x^k}\left|\frac{\partial x}{\partial y}\right|^{+1}\varepsilon^{ij\ldots k},$$

$$\varepsilon_{ij\ldots k}=\frac{\partial x^\alpha}{\partial y^i}\cdots\frac{\partial x^\gamma}{\partial y^k}\left|\frac{\partial x}{\partial y}\right|^{-1}\varepsilon_{\alpha\beta\ldots\gamma}.$$

Comparison with (1.2.38) indicates that we are dealing with a relative tensor of weight $+1$ in $\varepsilon^{\alpha\beta\ldots\gamma}$ and a tensor of weight -1 in $\varepsilon_{ij\ldots k}$. From (1.2.122) and its generalization follows that the generalized Kronecker-tensor has, as the product of two relative tensors of weight -1 and $+1$, the weight 0, this being an absolute tensor.

1.2.10 *Gauss' integral theorem*

This theorem is of fundamental significance in the realm of continuum mechanics. We shall, therefore, prove this theorem, enabling us to use it later in our calculation. We shall limit its use to orthogonal cartesian coordinates and set out to prove the theorem only as far as these special coordinates are concerned.

The theorem states:

$$\int_O A_i n_i \,\mathrm{d}O=\int_V A_{i,i}\,\mathrm{d}V, \tag{1.2.124}$$

when 0 is the surface of a three-dimensional domain of volume V, n_j is the outer unit-normal of 0 and A_i is a vector of class C^1 in V. Besides $A_{i,j}$ is an abbreviation for $\partial A_i/\partial x_j$.

Let us assume that $A_2=A_3=O$. Then (1.1.124) changes to

$$\int_O A_3 n_3\,\mathrm{d}O=\int_V\frac{\partial A_3}{\partial x_3}\,\mathrm{d}V. \tag{1.2.125}$$

Let us assume that area O is *regular*, i.e., that it is penetrated in two points P_1, P_2 by the line vertical to the x_1, x_2-plane.

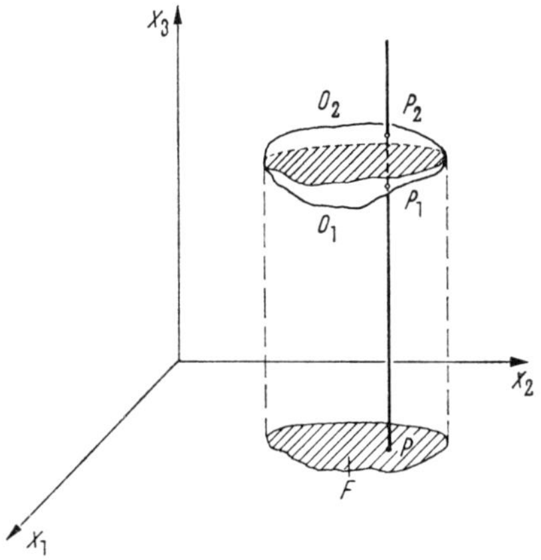

Figure 5

Through projection of O on plane x_1, x_2, we obtain area F, when point P in F represents the image of the penetration points P_1, P_2. By varying P in F, the corresponding points P_1, P_2 describe the areas O_1, O_2 on O, when

$$x_3 = f_1(x_1, x_2) \text{ for } P_1 \text{ on } O_1,$$
$$x_3 = f_2(x_1, x_2) \text{ for } P_2 \text{ on } O_2$$

is valid for the respective x_3-coordinate. Thus we write

$$\int_V \frac{\partial A_3}{\partial x_3} \mathrm{d}V = \int_F \left[\int_{f_1}^{f_2} \frac{\partial A_3}{\partial x_3} \mathrm{d}x_3\right] \mathrm{d}x_2 \mathrm{d}x_1 = \int_F [A_3(x_1, x_2, f_2) - A_3(x_1, x_2, f_1)] \mathrm{d}x_2 \mathrm{d}x_1 . \tag{1.2.126}$$

In addition,

$$\int_O A_3 n_3 \mathrm{d}O = \int_{O_1} A_3 n_3 \mathrm{d}O + \int_{O_2} A_3 n_3 \mathrm{d}O$$

is valid.

It is known that

$$\mathrm{d}O = \sqrt{1+(f_{1,1})^2+(f_{1,2})^2} \mathrm{d}x_1 \mathrm{d}x_2 ,$$

$$n_3 = \pm \frac{1}{\sqrt{1+(f_{1,1})^2+(f_{1,2})^2}} \begin{cases} \text{for } O_2, \\ \text{for } O_1. \end{cases}$$

From this is

$$n_3 \, \mathrm{d}O = \pm \mathrm{d}x_1 \, \mathrm{d}x_2 \begin{cases} \text{for } O_2 \\ \text{for } O_1, \end{cases}$$

which yields

$$\int_O A_3 n_3 \, \mathrm{d}O = \int_{O_2} A_3 \, \mathrm{d}x_1 \, \mathrm{d}x_2 - \int_{O_1} A_3 \, \mathrm{d}x_1 \, \mathrm{d}x_2 = \int_F [A_3(x_1, x_2, f_2) - A_3(x_1, x_2, f_1)] \, \mathrm{d}x_1 \, \mathrm{d}x_2. \tag{1.2.127}$$

The comparison of (1.2.127) and (1.1.126) indicates that (1.2.125) is correct. The proof of (1.2.125) is extended for the case of area O being only 'piecewise smooth' and being arbitrarily located, since the area is then at least 'piecewise regular'. From this we deduce that (1.1.125) applies generally.

Since the coordinate axes are completely equivalent to each other in this context, then (1.2.125) is also true, if index 3 is substituted by 1 or 2. Thus, the following is obtained:

$$\int_O A_i n_{(i)} \, \mathrm{d}O = \int_V A_{(i),i} \, \mathrm{d}V \quad i = 1, 2, 3,$$

and, by addition of these three formulae, (1.1.124). In this way, Gauss' integral theorem has been proven.

(1.2.124) is notated symbolically as

$$\int_O \boldsymbol{A} \boldsymbol{n} \, \mathrm{d}O = \int_V \operatorname{div} \boldsymbol{A} \, \mathrm{d}V. \tag{1.1.128}$$

This is useful since it enables a transition from (1.2.128) to the representation of the theorem in terms of more general coordinate systems. According to (1.1.9)

$$\nabla = \boldsymbol{k}^i \frac{\partial}{\partial x^i} \tag{1.2.129}$$

represents the Nabla-vector for cartesian coordinates x^i in component notation. Changing over to curvilinear coordinates $y^i = y^i(x^1, x^2, x^3)$, then,

$$\frac{\partial}{\partial x^i} = \frac{\partial y^j}{\partial x^i}\frac{\partial}{\partial y^j} \tag{1.2.130}$$

is valid. In addition, (1.2.109) is at our disposal. Therefore, (1.2.129) changes to

$$\nabla = \frac{\partial x^i}{\partial y^k}\frac{\partial y^j}{\partial x^i}\boldsymbol{b}^k\frac{\partial}{\partial y^j} = \delta^j_k\boldsymbol{b}^k\frac{\partial}{\partial y^j} = \boldsymbol{b}^j\frac{\partial}{\partial y^j}\,. \tag{1.2.131}$$

Using (1.2.131) on vector $\boldsymbol{A} = A^i\boldsymbol{b}_i$, then we obtain

$$\operatorname{div}\boldsymbol{A} = \nabla\boldsymbol{A} = \left(\boldsymbol{b}^j\frac{\partial}{\partial y^j}\right)(A^i\boldsymbol{b}_i),$$

which is

$$\operatorname{div}\boldsymbol{A} = \boldsymbol{b}^j\left(\frac{\partial A^i}{\partial y^j}\boldsymbol{b}_i + A^i\frac{\partial \boldsymbol{b}_i}{\partial y^j}\right). \tag{1.2.132}$$

By way of (1.2.105) we arrive from (1.2.132) at

$$\operatorname{div}\boldsymbol{A} = \frac{\partial A^i}{\partial y^j}\boldsymbol{b}^j\boldsymbol{b}_i + A^i\begin{Bmatrix}\alpha\\ i\,j\end{Bmatrix}\boldsymbol{b}^j\boldsymbol{b}_\alpha.$$

By renaming the indices in the second term on the right-hand side yields

$$\operatorname{div}\boldsymbol{A} = \left(\frac{\partial A^i}{\partial y^j} + \begin{Bmatrix}i\\ kj\end{Bmatrix}A^k\right)\boldsymbol{b}^j\boldsymbol{b}_i = A^i|_j\,\boldsymbol{b}^j\boldsymbol{b}_i\,,$$

which because of (1.2.101) is converted into

$$\operatorname{div}\boldsymbol{A} = A^i|_i\,. \tag{1.2.133}$$

It is

$$\mathrm{d}V = \left|\frac{\partial x^i}{\partial y^j}\right|\mathrm{d}y^1\mathrm{d}y^2\mathrm{d}y^3 \tag{1.2.134}$$

for the volume element $\mathrm{d}V$ in curvilinear coordinates. But it is

$$\left|\frac{\partial x^i}{\partial y^j}\right| \cdot \left|\frac{\partial x^i}{\partial y^j}\right| = \left|\frac{\partial x^\alpha}{\partial y^i}\frac{\partial x^\alpha}{\partial y^j}\right| = |g_{ij}| = g,$$

so that

$$\left|\frac{\partial x^i}{\partial y^j}\right| = \sqrt{g} \tag{1.2.135}$$

must be true. Inserting (1.2.135) in (1.2.134) yields

$$\mathrm{d}V = \sqrt{g}\,\mathrm{d}y^1\mathrm{d}y^2\mathrm{d}y^3. \tag{1.2.136}$$

Using (1.2.133) and (1.2.136), let us write the volume integral $\int \operatorname{div} \boldsymbol{A}\mathrm{d}V$ for any curvilinear coordinates. It is as follows:

$$\int_V \operatorname{div} \boldsymbol{A}\,\mathrm{d}V = \int_V A^i|_i\sqrt{g}\,\mathrm{d}y^1\mathrm{d}y^2\mathrm{d}y^3. \tag{1.2.137}$$

This equation can be further changed since

$$A^i|_i = \frac{\partial A^i}{\partial y^i} + \left\{\frac{i}{ik}\right\} A^k,$$

and therefore,

$$A^i|_i\sqrt{g} = \frac{\partial A^i}{\partial y^i}\sqrt{g} + \sqrt{g}\left\{\frac{i}{ik}\right\} A^k.$$

Because of (1.2.70) we can rewrite it into

$$A^i|_i\sqrt{g} = \frac{\partial A^i}{\partial y^i}\sqrt{g} + \frac{\partial\sqrt{g}}{\partial y^k} A^k,$$

and, after renaming the dummy indices, into

$$A^i|_i\sqrt{g} = \frac{\partial(A^i\sqrt{g})}{\partial y^i}. \tag{1.2.138}$$

Therefore, (1.2.137) changes to

$$\int_V \operatorname{div} \boldsymbol{A} \, \mathrm{d}V = \int_V \frac{\partial(A^i\sqrt{g})}{\partial y^i} \mathrm{d}y^1 \mathrm{d}y^2 \mathrm{d}y^3. \tag{1.2.139}$$

Let us assume $\boldsymbol{A} = A^i \boldsymbol{b}_i$ and $\boldsymbol{n} = n_j \boldsymbol{b}^j$ for the formation of

$$\boldsymbol{A}\boldsymbol{n} = A^i \boldsymbol{b}_i n_j \boldsymbol{b}^j$$

which because of (1.2.101) yields

$$\boldsymbol{A}\boldsymbol{n} = A^i n_i . \tag{1.2.140}$$

We are now able to notate for the area integral in (1.2.128)

$$\int_O \boldsymbol{A}\boldsymbol{n} \, \mathrm{d}O = \int_O A^i n_i \, \mathrm{d}O, \tag{1.2.140}$$

which from (1.2.128) through (1.2.139) and (1.2.140) yields Gauss' integral theorem for curvilinear coordinates expressed as:

$$\int_O A^i n_i \, \mathrm{d}O = \int_V \frac{\partial(A^i\sqrt{g})}{\partial y^i} \mathrm{d}y^1 \mathrm{d}y^2 \mathrm{d}y^3. \tag{1.2.141}$$

2

Physical fundamentals

2.1 Analysis of strain

2.1.1 *The deformable continuum*

In the study of deformable bodies, i.e., of bodies which are deformed following the effect of forces, several approaches may be used. Considering the experimentally proven fact of the atomic structure of matter, emphasizing the *microscopic approach* seems to lend itself best. It sets out to interpret the material as discontinuous and to consider its composition of discreet elements of atomic or molecular dimension. Thus a thorough, physically well-founded view at the phenomena occurring in stressed, deformed bodies, is obtained. Unfortunately, the evidence becomes so involved that any useful statement made in this way is concerned only with simple basic facts and is of very restricted value. In addition, the theory obtained from the microscopic approach is to this day, based on several uncertain assumptions. For the engineer, therefore, this method is yet too unfounded and unsafe.

We shall, therefore, use the *macroscopic approach*, which presupposes the existence of a *continuum*. It is assumed that the space taken up by a body is continuously filled with matter, and that all significant quantities pertaining to the deformation of the strained body are continuous field functions. Presupposing in addition that the matter is composed of identical volume elements, then it is a homogeneous continuum. Using the macroscopic approach, a *phenomenological theory* is set up. This theory describes the behaviour of the continuum following a certain stress state basing it on results of experiments. It uses relations which contain certain material constants. It is for the *atomic, microscopic theory* to explain in profound physical terms relations and constants such as these, taking the true structure of matter into account. This has been done to a certain extent, e.g., within the limit of the solid state body theory, metal physics theory and the dislocation theory [3, 4, 5].

Let the body keep a certain position within the space before being strained. Then a characterizing coordinate triple is assigned to the volume elements, which set up the body following the continuum-theory, and which we shall assume to be mass points. Owing to load, the volume elements will alter their position within the space, undergoing *displacements*, and corresponding to their new position, new coordinate triples are assigned to them. The difference between the old and the new coordinates give rise to the components of the particular *displacement vectors* of the mass points. The changes in the spatial arrangement of the body, following displacement of the volume elements, are divided into a 'rigid-body' movement (translation and rotation of the original body as a whole) and into *strain*. We shall only discuss the strain, leaving aside the rigid body movement.

In the case of strain, the body reacts in several different ways. It may be elastic or plastic, or it may flow. The whole range of reaction in homogeneous, isotropic bodies is described by *rheology* [6]. Let us keep to *elastic bodies*. These are such that strains caused by forces of finite magnitude do themselves remain finite (i.e., there is no flow) and disappear totally after removal of the load (i.e., no plastic, permanent strain remains).

It is the *theory of elasticity* which engages in the problem of elastic continua.

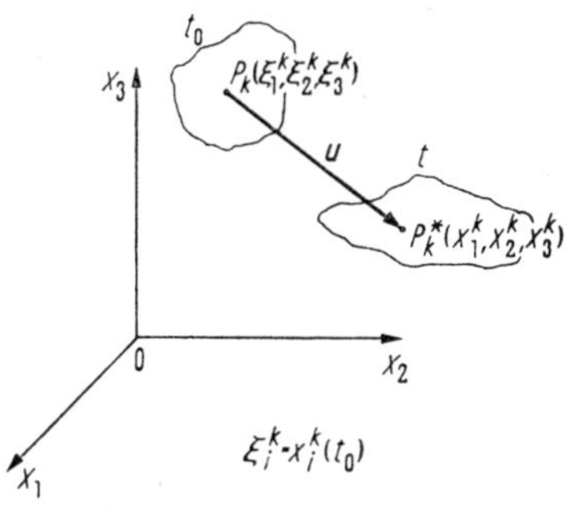

Figure 6

2.1.2 *Lagrange's and Euler's coordinates*

In an unloaded condition, i.e., at time t_0 let the elastic body be represented by point set $\{P_k\}$ and in a loaded condition, i.e., at time t, by point set $\{P_k^*\}$. The transition from point P_k to point P_k^* is indicated by the displacement vector $\boldsymbol{u}$ [Figure 6]. It is possible to choose between two ways of representing this deformation process. In the first case, the observer is closely connected with the k-th volume element of the body, describing all procedures from this point of view. This is the *material* or *substantial* or *Lagrange's approach.* The k-th volume element in question, which coincides at time t_0 with point P_k of the space, is given as an identification mark the coordinate triple $\xi_i^{(k)} = x_i^{(k)}(t_0)$, $i = 1, 2, 3$, which had been associated with point P_k. The ξ_i are called *Lagrange's coordinates.* During deformation, the k-th element incorporates other space points P_k^*, little by little. The related coordinate triples x_i^k, $i = 1, 2, 3$, are described from the substantial viewpoint by

$$x_i = x_i(\xi_1, \xi_2, \xi_3, t), \; i = 1, 2, 3. \tag{2.1.1}$$

The Lagrange approach thus uses Lagrange's coordinates as independent variables. A characteristic f, related to the matter of the body, is determined by

$$f = f(\xi_1, \xi_2, \xi_3, t). \tag{2.1.2}$$

Its derivative with respect to time is

$$\frac{\mathrm{d}f}{\mathrm{d}t} = \frac{\partial f}{\partial t/_{\mathscr{L}}}. \tag{2.1.3}$$

The index $/_{\mathscr{L}}$ on the right-hand side of (2.1.3) should indicate Lagrange's point of view during differentiation. For the displacement vector,

$$u_i = x_i(\xi_1, \xi_2, \xi_3 t) - \xi_i, \; i = 1, 2, 3, \tag{2.1.4}$$

is true, so that

$$\frac{\partial u_i}{\partial \xi_j} = \frac{\partial x_i}{\partial \xi_j} - \delta_{ij}, \; i, j = 1, 2, 3, \tag{2.1.5}$$

is valid.

In the second case, the observer is closely connected with point P_k of the space, which has the coordinates $x_i^{(k)}$, $i = 1, 2, 3$. The x_i are termed *Euler's coordinates*. The observer perceives the deformation of the body from point P_k as well as the change of the characteristics and conditions connected with the body. The volume elements, passing the observer's point in time, and which have been marked by Lagrange coordinates, are described by

$$\xi_i = \xi_i(x_1, x_2, x_3, t), \; i = 1, 2, 3. \tag{2.1.6}$$

In this kind of method, which is termed local or Euler method, the Euler coordinates are used as independent variables. Characteristics or conditions are indicated by

$$f = f(x_1, x_2, x_3, t), \tag{2.1.7}$$

and the derivative with respect to time is

$$\frac{\mathrm{d}f}{\mathrm{d}t} = \frac{\partial f}{\partial x_i}\dot{x}_i + \frac{\partial f}{\partial t}, \tag{2.1.8}$$

where $\mathrm{d}f/\mathrm{d}t$ is the *substantial*, $\partial f/\partial t$ is the *local* and $\dot{x}_i\,\partial f/\partial x_i$ is the *convective* derivative. Let us write for the displacement vector

$$u_i = x_i - \xi_i(x_1, x_2, x_3, t), \; i = 1, 2, 3, \tag{2.1.9}$$

and therefore

$$\frac{\partial u_i}{\partial x_j} = \delta_{ij} - \frac{\partial \xi_i}{\partial x_j}, \; i, j = 1, 2, 3 \tag{2.1.10}$$

is valid.

For the sake of generalization, we have assumed so far, that the deformation of a body is a procedure depending on time. This is the case, for example, when the body flows. Instead, let us look at elastic deformations only, and let us not consider time dependence. We shall discuss the configuration of a body given by $\{P_k\}$ *before* the loading and the configuration given by $\{P_k^*\}$ *after* loading. In the transition of $\{P_k\}$ to $\{P_k^*\}$ time $t - t_0$ will naturally pass, but this is of no importance to our discussion, which compares the conditions $\{P_k\}$ and $\{P_k^*\}$. We shall, therefore, abandon an inclusion of time in our mathematical formula, unless dynamic procedures in the continuum will be discussed.

Therefore, instead of (2.1.1) and (2.1.4)

$$x_i = x_i(\xi_1, \xi_2, \xi_3), \; i = 1, 2, 3, \tag{2.1.1*}$$

and

$$u_i = x_i(\xi_1, \xi_2, \xi_3) - \xi_i, \; i = 1, 2, 3, \tag{2.1.4*}$$

may be used, as well as instead of (2.1.6) and (2.1.9)

$$\xi_i = \xi_i(x_1, x_2, x_3), \; i = 1, 2, 3, \tag{2.1.6*}$$

and

$$u_i = x_i - \xi_i(x_1, x_2, x_3), \; i = 1, 2, 3. \tag{2.1.9*}$$

At the time t_0 the body is not deformed. Its volume elements, taken to be mass points, have been marked by the coordinates ξ_i. The Lagrange specification method, which uses the ξ_i as independent coordinates, is thus a specification containing the coordinates of the non-deformed body.

At time t, as a result of load, the body is deformed. Following displacements, its volume elements have arrived at different space points, which have coordinates x_i. Identifying the volume element with the momentarily occupied space-point, the coordinates x_i are those of the deformed body. The Euler specification method, which uses the x_i as independent coordinates, is thus a specification using the coordinates of the deformed body. Presupposing small deformations, both specification methods will differ so little from each other that they can be called identical, which we shall see later. Since in the theory of elasticity for the most part, small displacements and strain are assumed, then we need not think about the kind of specification and the choice of coordinates. We may choose any way, because, in this case, there is no difference between the two kinds of specification.

If dynamic problems of the theory of elasticity are discussed, then the Lagrange specification method is best used, since Newton's laws of motion must be written for the particular volume element, marked by Lagranges' coordinates. In stability problems, containing finite deformations, and in which the states of the body should be differentiated as non-deformed and deformed, the two specification methods must be kept apart, and one, preferably Lagrange's, should be chosen. We shall discuss this in 2.4.

2.1.3 *Affine mapping*

It has been mentioned in the previous paragraph that the deformation process can be interpreted as the change in Euclidean space of one set of point masses $\{P_k\}$ to another set of point masses $\{P_k^*\}$. In mathematical notation it is given by the mapping (2.1.1*). The mapping functions, x_i, should be continuous functions, since it is assumed, if not otherwise specified, that following deformation, all volume elements of the body will jointlessly fit together again. In addition, there should be a one-one and onto relation between ξ_i and x_i, making it possible to solve in (2.1.1*) for ξ_i, thus obtaining (2.1.6*). The simplest relation possible for (2.1.1*) is given by:

$$x_i = a_{i0}+(\delta_{ij}+a_{ij})\xi_j, \ i,j = 1, 2, 3, \tag{2.1.11}$$

through the constant coefficients a_{i0} and a_{ij}. This represents an *affine mapping*. Based on the assumptions valid for mapping (2.1.1*), of which (2.1.11) is a special case, then

$$\det(\delta_{ij}+a_{ij}) \neq 0$$

must be valid, i.e., the coefficient matrix of (2.1.11) is non-singular. Using suitable constant coefficients b_{i0} and b_{ij},

$$\xi_i = b_{i0}+(\delta_{ij}+b_{ij})x_j, \ i,j = 1, 2, 3, \tag{2.1.12}$$

is obtained from (2.1.11). The affine mapping has several known characteristics, which will be discussed in short.

The plane

$$A_0+A_j\xi_j = 0, \ j = 1, 2, 3, \tag{2.1.13}$$

is changed into a plane. Inserting (2.1.13) in (2.1.12), we obtain

$$A_0+A_jb_{j0}+A_j(\delta_{ji}+b_{ji})x_i = 0,$$

which, using the new expressions $A_0+A_jb_{j0} = A_0^*$, $A_j(\delta_{ji}+b_{ji}) = A_i^*$, changes into

$$A_0^* + A_i^* x_i = 0, \; i = 1, 2, 3,$$

representing indeed the equation of a plane.

A straight line is changed into a straight line by using affine mapping. This becomes obvious by imagining the original straight line as an intersection of two planes and by considering the representation characteristics set up for planes. Taking a vector to be a directed portion of a straight line, it is apparent that the affine mapping changes a vector into a vector. Let vector

$$K_i = \xi_i^{(2)}-\xi^{(1)}, \; i = 1, 2, 3, \tag{2.1.14}$$

be determined by directed distance $\overrightarrow{P_1P_2}$, and let vector

$$K_i^* = x_i^{(2)}-x_i^{(1)}, \; i = 1, 2, 3, \tag{2.1.15}$$

be determined by $\overrightarrow{P_1^*P_2^*}$. Let K_i^* be the mapping of K_i. The dependence between the coordinates of both these vectors is obtained from (2.1.15), if (2.1.11) is inserted. This yields:

$$K_i^* = (\delta_{ij}+a_{ij})(\xi_j^{(2)}-\xi_j^{(1)}),$$

which because of (2.1.14) changes to

$$K_i^* = (\delta_{ij}+a_{ij})K_j, \; i, j = 1, 2, 3. \tag{2.1.16}$$

Equation (2.1.16) represents a *linear vector function*. Its coefficients $\delta_{ij}+a_{ij}$ are components of a tensor bearing the term *affinor*.

What has been said of vectors so far, indicates that two equal vectors remain equal after mapping and that parallel vectors following mapping will be parallel again without any change in the relation of their respective lengths. We also deduce from this, that two identical and identically orientated polygons are transformed into two other identical and identically orientated polygons through affine mapping. The same applies to every geometric configuration as the limiting case of polygonal configuration. This means that geometrically equal parts of a body, regardless of their position, are mapped, i.e., deformed, in the same way. Affine mapping is also termed as *homogeneous* mapping.

Let us set the case that the two affine mappings:

$$x_i = a_{io} + (\delta_{ij} + a_{ij})\xi_j, \quad i, j = 1, 2, 3, \tag{2.1.11}$$

and

$$y_k = b_{ko} + (\delta_{ki} + b_{ki})x_i, \quad i, k = 1, 2, 3, \tag{2.1.17}$$

are carried out successively. To obtain the thus determined relation between ξ_i and y_i, (2.1.11) must be inserted in (2.1.17), thus yielding

$$\begin{aligned} y_k &= b_{ko} + (\delta_{ki} + b_{ki})[a_{io} + (\delta_{ij} + a_{ij})\xi_j] \\ &= bk_0 + (\delta_{ki} + b_{ki})(a_{io} + \xi_i + a_{ij}\xi_j). \end{aligned}$$

Disintegrating the brackets on the right-hand side, it yields

$$y_k = (b_{ko} + a_{ko}) + [\delta_{ki} + (a_{ki} + b_{ki})\xi_i + (b_{ki}a_{io} + b_{ki}a_{ij}\xi_j). \tag{2.1.18}$$

In addition, let us assume that coefficients $a_{io}, a_{ij}, b_{ko}, b_{ki}$ are small quantities, so that the products constructed together with these are permitted to be neglected. Mappings to which this applies are termed *infinitesimal affine mappings*. The brackets on the extreme right-hand side disappear in cases like these, and using

$$c_{ki} = a_{ki} + b_{ki}, \quad i = 0, 1, 2, 3, \quad k = 1, 2, 3, \tag{2.1.19}$$

there remains the relation

$$y_k = c_{ko} + (\delta_{ki} + c_{ki})\xi_i, \quad i, k = 1, 2, 3, \tag{2.1.20}$$

which is another infinitesimal affine mapping. (2.1.19) yields the fact usually only valid for infinitesimal affine mappings, that the coefficients of compound mapping (2.1.20) are obtained as the sum of the coefficients of mappings (2.1.11) and (2.1.17). Therefore, it is of no importance in which order the single infinitesimal mappings are joined to obtain the compound mapping. We have discussed infinitesimal mappings, since we shall only be using these later on.

Let us observe vector K_i which, because of an affine mapping (2.1.16), changes to K_i^*. Following mapping, the following change has occurred:

$$\delta K_i = K_i^* - K_i. \tag{2.1.21}$$

From (2.1.16) we read off for (2.1.21)

$$\delta K_i = a_{ij} K_j. \tag{2.1.22}$$

The tensor a_{ij} in (2.1.22) can be divided into a symmetric and antisymmetric part, i.e., into ε_{ij} and ω_{ij}, so that

$$\begin{aligned} a_{ij} &= \tfrac{1}{2}(a_{ij}+a_{ji})+\tfrac{1}{2}(a_{ij}-a_{ji}) = \varepsilon_{ij}+\omega_{ij}, \\ \varepsilon_{ij} &= \tfrac{1}{2}(a_{ij}+a_{ji}), \quad \omega_{ij} = \tfrac{1}{2}(a_{ij}-a_{ji}) \end{aligned} \tag{2.1.23}$$

is valid. Thus, (2.1.22) changes to

$$\delta K_i = \omega_{ij} K_j + \varepsilon_{ij} K_j = \delta K_{i1} + \delta K_{i2}. \tag{2.1.24}$$

Let us examine the significance of part

$$\delta K_{i1} = \omega_{ij} K_j. \tag{2.1.25}$$

Let us apply the fact that corresponding to antisymmetric tensor ω_{ij}, by using tensor ε_{ijk} ,the 'vector of antisymmetric tensor'

$$\omega_k = \tfrac{1}{2}\varepsilon_{ikj}\omega_{ij} \tag{2.1.26}$$

can be formed. By its use, tensor ω_{ij} is written as

$$\omega_{ij} = \varepsilon_{ikj}\omega_k. \tag{2.1.27}$$

Inserting (2.1.27) in (2.1.25) yields

$$\delta K_{i1} = \varepsilon_{ikj}\omega_k K_j, \tag{2.1.28}$$

from which, because of the known characteristic of the ε-tensor, we can read off that δK_{i1} represents the vector product of vectors ω_i and K_i. This fact indicates that δK_{i1} corresponds to a *rotation.* As a result of δK_{i1}, the body performs a rigid-body motion in the transition of $\{P_k\}$ to $\{P_k^*\}$. This becomes obvious, since vector K_i in the mapping through ω_{ij}, i.e., by

$$K_i^* = (\delta_{ij}+\omega_{ij})K_j, \tag{2.1.29}$$

does not alter its length. Since this is not the identical mapping, then the mapping given by (2.1.29) could only be a rotation, if the vector keeps its length, despite mapping. No change of length does actually occur. Because of (2.1.29)

$$K_i^* - K_i = \delta K_{i1} = \omega_{ij} K_j. \tag{2.1.25}$$

is valid.

The square of the vector's length is K_i^2. Through differentiation we obtain $2K_i\delta K_i$ which must be nought for a vector keeping its length. Let us verify this fact if for δK_i the value δK_{i1}, especially determined by representation (2.1.29), is inserted. It is valid

$$K_i\delta K_{i1} = K_i\omega_{ij}K_j. \tag{2.1.30}$$

Written in full, it is

$$K_i K_{i1} = \omega_{11} K_1^2+\omega_{22} K_2^2+\omega_{33} K_3^2+(\omega_{12}+\omega_{21})K_1 K_2 +(\omega_{23}+\omega_{32})K_2 K_3+(\omega_{31}+\omega_{13})K_3 K_1.$$

This is identically equal to nought; since in accordance with the characteristics of the antisymmetric tensor ω_{ij},

$$\omega_{11} = \omega_{22} = \omega_{33} = \omega_{12}+\omega_{21} = \omega_{23}+\omega_{32} = \omega_{31}+\omega_{13} = 0$$

is valid. Thus, it has been proven that following mapping (2.1.29), the vector does not alter its length and thus through ω_{ij} a rotation results.

Using (2.1.23), mapping (2.1.11) is rewritten as

$$x_i = a_{io}+(\delta_{ij}+\omega_{ij}+\varepsilon_{ij})\xi_j.$$

Because of (2.1.4*) this is followed by

$$u_i = x_i-\xi_i = a_{io}+\omega_{ij}\xi_j+\varepsilon_{ij}\xi_j. \tag{2.1.31}$$

The displacement u_i which transforms the points of mass $\{P_k\}$ to the points of mass $\{P_k^*\}$ (Figure 7) is composed of part

$$u_{i1} = a_{io}+\omega_{ij}\xi_j, \tag{2.1.32}$$

and of

$$u_{i2} = \varepsilon_{ij}\xi_j.$$

Part u_{i1} is obviously caused by the rigid body motion, since a_{io} corresponds to a *translation*, and as we have seen, $\omega_{ij}\xi_j$ has been obtained from a *rotation*. Part u_{i2} must represent the strain resulting from the affine mapping.

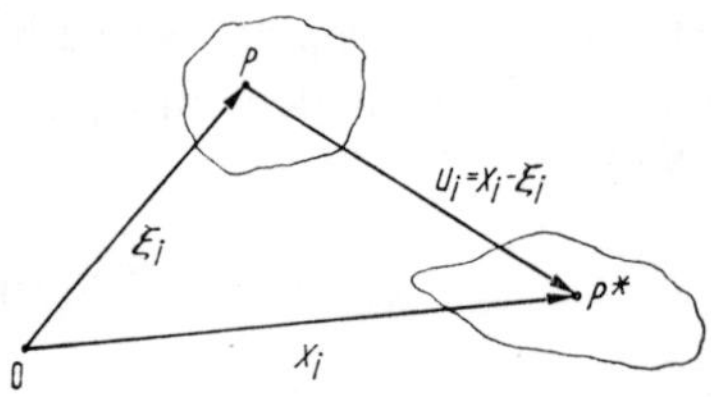

Figure 7

Henceforth only pure strain shall be discussed, the rigid body movement will be totally disregarded, and hence,

$$u_i = \varepsilon_{ij}\xi_j \tag{2.1.33}$$

will be written for the displacement. However, we must always remember that the displacements expressed in *this way* are yet to be completed by terms originating from the rigid body motion. But these are known in special problems because of the presupposed constraints (conditions of support) of the considered body, so that, as the case may be, supplementing (2.1.33) to a complete specification of displacement is quite possible.

2.1.4 *Strain tensor*

Through deformation the body has reached configuration $\{P_k^*\}$ from configuration $\{P_k\}$. The material point of the body, which coincided with the origin of coordinates, O, before deformation, has arrived at O^* through displacement vector u_{i0}. The material point P with coordinates ξ_i and situated close to O has been changed to P^* as a result of u_i. The displacement vector u_i is generally different from u_{i0}, since the components of displacement vectors are functions of Lagrange's coordinates ξ_i. Following deformation, the coordinates x_i of the material points of the deformed body are to be measured from O, since the origin of coordinates, O, is fixed in space.

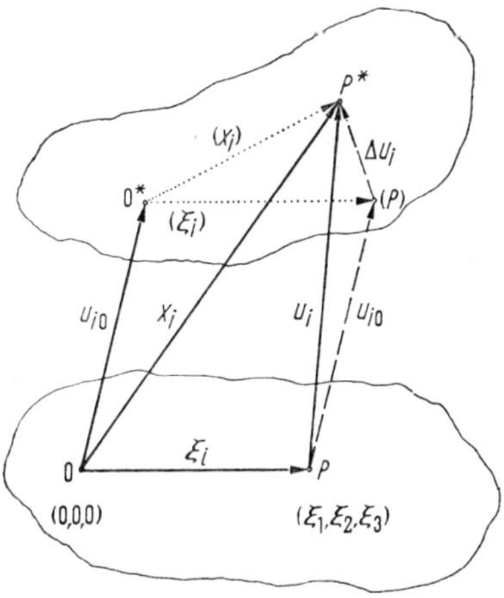

Figure 8

From Figure 8 we read off

$$u_i = u_{i0} + \Delta u_i.$$

Using a Taylor-expansion, we keep the two first terms of this expansion, i.e.

$$u_i = u_{i0} + \frac{\partial u_i}{\partial \xi_j} \xi_j. \tag{2.1.34}$$

In this expansion, the terms of higher order can indeed be neglected, since it was assumed that only the deformation of the nearest surrounding of O would be discussed and hence the ξ_i are taken to be very small.

In the same way from Figure 8

$$x_i = \xi_i + u_i,$$

which because of (2.1.34) (or by reading from Figure 8) is represented as

$$x_i = u_{io} + \frac{\partial u_i}{\partial \xi_j} \xi_j + \xi_i .$$

Using the Kronecker-symbol it is

$$x_i = u_{io} + \left(\delta_{ij} + \frac{\partial u_i}{\partial \xi_j}\right) \xi_j . \tag{2.1.35}$$

Comparing (2.1.35) and (2.1.11), it becomes evident that, in considering the deformation of the body, we are actually dealing with an affine transformation. If not otherwise specified, let us always assume that this transformation is also infinitesimal, i.e., u_{i0} and $\partial u_i/\partial \xi_j$ are also small quantities.

The mapping (2.1.35) contains the translation u_{i0}, which is taken to be the displacement of the origin of coordinates O towards O^*. We eliminate this translation by measuring, as from now, the coordinates from O^*. Then, using the bracketed data, we read from Figure 8

$$x_i = \xi_i + \Delta u_i .$$

But, as the Taylor-expansion indicated,

$$\Delta u_i = \frac{\partial u_i}{\partial \xi_j} \xi_j .$$

Hence, we are permitted to write

$$x_i = \xi_i + \frac{\partial u_i}{\partial \xi_j} \xi_j . \tag{2.1.36}$$

For this consideration, from point O^*, let us put $x_i - \xi_i = u_i$. In this sense, point O^* does not have any displacement, because the translation had previously been omitted. For the displacement of P we obtain, either from Figure 8 or by changing (2.1.36),

$$x_i - \xi_i = u_i = \frac{\partial u_i}{\partial \xi_j} \xi_j. \tag{2.1.37}$$

As discussed in (2.1.3), the displacement given by (2.1.37) is divided into rotation

$$u_{i1} = \frac{1}{2}\left(\frac{\partial u_i}{\partial \xi_j} - \frac{\partial u_j}{\partial \xi_i}\right) \xi_j$$

and into strain

$$u_{i2} = \frac{1}{2}\left(\frac{\partial u_i}{\partial \xi_j} + \frac{\partial u_j}{\partial \xi_i}\right) \xi_j.$$

Furthermore, instead of tensor a_{ij} used previously, we now split the tensor $u_{ij} \equiv \partial u_i / \partial \xi_j$ into its antisymmetric part

$$\omega_{ij} = \frac{1}{2}\left(\frac{\partial u_i}{\partial \xi_j} - \frac{\partial u_i}{\partial \xi_i}\right)$$

and into its symmetric part

$$\varepsilon_{ij} = \frac{1}{2}\left(\frac{\partial u_i}{\partial \xi_j} + \frac{\partial u_j}{\partial \xi_i}\right). \tag{2.1.38}$$

The symmetric tensor, defined by (2.1.38), is termed *strain tensor*. In future, by neglecting all rigid body motions, we shall only discuss that part of the displacement given by (2.1.38) and set down

$$u_i = \varepsilon_{ij} \xi_j, \tag{2.1.33}$$

as has already been discussed in Section 2.1.3. The additional knowledge we have thus obtained is that tensor ε_{ij}, only formally introduced in 2.1.3.

has attained physical significance through definition (2.1.38). The identity of ε_{ij} as a tensor can be checked by (2.1.33) using the quotient law.

To explain in detail the significance of strain tensor (2.1.38), let us examine into what kind of geometric figure a small material sphere, having its centre in O^*, changes following strain.

Let us look at unit sphere

$$\xi_i \xi_i = 1. \tag{2.1.39}$$

Let $x_i = \xi_i + u_i$ be valid. Since we shall only consider the strain, we use (2.1.33). This yields

$$x_i = \xi_i + \varepsilon_{ij}\xi_j = (\delta_{ij} + \varepsilon_{ij})\xi_j. \tag{2.1.40}$$

To abbreviate, let us introduce the expression

$$\varphi_{ij} = \delta_{ij} + \varepsilon_{ij}.$$

Thus (2.1.40) changes to

$$x_i = \varphi_{ij}\xi_j. \tag{2.1.41}$$

Through reversal of (2.1.41),

$$\xi_i = \varphi_{ij}^{-1} x_j \tag{2.1.42}$$

is obtained, where the components of the inverse tensor φ_{ij}^{-1}, because of symmetry of φ_{ij}, read

$$\varphi_{ij}^{-1} = \frac{\text{co}\,(\varphi_{ij})}{|\varphi_{ij}|}. \tag{2.1.43}$$

In (2.1.43), co (φ_{ij}) is the *cofactor* of element φ_{ij}, and $|\varphi_{ij}|$ is the determinant belonging to the matrix (φ_{ij}).

Inserting (2.1.42) in (2.1.39) yields

$$\varphi_{ij}^{-1}\varphi_{ik}^{-1} x_j x_k = 1, \tag{2.1.44}$$

this being an ellipsoid.

The unit-sphere about O^*, i.e., (2.1.39), changes into the ellipsoid (2.1.44) because of the strain applied to it. This ellipsoid is termed *measure-ellipsoid,* since the radius vectors of the ellipsoid associated with the radii of the unit-sphere, are interpreted as scales for the mapping of any vectors ξ_i into their images, i.e., the vectors x_i.

Let t_j be a vector in the tangential plane of the measure-ellipsoid, then the tangential plane of the ellipsoid in point x_k is represented by

$$\varphi_{ij}^{-1}\varphi_{ik}^{-1}t_j x_k = 1. \tag{2.1.45}$$

From (2.1.44) following differentiation,

$$\varphi_{ij}^{-1}\varphi_{ik}^{-1}\mathrm{d}x_j x_k + \varphi_{ij}^{-1}\varphi_{ik}^{-1}x_k \mathrm{d}x_k = 2\varphi_{ij}^{-1}\varphi_{ik}^{-1}\mathrm{d}x_j x_k = 0 \tag{2.1.46}$$

is valid. We read off (2.1.46) that vector $\varphi_{ij}^{-1}\varphi_{ik}^{-1}x_k$ is vertical to vector $\mathrm{d}x_j$ in the tangential plane. Thus, it is proportional to the normal vector n_j of the ellipsoid at point x_k. Terming the proportionality factor p, then the following is true for the normal unit vector:

$$n_j = p\varphi_{ij}^{-1}\varphi_{ik}^{-1}x_k. \tag{2.1.47}$$

Let

$$n_j x_j = p\varphi_{ij}^{-1}\varphi_{ik}^{-1}x_j x_k$$

be true, and because of (2.1.44), let

$$n_j x_j = p \tag{2.1.48}$$

be valid. It yields that p represents the distance of the ellipsoid's tangential plane from the ellipsoid's centre. This fact and another, which we shall discuss below, permits to indicate for any radius ξ_i of the unit-sphere its image x_i, which results from strain, by using the measure-ellipsoid.

If the strain is determined by specification of ε_{ij}, then the φ_{ij} as well as the φ_{ij}^{-1} are known, and in addition to the unit-sphere (2.1.39), the measure-ellipsoid (2.1.44) can be drawn (see Figure 9).

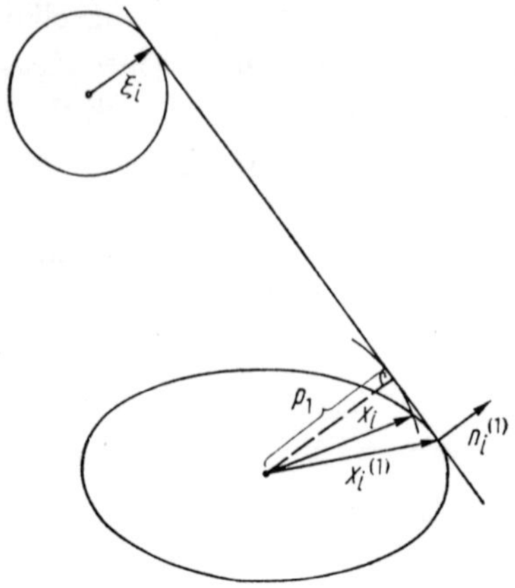

Figure 9

Let us select radius-vector $x_i^{(1)}$ of the measure-ellipsoid. It determines a tangential plane of distance p_1 from the centre of the ellipsoid and having normal vector

$$n_i^{(1)} = p_1 \varphi_{ji}^{-1} \varphi_{jk}^{-1} x_k^{(1)}. \tag{2.1.49}$$

Let us find radius ξ_i of the unit-sphere which is parallel to the normal vector $n_i^{(1)}$. Then,

$$\xi_i = n_i^{(1)} \tag{2.1.50}$$

must be valid for this radius. Inserting (2.1.50) in (2.1.41) yields

$$x_i = \varphi_{ij} n_j^{(1)}.$$

Through (2.1.49) it changes to

$$x_i = p_1 \varphi_{mj}^{-1} \varphi_{ij} \varphi_{mk}^{-1} x_k^{(1)}. \tag{2.1.51}$$

But it is $\varphi_{ij}\varphi_{mj}^{-1} = \delta_{im}$, so that (2.1.51) changes to

$$x_i = p_1 \varphi_{ik}^{-1} x_k^{(1)},$$

and because of (2.1.42) finally into

$$x_i = p_1 \xi_i^{(1)}. \tag{2.1.52}$$

For the magnitude of radius vector x_i, which is the mapping of the sphere's radius, ξ_i, we obtain from (2.1.52)

$$|x_i| = p_1 |\xi_i^{(1)}| \equiv p_1 \,. \tag{2.1.53}$$

The affine mapping, which corresponds to the strain, thus attaches to every radius ξ_i of the unit-sphere after (2.1.53) a radius vector x_i of the measure-ellipsoid, whose absolute value is equal to the distance of the centre of that ellipsoid's tangential plane, which stands vertical upon radius ξ_i of the unit-sphere. This fact yields a simple construction of image x_i of ξ_i by using the measure-ellipsoid (Figure 9).

There are further methods of explaining a state of strain. One of these is, for example, Cauchy's strain-surface[1]

$$\varepsilon_{ij}\xi_i\xi_j = \pm k^2, \tag{2.1.54}$$

where k is a constant and the plus or minus sign on the right-hand side is chosen in such a way that (2.1.54) yields a real surface. Details of this are given in references [7, 8].

By using the measure-ellipsoid further, let us determine the significance of the strain-tensor's components. To begin with, let us look at sphere radius $\boldsymbol{r}$, which has the direction of the ξ_1 axis. It is valid for its components

$$\xi_1 = 1, \; \xi_2 = \xi_3 = 0. \tag{2.1.55}$$

Following strain, it is changed into a radius-vector $\boldsymbol{r}^*$ of the measure-ellipsoid, and because of (2.1.41) we obtain for its components

$$x_1 = \varphi_{11} = 1+\varepsilon_{11}, \; x_2 = \varphi_{21} = \varepsilon_{21}, \; x_3 = \varphi_{31} = \varepsilon_{31}. \tag{2.1.56}$$

The elongation E_1, which affects $\boldsymbol{r}$ when it changes to $\boldsymbol{r}^*$, is defined as a relative change of length. Thus

[1] M. Lagally (see [7]), calls this surface the *tensor-surface*, or *dilatation-surface*, in contrast to the measure-ellipsoid.

$$E_1 = \frac{|\boldsymbol{r}^*| - |\boldsymbol{r}|}{|\boldsymbol{r}|}. \tag{2.1.57}$$

From (2.1.55) and (2.1.56), we obtain

$$|\boldsymbol{r}^*| = \sqrt{1+2\varepsilon_{11}+\varepsilon_{11}^2+\varepsilon_{21}^2+\varepsilon_{31}^2}, \; |\boldsymbol{r}| = 1. \tag{2.1.57a}$$

We use the fact that ε_{11}, ε_{21} and ε_{31} are small quantities, since the affine representation, found through ε_{ij}, is supposed to be infinitesimal. Then ε_{11}^2, ε_{21}^2 and ε_{31}^2 may be neglected as quantities which in higher order are small, and by using an approximate expression for the root, we obtain

$$|\boldsymbol{r}^*| \approx \sqrt{1+2\varepsilon_{11}} \approx 1+\varepsilon_{11}, \; |\boldsymbol{r}|=1. \tag{2.1.58}$$

Inserting (2.1.58) into (2.1.57) yields

$$E_1 = \varepsilon_{11},$$

which indicates that ε_{11} signifies *elongation*, or *normal strain*. The same significance is obtained by corresponding calculation for the remaining diagonal terms ε_{22} and ε_{33} of the strain-tensor.

In the second case let us look at a sphere-radius $\boldsymbol{r}^{(1)}$, having the direction of the ξ_1 axis, as well as at a sphere-radius $\boldsymbol{r}^{(2)}$, having the direction of the ξ_2 axis. Both radii form an angle of 90°. Following the strain, both sphere radii have changed into radius-vectors $\boldsymbol{r}^{(1)*}$, $\boldsymbol{r}^{(2)*}$ of the measure-ellipsoid, and include angle ϑ. The components of $\boldsymbol{r}^{(1)}$, $\boldsymbol{r}^{(2)}$ are:

$$\xi_1^{(1)} = 1, \; \xi_2^{(1)} = \xi_3^{(1)} = 0, \; \xi_1^{(2)} = 0, \; \xi_2^{(2)} = 1, \; \xi_3^{(2)} = 0,$$

and the components of $\boldsymbol{r}^{(1)*}$ and $\boldsymbol{r}^{(2)*}$ are obtained according to (2.1.41) as

$$x_1^{(1)} = 1+\varepsilon_{11}, \; x_2^{(1)} = \varepsilon_{21}, \; x_3^{(1)} = \varepsilon_{31},$$

and

$$x_1^{(2)} = \varepsilon_{12}, \; x_2^{(2)} = 1+\varepsilon_{22}, \; x_3^{(2)} = \varepsilon_{32}$$

respectively.

Using the scalar product of vectors $\boldsymbol{r}^{(1)*}$ and $\boldsymbol{r}^{(2)*}$, the cosine of angle ϑ is calculated.

We find

$$\cos\vartheta = \frac{\boldsymbol{r}^{(1)*}\cdot\boldsymbol{r}^{(2)*}}{|\boldsymbol{r}^{(1)*}||\boldsymbol{r}^{(2)*}|}. \tag{2.1.59}$$

Using the components of vectors as indicated above,

$$\boldsymbol{r}^{(1)*}\boldsymbol{r}^{(2)*} = (1+\varepsilon_{11})\varepsilon_{12}+\varepsilon_{21}(1+\varepsilon_{22})+\varepsilon_{31}\varepsilon_{32}$$

is obtained. Because of the smallness of components ε_{ij} we are permitted to neglect their products, hence we write

$$\boldsymbol{r}^{(1)*}\boldsymbol{r}^{(2)*} = \varepsilon_{12}+\varepsilon_{21} = 2\varepsilon_{12} \tag{2.1.60}$$

as an approximation. In the same way, as an approximation, and according to (2.1.58), $|\boldsymbol{r}^{(1)*}| = 1+\varepsilon_{11}$ is valid, and similarly we obtain $|\boldsymbol{r}^{(2)*}| = 1+\varepsilon_{22}$. Accordingly, neglecting product $\varepsilon_{11}\varepsilon_{22}$ again,

$$|\boldsymbol{r}^{(1)*}||\boldsymbol{r}^{(2)*}| = 1+\varepsilon_{11}+\varepsilon_{22}.$$

In addition, let us neglect ε_{11} and ε_{22} in relation to 1, so that

$$|\boldsymbol{r}^{(1)*}||\boldsymbol{r}^{(2)*}| = 1 \tag{2.1.61}$$

is fairly accurate. Inserting (2.1.60) and (2.1.61) in (2.1.59), we obtain

$$\cos\vartheta = 2\varepsilon_{12}. \tag{2.1.62}$$

Introducing angle γ_{12}, through which the original right angle between $\boldsymbol{r}^{(1)}$ and $\boldsymbol{r}^{(2)}$ is changed during the strain process, it is

$$\cos\vartheta = \cos\left(\frac{\pi}{2}-\gamma_{12}\right) = \sin\gamma_{12} = 2\varepsilon_{12} \tag{2.1.63}$$

Because of the smallness of γ_{12} we are permitted to write $\sin\gamma_{12} = \gamma_{12}$ in approximation, with which (2.1.63) finally changes to

$$\varepsilon_{12} = \tfrac{1}{2}\gamma_{12}. \tag{2.1.64}$$

Similarly

$$\varepsilon_{23} = \tfrac{1}{2}\gamma_{23}, \ \varepsilon_{31} = \tfrac{1}{2}\gamma_{31}$$

is calculated, and we can write

$$\varepsilon_{ij} = \tfrac{1}{2}\gamma_{ij}, \ i \neq \mathrm{j}. \tag{2.1.65}$$

From (2.1.65) we read off the explicit significance of the strain-tensor's components beyond the diagonal: these components represent half of the change of angle of the original right angles between the ξ_i and ξ_j axes. The ε_{ij} are, therefore, termed *shearing strains*. Let us find the relative change of a volume which is obtained after infinitesimal strain. Let

$$V = l_1 l_2 l_3 \tag{2.1.66}$$

be the original volume of a block having the edges l_1, l_2, l_3 pointing in the principal directions. Through strain, the volume changes into

$$V+\delta V = l_1(1+\varepsilon_{11})l_2(1+\varepsilon_{22})l_3(1+\varepsilon_{33}),$$

which except for quantities of higher order equals

$$V+\delta V = l_1 l_2 l_3(1+\varepsilon_{11}+\varepsilon_{22}+\varepsilon_{33}) = V(1+\varepsilon_{ii}). \tag{2.1.67}$$

By *cubical dilatation* the relative change of volume $\Theta = \delta V/V$ is understood, which, through (2.1.66) and (2.1.67), becomes

$$\Theta = \frac{\delta V}{V} = \frac{(V+\delta V)-V}{V} = \frac{V+\delta V}{V} - 1 = \varepsilon_{ii}.$$

It will become evident later on that through

$$\Theta = \varepsilon_{ii}, \tag{2.1.68}$$

the physical significance of the first invariant of the strain tensor has been obtained.

Image x_i of radius ξ_i of the unit-sphere does not generally have the direction of ξ_i. This fact is indicated in Figure 9. Let us examine the special cases, in which, as an exception, the direction of both vectors is the same. Then

$$x_i = k\xi_i \tag{2.1.69}$$

is true. Let $k = \varepsilon+1$ be the proportional factor, yet to be determined, of both vectors x_i and ξ_i, which by assumption are parallel. Because of $\xi_i = \delta_{ij}\xi_j$, and following (2.1.41), it can be written for (2.1.69)

$$\varphi_{ij}\xi_j = k\delta_{ij}\xi_j,$$

and following further change,

$$(\varphi_{ij}-k\delta_{ij})\xi_j = 0. \tag{2.1.70}$$

Considering $\varphi_{ij} = \delta_{ij}+\varepsilon_{ij}$ and $k = \varepsilon+1$, then (2.1.70) changes to

$$(\varepsilon_{ij}-\varepsilon\delta_{ij})\xi_j = 0. \tag{2.1.71}$$

This represents a homogeneous system of equations which gives solutions only for certain values of ε. From (2.1.69) it follows

$$\varepsilon = \frac{x_i-\xi_i}{\xi_i},$$

which indicates that ε represents the normal strain of the particular marked sphere-radii. They are termed *principal normal strains.* In particular, *those* directions are yielded from (2.1.71) and from another, yet to be named relation, to which the marked radii, together with their image x_i, are parallel. They are termed *principal directions.*

The construction of the measure-ellipsoid shown in Figure 9 indicates that ξ_i and the respective image x_i can both have only the same direction, if the radius-vector x_i of the measure-ellipsoid has the direction of that normal vector n_i of the measure-ellipsoid which is determined by the respective sphere-radius. From this follows the explicit significance of the specific x_i in consideration and the corresponding principal directions:

The particular x_i are the principal-axes segments of the measure-ellipsoid, and the principal directions are the directions corresponding to these principal axes segments. Considering this explicit significance, it enables us to determine without proof, that there will always be three principal directions, which in addition are vertical to each other. They represent the *orthogonal principal axes system* of the measure-ellipsoid, and, respectively, the orthogonal principal axes system of the strain tensor ε_{ij}. In the same way there will always be three real values for the *principal normal strains* ε, which naturally do not always have to differ from each other.

Let us calculate the principal normal strains and the principal directions. The system of equations (2.1.71) is homogeneous. For non-trivial solutions of the ξ_j,

$$\det(\varepsilon_{ij}-\varepsilon\delta_{ij}) = 0 \tag{2.1.72}$$

is required. (2.1.72) is termed *characteristic equation* of the strain tensor ε_{ij}. This is a cubic equation, which can be written as

$$\varepsilon^3 - J_{1V}\varepsilon^2 + J_{2V}\varepsilon - J_{3V} = 0 \tag{2.1.73}$$

after expansion of the determinant. The coefficients J_{iV}, $i = 1, 2, 3$, of this algebraic equation are the invariants of the strain tensor; roots ε_i, $i = 1, 2, 3$, of (2.173) are the sought principal normal strains.

Based on Vieta's root theorem, the relations

$$\begin{aligned} J_{1V} &= \varepsilon_1+\varepsilon_2+\varepsilon_3\,, \\ J_{2V} &= \varepsilon_1\varepsilon_2+\varepsilon_2\varepsilon_3+\varepsilon_3\varepsilon_1\,, \\ J_{3V} &= \varepsilon_1\varepsilon_2\varepsilon_3 \end{aligned} \tag{2.1.74}$$

are obtained for the invariants ε_i. These relations determine the connection between the invariants and the principal normal strains. From the expansion of determinant (2.1.72) further relations follow:

$$\begin{aligned} J_{1V} &= \varepsilon_{ii} = \Theta, \\ J_{2V} &= \frac{1}{2!}\,\delta^{ij}_{pq}\,\varepsilon_{pi}\,\varepsilon_{qj}\,, \\ J_{3V} &= \frac{1}{3!}\,\delta^{ijk}_{pqr}\,\varepsilon_{pi}\,\varepsilon_{qj}\,\varepsilon_{rk}\,, \end{aligned} \tag{2.1.75}$$

which, when written using the *generalized Kronecker symbol* δ^{ijk}_{pqr}, represent the connection of the invariants with the general components of the strain-tensor.

After calculating the three (not always different) roots ε_i of the characteristic equations (2.1.72), (2.1.73), then these can be used in calculating the three principal directions. The principal directions are directly determined by the components of vectors ξ_i, which are obtained from the equation system (2.1.71) and the additional equation

$$\xi_i \xi_i = 1. \tag{2.1.76}$$

This is so because the ξ_i have been determined as unit vectors and their components are, therefore, direction cosines. This also justifies the introduction of (2.1.76).

If, for example, the main value ε_k, having a fixed index k, is inserted in

$$\begin{aligned} (\varepsilon_{ij} - \varepsilon_k \delta_{ij})\xi_j^{(k)} &= 0, \\ \xi_i^{(k)} \xi_i^{(k)} &= 1, \end{aligned} \tag{2.1.77}$$

then (2.1.77) yields the unit vector $\xi_j^{(k)}$, k fixed, $j = 1, 2, 3$, which determines the principal direction belonging to the principal strain, ε_k, k fixed. If the ε_k are all different, then three different orthogonal principal directions are obtained. If two principal strains are equal to each other, then the third different one, ε_k, from (2.1.77) has a principal direction, and the other two missing principal directions can be determined by setting up, in addition to the already known direction, an orthogonal right-hand system. However, this cannot be done in a unique way. If all principal strains are equal to each other, then any orthogonal right-hand system can be chosen as the principal direction system. For the following reasons we cannot work with only the system (2.1.71), but must use the equation system (2.1.77), which has been supplemented by (2.1.76), for the determination of the principal directions: Only two equations are independent from the three equations (2.1.71)! To calculate the three unknown direction cosines $\xi_j^{(k)}$, k fixed, $j = 1, 2, 3$, a third independent equation, which is (2.1.76), is needed.

Using the cubical dilatation Θ, the spherical tensor $\Theta/3 \cdot \delta_{ij}$ is found. It can be used to represent the strain tensor ε_{ij} as

$$\varepsilon_{ij} = \tfrac{1}{3}\Theta\delta_{ij}+\varepsilon_{ij}^{(0)}. \tag{2.1.78}$$

The strain tensor has been split into two parts in (2.1.78). The first part is the said spherical tensor which yields the true dilatation. The second part is the *strain deviator*

$$\varepsilon_{ij}^{(0)} = \varepsilon_{ij}-\tfrac{1}{3}\Theta\delta_{ij}, \tag{2.1.79}$$

which indicates the 'deviation' of the strain tensor from the spherical tensor and which corresponds thus to the shear deformation, for which the cubical dilatation vanishes. Its components are termed *strain deviations*. The deviator has the special characteristic that its first invariant $J_{1\text{v}^{\text{o}}}$ is equal to nought, since because of $\delta_{ii} = 3$ [1] and $\Theta = \varepsilon_{ii}$,

$$J_{1\text{v}^{\text{o}}} = \varepsilon_{ii}^{(0)} = \varepsilon_{ii}-\tfrac{1}{3}\Theta\cdot 3 = \varepsilon_{ii}-\varepsilon_{ii} = 0 \tag{2.1.80}$$

is valid.

In addition, as shall be seen, it plays a special part in the setting up of constitutive equations. This is the reason why already here the deviator is being mentioned.

For the modelling of 2-order tensors (dyadics), to which belongs the strain tensor, O. Mohr has indicated a special graphical procedure. It is termed method of Mohr's circles, and can be used for strain tensors. It will not be discussed at this point, but later on, in connection with stress tensors, which too are dyadics. What shall be said then can also be rationally applied to strain tensors.

2.1.5 *Compatibility conditions*

In an unstrained continuum, the following is valid for the square of the line element, ds,

$$ds^2 = d\xi_i d\xi_i. \tag{2.1.81}$$

---

[1] Here, δ_{ij} is to be taken as a unit vector and not as Kronecker-symbol. Hence, $\delta_{ii} = \delta_{11}+\delta_{22}+\delta_{33} = 1+1+1 = 3$.

For the strained continuum, there is correspondingly for the line element, ds^*,

$$ds^{*2} = dx_i dx_i. \tag{2.1.82}$$

But because of (2.1.1*),

$$dx_i dx_i = \frac{\partial x_i}{\partial \xi_\alpha}\frac{\partial x_i}{\partial \xi_\beta} d\xi_\alpha d\xi_\beta = g_{\alpha\beta} d\xi_\alpha d\xi_\beta. \tag{2.1.83}$$

Then the metric tensor

$$g_{\alpha\beta} = \frac{\partial x_i}{\partial \xi_\alpha}\frac{\partial x_i}{\partial \xi_\beta},$$

which determines the metric of the strained continuum, has been used. The strain tensor ε_{ij} is defined by condition

$$ds^{*2} - ds^2 = 2\varepsilon_{ij} d\xi_i d\xi_j. \tag{2.1.84}$$

We shall discuss this in detail. According to (2.1.81), $ds^2 = \delta_{ij} d\xi_i d\xi_j$, and according to (2.1.82) as well as (2.1.83), $ds^{*2} = g_{ij} d\xi_i d\xi_j$. Inserting these relations in (2.1.84), then

$$(g_{ij} - \delta_{ij}) d\xi_i d\xi_j = 2\varepsilon_{ij} d\xi_i d\xi_j$$

is obtained, and from this

$$g_{ij} = 2\varepsilon_{ij} + \delta_{ij}, \tag{2.1.85}$$

through which the connection between the metric tensor of the strained continuum and the strain tensor has been determined. The unstrained, as well as the strained continuum, are parts of an Euclidean space, for which Riemann's curvature tensor R_{ijkl} must disappear. It is

$$\begin{aligned} R_{ijkl} = \frac{1}{2}\left(\frac{\partial^2 g_{il}}{\partial \xi_j \partial \xi_k} - \frac{\partial^2 g_{jl}}{\partial \xi_i \partial \xi_k} - \frac{\partial^2 g_{ik}}{\partial \xi_j \partial \xi_l} + \frac{\partial^2 g_{jk}}{\partial \xi_i \partial \xi_l}\right) \\ + g^{\alpha\beta}([jk, \beta][il, \alpha] - [jl, \beta][ik, \alpha]) \end{aligned} \tag{2.1.86}$$

for the strained continuum.

Using (2.1.85) for R_{ijkl} and making the tensor equal to nought, then by neglecting the terms following from the second round bracket in (2.1.86), which are small by a higher order, the equations

$$\frac{\partial^2 \varepsilon_{il}}{\partial \xi_j \partial \xi_k} + \frac{\partial^2 \varepsilon_{jk}}{\partial \xi_i \partial \xi_l} - \frac{\partial^2 \varepsilon_{jl}}{\partial \xi_i \partial \xi_k} - \frac{\partial^2 \varepsilon_{ik}}{\partial \xi_j \partial \xi_l} = 0 \tag{2.1.87}$$

are obtained. These are the *compatibility conditions* by St. Vénant, which must be satisfied by the components of the strain tensor. Writing (2.1.87) in full, then from 81 it is possible, to arrive at six essential equations, which are:

$$2\,\frac{\partial^2 \varepsilon_{12}}{\partial \xi_1 \partial \xi_2} = \frac{\partial^2 \varepsilon_{11}}{\partial \xi_2^2} + \frac{\partial^2 \varepsilon_{22}}{\partial \xi_1^2},$$

(a further two by cyclic exchanges of indices), (2.1.88)

$$\frac{\partial^2 \varepsilon_{11}}{\partial \xi_2 \partial \xi_3} = \frac{\partial}{\partial \xi_1}\left(\frac{\partial \varepsilon_{12}}{\partial \xi_3} + \frac{\partial \varepsilon_{31}}{\partial \xi_2} - \frac{\partial \varepsilon_{23}}{\partial \xi_1}\right),$$

(a further two by cyclic exchange of indices).

The compatibility equations are necessary and sufficient for the fact that, from the six components of a given strain tensor ε_{ij}, the three components of displacement vector u_i can be calculated. Hence, they will be found as integrability conditions: It is

$$\varepsilon_{ij} = \frac{1}{2}\left(\frac{\partial u_i}{\partial \xi_j} + \frac{\partial u_j}{\partial \xi_i}\right).$$

From this, by simple calculation, we obtain

$$\frac{\partial \varepsilon_{ik}}{\partial \xi_j} + \frac{\partial \varepsilon_{jk}}{\partial \xi_i} - \frac{\partial \varepsilon_{ij}}{\partial \xi_k} = \frac{\partial^2 u_k}{\partial \xi_i \partial \xi_j}. \tag{2.1.89}$$

This represents a differential equation system to determine the u_i. From the known requirement

$$\frac{\partial}{\partial \xi_i}\,\frac{\partial^2 u_k}{\partial \xi_j \partial \xi_h} = \frac{\partial}{\partial \xi_h}\,\frac{\partial^2 u_k}{\partial \xi_i \partial \xi_j}$$

for the integrability of (2.1.89), is yielded from the left-hand sides of (2.1.89), as an integrability condition for this system,

$$\frac{\partial}{\partial \xi_i}\left(\frac{\partial \varepsilon_{jk}}{\partial \xi_h} + \frac{\partial \varepsilon_{hk}}{\partial \xi_j} - \frac{\partial \varepsilon_{jh}}{\partial \xi_k}\right) = \frac{\partial}{\partial \xi_h}\left(\frac{\partial \varepsilon_{ik}}{\partial \xi_j} + \frac{\partial \varepsilon_{jk}}{\partial \xi_i} - \frac{\partial \varepsilon_{ij}}{\partial \xi_k}\right),$$

which is easily recalculated into (2.1.87) by renaming the indices. Thus the significance of the compatibility conditions as integrability conditions has been proven.

Finally, let us establish that not any symmetric 2-order tensor can be given as a strain tensor, but that only a tensor, which additional to the symmetry satisfies the conditions (2.1.87), is eligible. We shall subsequently bear this in mind. There is another fact which is important and which has been indicated by E. Beltrami. Only three of the compatibility conditions are independent of each other. This is easy to prove. Let us start by introducing notations

$$-\frac{\partial^2 \varepsilon_{11}}{\partial \xi_2^2} - \frac{\partial^2 \varepsilon_{22}}{\partial \xi_1^2} + 2\frac{\partial^2 \varepsilon_{12}}{\partial \xi_1 \partial \xi_2} = A_{12}, \text{ etc.,}$$

$$\frac{\partial^2 \varepsilon_{11}}{\partial \xi_2 \partial \xi_3} - \frac{\partial}{\partial \xi_1}\left(\frac{\partial \varepsilon_{12}}{\partial \xi_3} + \frac{\partial \varepsilon_{31}}{\partial \xi_2} - \frac{\partial \varepsilon_{23}}{\partial \xi_1}\right) = B_{23}, \text{ etc.}$$

Using these, the identities

$$\begin{aligned} &\frac{\partial A_{12}}{\partial \xi_3} + \frac{\partial B_{23}}{\partial \xi_2} + \frac{\partial B_{13}}{\partial \xi_1} = 0, \\ &\frac{\partial A_{23}}{\partial \xi_1} + \frac{\partial B_{31}}{\partial \xi_3} + \frac{\partial B_{21}}{\partial \xi_2} = 0, \\ &\frac{\partial A_{31}}{\partial \xi_2} + \frac{\partial B_{12}}{\partial \xi_1} + \frac{\partial B_{32}}{\partial \xi_3} = 0 \end{aligned} \tag{2.1.90}$$

are written, by which the six equations (2.1.88) are reduced to only three independent equations.

2.2 Analysis of stress

2.2.1 *Forces acting on and in a continuum*

The forces affecting a deformable continuum are divided into two classes. One class includes the spatially distributed *volume forces*. The other class incorporates *stresses* which are distributed on a plane and which are exerted on the surface element from the adjacent volume elements, the surface element serving as the reference plane for the particular stress.

The volume forces will, henceforth, be known as $\boldsymbol{K}\mathrm{d}V$, $K_i \mathrm{d}V$, respectively. Components K_i thus represent a force per unit volume.

Similarly, the surface forces will be termed by $\boldsymbol{\sigma}_n \mathrm{d}F$, $\sigma_{ni} \mathrm{d}F$, respectively. Index n indicates that the reference plane, i.e., the surface element $\mathrm{d}F$, has the normal $\boldsymbol{n}$ pointing outside. This normal characterizes the plane. To differentiate, let us call $\boldsymbol{\sigma}_n$ a *stress vector* and its components, which have the dimension of a force per unit area, *stresses*. Thus, stresses do not represent a vector, but the modulus of vector components. Let us say the following about stresses: we have assumed that only true forces are transferred by the surrounding to the surface element concerned. It is not impossible that 'forces' in a more general meaning, i.e., moments, are exerted by the surrounding. These would then give rise to *couple stresses*. Considering couple stresses, in addition to stresses in their proper sense, leads to a special theory which has been set up by E. and F. Cosserat, and which has been extended in the meantime. It has found a certain significance for the *dislocation theory*, see [9]. We shall disregard couple stresses in the following, and remain in the realm of classic conditions, i.e., permit stresses brought up by proximity forces only. As we shall see, this will be consequential for the structure of stress tensors.

Both, volume forces and stresses, would occur as *external* or *internal forces*. External forces are those, which affect the continuum from outside. Internal forces act between the volume elements of the continuum. Gravity is, for example, an external force, and the gravitation force between the volume elements of the continuum is an internal volume force. Internal stresses become apparent if a volume element is separated from the continuum and is considered with regard to the forces transferred by the neighbouring volume elements. External stresses result from external working loads, which are exerted on the area of contact of the continuum.

We used Lagrange's approach as the most natural approach in dealing with strain. It will now be easier and more natural to describe stresses for the deformed element and to use the coordinates x_i of the deformed body. We shall, therefore, prefer to use the Euler approach when dealing with stresses. Moreover, let us assume the volume forces as being continuous, having continuous first derivations, and let the surface forces be piecewise continuous functions of coordinates x_i.

2.2.2 *The stress tensor*

In Figure 10, let us consider a surface element dF of the continuum, which is situated in the interior or at the boundary of the same. Let us determine an 'exterior' and 'interior' side of dF marking the exterior side by depositing on it the 'positive' counting 'outside' pointing normal $\boldsymbol{n}$ of the surface element. In dF operates force $\boldsymbol{\sigma}_n^*$ which is relative to the unit surface. By setting up a limit value, we obtain

$$\boldsymbol{\sigma}_n = \lim_{\mathrm{d}F \to 0} \frac{\boldsymbol{\sigma}_n^* \mathrm{d}F}{\mathrm{d}F} \tag{2.2.1}$$

The $\boldsymbol{\sigma}_n$ is a stress vector of absolute value $|\boldsymbol{\sigma}_n|$, and $|\boldsymbol{\sigma}_n|$ *is the stress which is exerted from that part of the continuum, which is situated on the 'exterior' side of* dF, *to this part of the continuum lying on the 'interior' side. Tensile stresses are counted as being positive.*
Based on the third law by J. Newton (actio-reactio), the stress vector $-\boldsymbol{\sigma}_n$

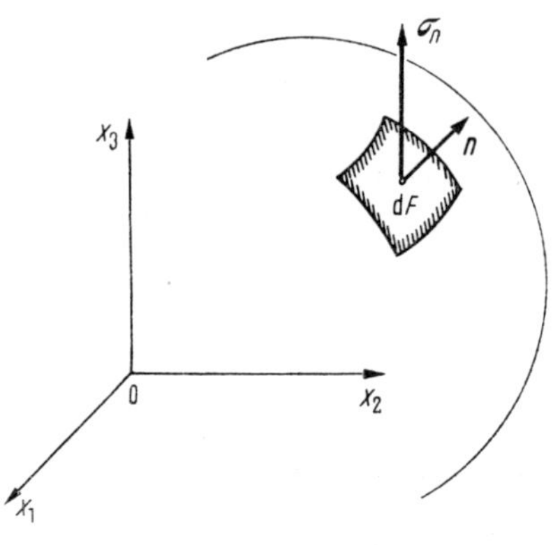

Figure 10

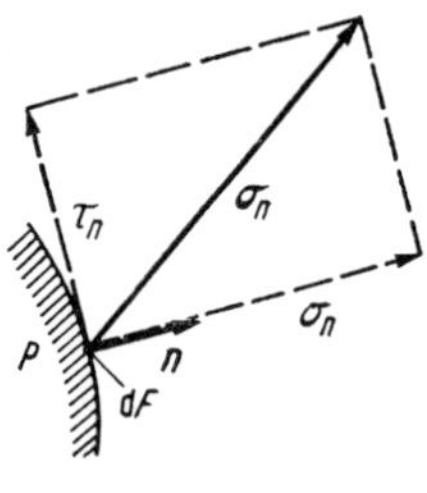

Figure 11

is exerted through the surface element dF to that part of the continuum situated on the exterior side of dF.

Working at a profile of surface element dF, which has been placed through point P of the continuum containing normal $\boldsymbol{n}$ and stress vector $\boldsymbol{\sigma}_n$ (Figure 11), it becomes evident that $\boldsymbol{\sigma}_n$ can be divided into two components, σ_n and τ_n, parallel and vertical to $\boldsymbol{n}$. σ_n is termed *normal stress*, it is the magnitude of a stress component in the direction of surface normal $\boldsymbol{n}$; and τ_n is termed *tangential stress*, since it is the magnitude of the stress component which is tangential to dF. We read from Figure 11 that

$$\sigma_n = \boldsymbol{\sigma}_n \boldsymbol{n} \tag{2.2.2}$$

and

$$\tau_n = \sqrt{|\sigma_n|^2 - \sigma_n^2} \tag{2.2.3}$$

are valid.

When describing the state of stress of the continuum in any point P, *one* stress vector $\boldsymbol{\sigma}_n$ is not sufficient, since through P infinitely numerous surface elements, having differently orientated normals $\boldsymbol{n}$, may be passed, thus specifying infinitely different vectors $\boldsymbol{\sigma}_n$, and every one being justified. But the state of stress is uniquely determined in point P, when three stress vectors related to three surface elements through P are given, whose normals are not permitted to be coplanar. Proof of this statement is given if it can be shown that it is possible to calculate the stress on any surface element through P, orientated in any way, by using these particulars.

It is best to choose, as reference surfaces, the three coordinate planes passing through P and being orthogonal to each other (Figure 12). They are spanned by two coordinate axes, and their normal is the third respective coordinate axis. Let us consider the three stress vectors

$$\boldsymbol{\sigma}_i = (\sigma_{ij}),\ i, j = (1, 2, 3) \tag{2.2.4}$$

with components σ_{ij}, which are related to the unit surface. Let us assemble σ_{ij} schematically into a matrix. It becomes evident soon that this matrix describes also a physical quantity. This is *stress tensor* σ_{ij}, which determines completely the stress condition in P.

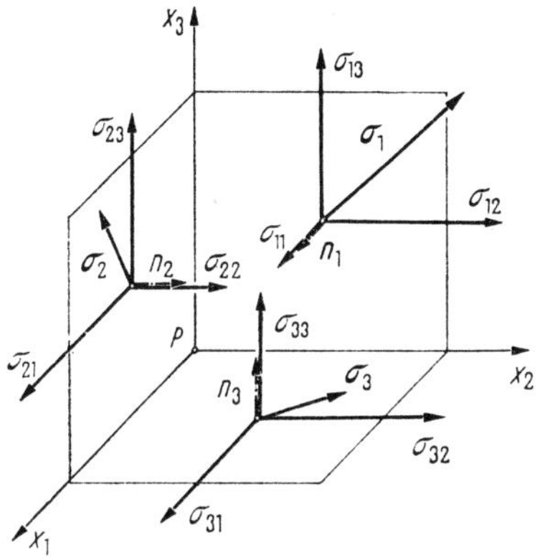

Figure 12

In the components σ_{ij} (i, j constant) of the stress tensor, the first index i indicates the normal of the reference plane and the second index j indicates the direction in which the stress-component-vector belonging to σ_{ij} (i, j constant), operates. If σ_{ij} is relative to a surface element having normal $\boldsymbol{n}_i$ (i constant), which has the direction of the *positive coordinate axis* x_i (i constant), then σ_{ij} is counted as *being positive*, if the related *stress component vector* has the direction of the *positive* x_j*-axis* (j constant). If, however, σ_{ij} is related to a surface element with normal $\boldsymbol{n}_i$ (i constant) which has the direction of the *negative* x_i*-axis* (i constant), then σ_{ij} is counted as *positive*, if the related *stress component vector* indicates the direction of *negative* x_j*-axis* (j constant). In the sense of this definition, all components σ_{ij} in Figure 12, are positive. Let as say that stresses having two equal indices are *normal stresses* and having two different indices are *shearing stresses*.

Let us show that using the components σ_{ij} of the stress tensor, the stress vector $\boldsymbol{\sigma}_n$ can actually be calculated in relation to any surface with normal vector $\boldsymbol{n}$ and passing through P. Let us consider the tetrahedron in Figure 13. The inclined surface $\mathrm{d}F$ with normal $\boldsymbol{n}$ does not pass through P, but the tetrahedron is assumed as being so small that everything will take place near to P. Then $\boldsymbol{\sigma}_n$ can be taken as being practically a stress vector of an inclined surface passing through P.

If the inclined tetrahedron surface has the area $\mathrm{d}F$, then the remaining surfaces of the tetrahedron, which coincide with the coordinate planes, obviously have the area $\mathrm{d}Fn_i$, $i = 1, 2, 3$, when n_i are the components of $\boldsymbol{n}$. Furthermore, let the stress vector $\boldsymbol{\sigma}_n$, which affects the inclined tetra-

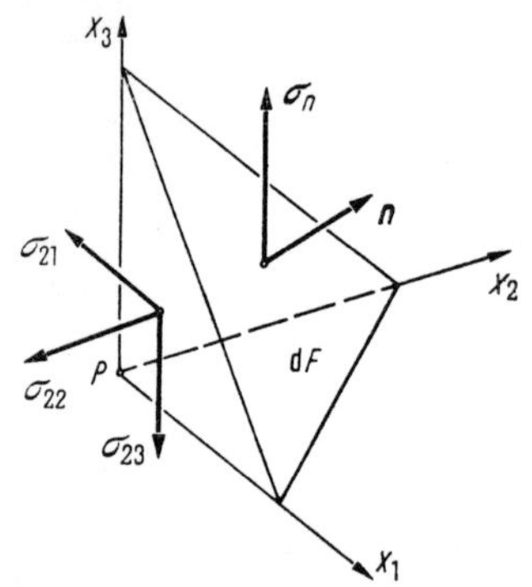

Figure 13

hedron surface, have components s_{in}, $i = 1, 2, 3$. Then by consideration of the equilibrium of forces in the tetrahedron in the direction of the x_i axis

$$s_{in}\mathrm{d}F = \sigma_{ji}\mathrm{d}Fn_j$$

is obtained which yields instantly

$$s_{in} = \sigma_{ji}n_j. \tag{2.2.5}$$

The relation (2.2.5) permits calculation of the components s_{in} (and hence implicitly of (σ_n)) for any area through P having normal $\boldsymbol{n} = (n_i)$ from the components σ_{ji} of the given stress tensor. Thus, it has been proven that the stress tensor, as has been said, describes in full the stress condition in P.

Let a deformable body, which is in equilibrium, be subdivided arbitrarily. Let one of these divisions fill out volume V, which is closed off by surface O. Then the division, the same as the entire body, is in equilibrium. Consequently, it must be valid for the equilibrium of all forces acting in the x_i direction,

$$\int_V K_i \mathrm{d}V + \int_O s_{in}\,\mathrm{d}O = 0. \tag{2.2.6}$$

Inserting the relation (2.2.5) in (2.2.6), we obtain

$$\int_V K_i \mathrm{d}V + \int_O \sigma_{ji}n_j\,\mathrm{d}O = 0. \tag{2.2.7}$$

This expression can be changed by using Gauss' integral theorem, since

$$\int_O \sigma_{ji} n_j \mathrm{d}O = \int_V \frac{\partial \sigma_{ji}}{\partial x_j} \mathrm{d}V$$

so that (2.2.7) changes to

$$\int_V \left(K_i + \frac{\partial \sigma_{ji}}{\partial x_j}\right) \mathrm{d}V = 0. \tag{2.2.8}$$

Since volume V, over which we are integrating, represents an arbitrary part of the body's total volume, and since the integrand of (2.2.8) is assumed to be continuous, then (2.2.8) can only be satisfied, if

$$\frac{\partial \sigma_{ji}}{\partial x_j} = -K_i \tag{2.2.9}$$

is valid for every point of the body. In this way we have obtained the *equilibrium conditions* valid for an elastic continuum. Let us add the *boundary conditions*. They are obtained, according to (2.2.5), as

$$p_i = \sigma_{ji} n_j, \tag{2.2.10}$$

where vector p_i has substituted s_{in} in (2.2.5). This vector represents the surface force per unit area operating on the surface element having the normal n_j.

If we want to know about the equilibrium of all moments, we are led to

$$M_i = \int_V \varepsilon_{ijk} x_j K_k \mathrm{d}V + \int_O \varepsilon_{ijk} x_j s_{kn} \mathrm{d}O = 0. \tag{2.2.11}$$

Let us convert the second integral, i.e.,

$$\int_O \varepsilon_{ijk} x_j s_{kn} \mathrm{d}O.$$

Through (2.2.5) we obtain

$$\int_O \varepsilon_{ijk} x_j s_{kn} \mathrm{d}O = \int_O \varepsilon_{ijk} x_j \sigma_{lk} n_l \mathrm{d}O,$$

which when applying Gauss' integral theorem changes to

$$\int_O \varepsilon_{ijk} x_j s_{kn} \mathrm{d}O = \int_V \frac{\partial}{\partial x_l}(\varepsilon_{ijk} x_j \sigma_{lk}) \mathrm{d}V = \int_V \varepsilon_{ijk} \left(\delta_{jl}\sigma_{lk} + x_j \frac{\partial \sigma_{lk}}{\partial x_l}\right) \mathrm{d}V. \tag{2.2.12}$$

But it is $\delta_{jl}\sigma_{lk} = \sigma_{jk}$ and $\partial\sigma_{lk}/\partial x_l = -K_k$, which yields

$$\int_O \varepsilon_{ijk} x_j s_{kn} \mathrm{d}O = \int_V \varepsilon_{ijk}(\sigma_{jk} - x_j K_k) \mathrm{d}V$$

for (2.2.12).

Inserting this relation in (2.2.11), then it is

$$\int_V \varepsilon_{ijk} \sigma_{jk} \, \mathrm{d}V = 0.$$

Again this is only possible if

$$\underset{V}{\varepsilon_{ijk}\sigma_{jk}} = 0$$

is valid. With respect to the special characteristic of the ε-tensor, this leads to the condition

$$\sigma_{jk} = \sigma_{kj}, \tag{2.2.13}$$

i.e., the stress tensor is *symmetric*. Let us note that the symmetry of the stress tensor is not given at any conditions at all. Instead, it is linked to the condition that no couple stresses occur in the surface forces. Otherwise, the equilibrium condition (2.2.11) had contained more terms and we would not have arrived at (2.2.13). When using the theory of the Cosserat-continuum, we calculate indeed with a non-symmetric stress tensor.

Let us show that we are really dealing with a tensor. To do this, let us set up stress vector $\boldsymbol{\sigma}_n$ at one time in the cartesian system of axes x_i and at another time in the rotated system of axes x_i' (Figure 14).

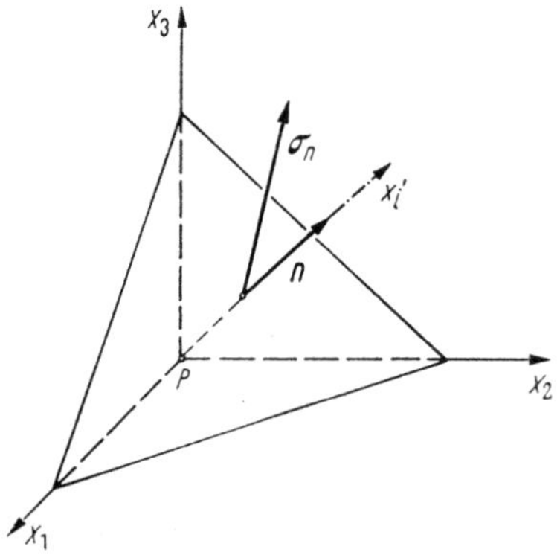

Figure 14

Let the latter be in such a position that one of its axes coincides with normal $\boldsymbol{n}$ of the plane on which σ_n is acting. Relative to the x_i'-system, $\boldsymbol{\sigma}_n$ contains components σ_{ij}', because they belong to stress component vectors operating on a surface which has a normal in the x_i'-direction, and the stress component vectors indicate the respective direction of one of the axes x_j' of the dashed (rotated) system of axes. At the same time, σ_{ij}' is also the projection of $\boldsymbol{\sigma}_n$ on the x_j'-direction. Thus

$$\sigma_{ij}' = \boldsymbol{\sigma}_n \boldsymbol{e}_j' \tag{2.2.14}$$

is valid, where e_j' is the unit vector of the x_j'-direction. Relative to the x_i-system, there is the component representation $\boldsymbol{\sigma}_n = (s_{kn})$ and $\boldsymbol{e}_j' = (a_{jk})$, where as usual, $a_{jk} = \cos(x_j', x_k)$. From (2.2.14),

$$\sigma_{ij}' = s_{kn} a_{jk}. \tag{2.2.15}$$

Following (2.2.5), relative to the x_i-system there is also

$$s_{kn} = \sigma_{lk} n_l. \tag{2.2.16}$$

In our case, the normal $\boldsymbol{n} = (n_l)$ coincides with the x_i'-axis of the rotated system. Thus, in particular,

$$n_l = \cos(x_i', x_l) = a_{il} \tag{2.2.17}$$

is valid. Using (2.2.16) and (2.2.17), (2.2.15) changes to

$$\sigma'_{ij} = \sigma_{lk} a_{il} a_{jk}, \tag{2.2.18}$$

which represents the transformation law of a cartesian tensor. Thus, we are justified to speak of the stress *tensor*.

Similarly, as in the strain tensor through the measure-ellipsoid, we can give a model representation for the stress tensor at the suggestion of G. Lamé using the *stress-ellipsoid*: let us put a surface-element, with normal $\boldsymbol{n}$, through point P of the continuum. Relative to the surface element operates stress vector $\boldsymbol{\sigma}_n$. If we imagine an s_{in}-system of axes set up in P, then we are permitted to use the components s_{in} of a specific stress vector $\boldsymbol{\sigma}_n$ to determine point P^* in the s_{in}-space (Figure 15). Changing the orientation of the surface-element in P, then little by little, we obtain different points P^* which fill up the surface of an ellipsoid. To determine this ellipsoid mathematically, we start from

$$n_i n_i = 1. \tag{2.2.19}$$

Then, from (2.2.5), we set up the relation

$$n_i = \sigma_{ij}^{-1} s_{jn}. \tag{2.2.20}$$

Analogous to (2.1.43),

$$\sigma_{ij}^{-1} = \frac{\text{co}\,(\sigma_{ij})}{|\sigma_{ij}|}$$

is valid. By inserting (2.2.20) in (2.2.19), we obtain

$$\sigma_{ij}^{-1} \sigma_{ik}^{-1} s_{jn}\, s_{kn} = 1 \text{ (do not add with respect to } n!). \tag{2.2.21}$$

This is the equation of an ellipsoid in the s_{in}-space. The radius vectors of this stress-ellipsoid indicate clearly magnitude and direction of all possible stress vectors $\boldsymbol{\sigma}_n$ at the reference point P. Thus, the stress-ellipsoid is a representation of the state of stress in P.

The radius vector s_{in} of the stress-ellipsoid does not generally coincide with the direction of normal n_i of its reference surface. However, there will be special directions through P for which, as an exception, this is true.

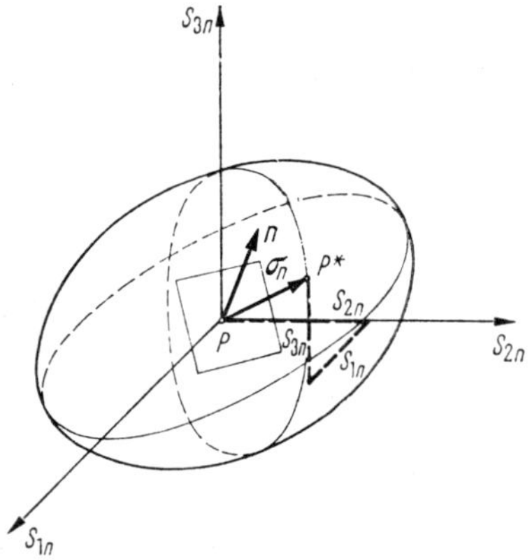

Figure 15

Then, s_{in} does not only have the same direction as normal n_i of the reference surface, but also as the normal of the tangential plane of the strain ellipsoid in that image point P^*, which is determined on the surface of the ellipsoid by s_{in}. Then s_{in} and n_i are proportional, which can be expressed using the proportionality factor σ by

$$s_{in} = \sigma n_i. \tag{2.2.22}$$

Using (2.2.5) and relation $n_i = \delta_{ji} n_j$, then (2.2.22) changes to

$$(\sigma_{ji} - \sigma\delta_{ji})n_j = 0. \tag{2.2.23}$$

This is a system of equations to determine the specific directions, marked by n_j, and which represent the *principal directions* of the stress tensor.

As in 2.1.4, we require that

$$\det\,(\sigma_{ji} - \sigma\delta_{ji}) = 0. \tag{2.2.24}$$

From this characteristic equation, which again is cubic and can be represented as

$$\sigma^3 - J_{1s}\sigma^2 + J_{2s}\sigma - J_{3s} = 0, \tag{2.2.25}$$

we obtain the three *principal stresses*, σ_i, $i = 1, 2, 3$, which may be equal to each other in special cases. The coefficients J_{is}, $i = 1, 2, 3$, from (2.2.25) are the invariants of the stress tensor for which corresponding to (2.1.74) and (2.1.75),

$$
\begin{aligned}
J_{1s} &= \sigma_1+\sigma_2+\sigma_3 = \sigma_{ii} = 3\sigma_M\,,\\
J_{2s} &= \sigma_1\sigma_2+\sigma_2\sigma_3+\sigma_3\sigma_1 = \frac{1}{2!}\delta^{ij}_{pq}\sigma_{pi}\sigma_{qj}\,,\\
J_{3s} &= \sigma_1\sigma_2\sigma_3 = \frac{1}{3!}\delta^{ijk}_{pqr}\sigma_{pi}\sigma_{qj}\sigma_{rk}
\end{aligned}
\tag{2.2.26}
$$

is valid. The *mean normal stress* $\sigma_M = \frac{1}{3}\sigma_{ii}$ has been used in this representation of J_{1s}.

The roots σ_i of (2.2.25), i.e., the principal stresses, are used to calculate the unit vector $n_j^{(k)}$ (k constant), $j = 1, 2, 3$, from

$$
\begin{aligned}
(\sigma_{ji}-\sigma_k\delta_{ji})n_j^{(k)} &= 0,\\
n_i^{(k)}n_i^{(k)} &= 1.
\end{aligned}
\tag{2.2.27}
$$

The unit vectors determine the principal direction belonging to the principal stress σ_k (k constant). Again, only three equations are independent in (2.2.27).

The principal directions of the stress tensor are the directions of the principal axes of the stress ellipsoid. The principal stresses correspond to the magnitudes of the semi-axes of the ellipsoid. Considering this, we can deduce that there are three principal stresses, of which two in relation to all possible moduli of stress vectors $\boldsymbol{\sigma}_n$ are extremal in P. There are also three principal directions. If two of the principal stresses are equal, then the stress ellipsoid becomes rotational-symmetrical. Then the equation (2.2.7) yields only one principal direction. The other two principal directions can be any pair of orthogonal directions through reference point P, and which set up an orthogonal right-hand system, together with the first principal direction. If all three principal directions are equal, then the stress ellipsoid degenerates into a sphere and the stress tensor into the spherical tensor

$$\sigma_{ij} = \sigma_M\delta_{ij}\,. \tag{2.2.28}$$

The principal directions are given by any orthogonal direction-triple, which corresponds to a right-hand system in P. Similarly, as in the case of the strain tensor, we can work with Cauchy's stress quadric

$$\sigma_{ij} x_i x_j = \pm k^2 \tag{2.2.29}$$

for representing the stress tensor. We shall not discuss this in detail, but point out [7, 8]. Corresponding to (2.1.78), the stress tensor can be represented by

$$\sigma_{ij} = \sigma_{\mathrm{M}} \delta_{ij} + \sigma_{ij}^{(0)}, \tag{2.2.30}$$

and thus be divided into two parts. The first part is the *spherical tensor* (2.2.28). The second part is the *stress-deviator*

$$\sigma_{ij}^{(0)} = \sigma_{ij} - \sigma_{\mathrm{M}} \delta_{ij}. \tag{2.2.31}$$

Its components are the *stress-deviations*, and the first invariant J_{1s^0} of the stress-deviator is equal to nought, since, because of $\delta_{ii} = 3$ and $\sigma_{\mathrm{M}} = \sigma_{ii}/3$,

$$\sigma_{ii}^{(0)} = \sigma_{ii} - \tfrac{1}{3}\sigma_{ii} \cdot 3 = \sigma_{ii} - \sigma_{ii} = 0 \tag{2.2.32}$$

is valid.

2.2.3 *The method of Mohr's circles*

Using the principal axes system of the stress tensor as the reference system, then the following is valid for the components s_{in} of stress vectors $\boldsymbol{\sigma}$ at a point P

$$s_{in} = \sigma_i n_i, \quad i = 1, 2, 3, \quad \text{(do not add with respect to } i\text{!)}. \tag{2.2.33}$$

Hence, it follows from (2.2.2)

$$\sigma_n = s_{in} n_i = \sigma_{(i)} n_i n_i = \sigma_1 n_1^2 + \sigma_2 n_2^2 + \sigma_3 n_3^2. \tag{2.2.34}$$

For the magnitude of the stress vector, we obtain from (2.2.3)

$$|\boldsymbol{\sigma}_n|^2 = \sigma_n^2 + \tau_n^2. \tag{2.2.35}$$

But since $|\boldsymbol{\sigma}_n|^2 = s_{in} s_{i(n)}$, then using (2.2.33) it follows from (2.2.35) that

$$|\sigma_n|^2 = \sigma_n^2 + \tau_n^2 = (\sigma_1 n_1)^2 + (\sigma_2 n_2)^2 + (\sigma_3 n_3)^2. \tag{2.2.36}$$

Let us consider identity

$$\left(\sigma_n - \frac{\sigma_2+\sigma_3}{2}\right)^2 + \tau_n^2 = -\sigma_n(\sigma_2+\sigma_3) + \left(\frac{\sigma_2+\sigma_3}{2}\right)^2 + (\sigma_n^2+\tau_n^2). \tag{2.2.37}$$

We change the right-hand side of (2.2.37) by inserting for σ_n (2.2.34) and for $\sigma_n^2+\tau_n^2$ the right-hand side of (2.2.36). Following this, we obtain after simple calculations, the relation

$$\left(\sigma_n - \frac{\sigma_2+\sigma_3}{2}\right)^2 + \tau_n^2 = n_1^2(\sigma_1-\sigma_2)(\sigma_1-\sigma_3) + \left(\frac{\sigma_2-\sigma_3}{2}\right)^2. \tag{2.2.38}$$

Through (2.2.38), a circle is established in the σ_n, τ_n-plane, having centre M_1 at $(\sigma_2+\sigma_3)/2$ on the σ_n-axis, and having the variable radius

$$R(n_1) = \left[n_1^2(\sigma_1-\sigma_2)(\sigma_1-\sigma_3) + \left(\frac{\sigma_2-\sigma_3}{2}\right)\right]^{\frac{1}{2}} \tag{2.2.39}$$

(see Figure 16). Then $n_1 = \cos\alpha_1$, and α_1 is the angle between the normal $\boldsymbol{n}$ and the x_1-axis. The radius R can, therefore, be given as

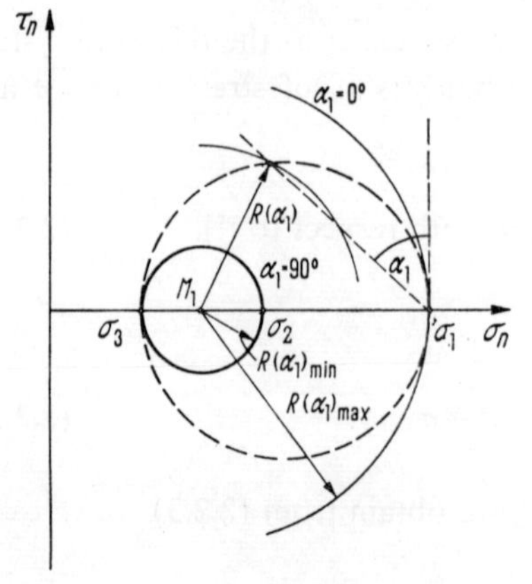

Figure 16

$$R = R(\alpha_1),$$

i.e., as the function of angle α_1.

We obtain a circle for any n_1 and thus any α_1, which runs concentric between the two for $\alpha_1 = 90°$ and $\alpha_1 = 0°$ valid limiting circles. The radii of the limiting circles are

$$R(\alpha_1)_{\min} = R(\alpha_1 = 90°) = \frac{\sigma_2 - \sigma_3}{2},$$

$$R(\alpha_1)_{\max} = R(\alpha_1 = 0°) = \sigma_1 - \frac{\sigma_2 + \sigma_3}{2},$$

which can be calculated for $\alpha_1 = 90°$ $(n_1 = 0)$ and $\alpha_1 = 0°$ $(n_1 = 1)$ from (2.2.39). The determination of $R(\alpha_1)$ for $0° < \alpha < 90°$, can be carried out graphically. Carrying out the construction when the assumed values $\sigma_1, \sigma_2, \sigma_3$ (these can always be assumed as being $\sigma_1 > \sigma_2 > \sigma_3$, without restriction of generality) as well as n_1 (α_1, respectively), can be taken from Figure 16 without special explanation.

As a first geometric location for the image point (σ_n, τ_n), which characterizes the stress vector $\boldsymbol{\sigma}_n$ in the σ_n, τ_n-plane, we have found a circle with centre M_1, and radius $R(\alpha_1)$. To determine the image point for σ_n definitely, at least one more geometric location has to be known. Two more are obtained by changing, in the same way as for (2.2.37), the two expressions:

$$\left(\sigma_n - \frac{\sigma_3 + \sigma_1}{2}\right)^2 + \tau_n^2 \text{ and } \left(\sigma_n - \frac{\sigma_1 + \sigma_2}{2}\right)^2 + \tau_n^2.$$

Thus, we obtain the two additional circles of the σ_n, τ_n-plane, having as their centre M_2 and M_3, respectively, at $(\sigma_3 + \sigma_1)/2$, i.e., $(\sigma_1 + \sigma_2)/2$, on the σ_n-axis, as well as radii $R(\alpha_2)$ and $R(\alpha_3)$, respectively, (Figure 17). These circles intersect with the first circle about M_1 in the sought point (σ_n, τ_n). To determine this point, only two circles are needed. The third may be omitted or else serve to check the accuracy of the procedure. To keep a clear picture, only two circles have been drawn in Figure 17, i.e., those having $R(\alpha_1)$ and $R(\alpha_3)$. In case of $R(\alpha_2)$, only the two limiting circles for $\alpha_2 = 0°$ and $\alpha_2 = 90°$ have been represented. We take from Figure 17 that the image points fill out the shaded 'triangle' for possible stress

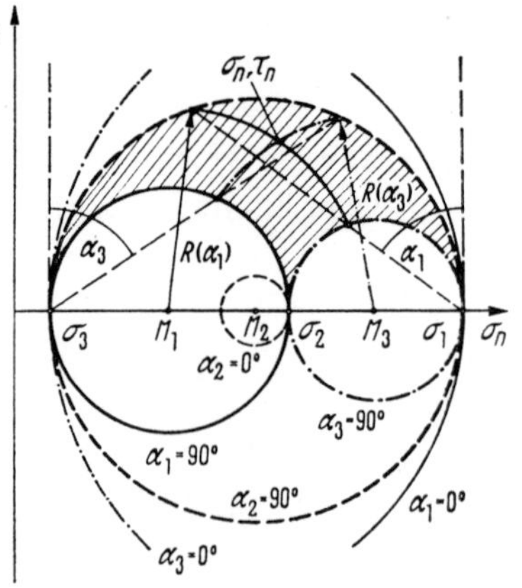

Figure 17

vectors $\boldsymbol{\sigma}_n$. This 'triangle' is formed by the limiting circles, which are set up for $\alpha_i = 90°$, $i = 1, 2, 3$.

We have assumed up to this point that the state of stress has been determined for reference point P by specification of principal stresses $\sigma_1, \sigma_2, \sigma_3$ and that for a reference surface passing through P and having the presupposed normal $\boldsymbol{n} = (n_1, n_2, n_3)$, the stress vector $\boldsymbol{\sigma}_n$ should be determined. Through specification of n_1, n_2, n_3 the angles $\alpha_1, \alpha_2, \alpha_3$ are known, which forms normal $\boldsymbol{n}$ with the principal axes x_1, x_2, x_3. To find σ_n, τ_n, i.e., the components which determine the vector $\boldsymbol{\sigma}_n$, by using Figure 17, to begin with, let us mark σ_1, σ_2 and σ_3 on the σ_n-axis. By dividing the intervals $\overline{\sigma_2\sigma_3}$, $\overline{\sigma_1\sigma_3}$ and $\overline{\sigma_1\sigma_2}$ into halves, the centres M_1, M_2 and M_3 are found. The limiting circles for $\alpha_1 = 90°$ ($\alpha_2 = 90°$, $\alpha_3 = 90°$) respectively are obtained, by applying the compass in points M_1 (M_2 and M_3, respectively), and drawing the circles with radii $(\sigma_2-\sigma_3)/2$, $(\sigma_1-\sigma_3)/2$ and $(\sigma_1-\sigma_2)/2$, respectively. Let us find the circles which pass through the wanted point σ_n, τ_n. Let us mark off on the vertical through σ_1 (σ_2, σ_3, respectively) the given angles α_1 (α_2, α_3, respectively). The respective sloped side of the angles α_1 (α_2, α_3) intersects at the limiting circles $\alpha_3 = 90°$, $\alpha_2 = 90°$ ($\alpha_1 = 90°$, $\alpha_3 = 90°$; $\alpha_1 = 90°$, $\alpha_2 = 90°$, respectively). Thus two respective points are obtained, which together with the respective centre M_1 (M_2, M_3) determines the determining circles passing through σ_n, τ_n. As stated, this construction has only been carried out for α_1 and α_3 in Figure 17. The graphic method of Mohr's circles can also be used in other problems. Let us show this by referring to a special case, which is of great practical significance.

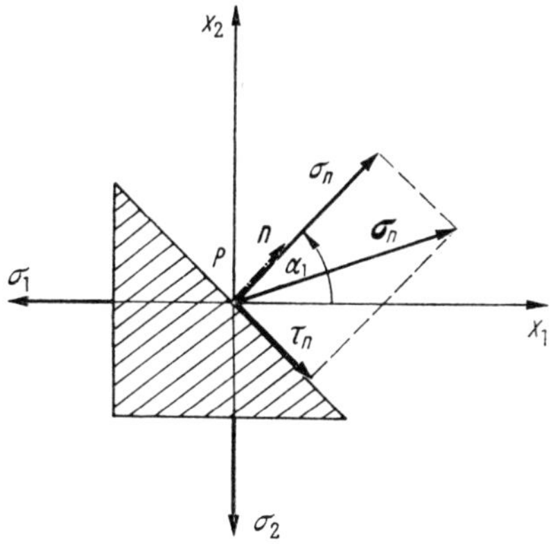

Figure 18

Let the following calculation be based on a *plane state of stress* for which $\sigma_3 \equiv 0$. Furthermore, let the stress vectors in P relative to such sections be determined, which are vertical to the stress-free x_1, x_2-plane (Figure 18). In such a case it is from the start $\alpha_3 = 90°$, so that the image point (σ_n, τ_n) must definitely be on the limiting circle passing through σ_1, σ_2 about M_3. The general construction of Figure 17 changes to the special construction of Figure 19. Let us mark off σ_1 and σ_2 on the σ_n-axis, let us divide interval $\overline{\sigma_1 \sigma_2}$ and find M_3. Let us draw a circle of radius $(\sigma_1 - \sigma_2)/2$ about M_3. The given angle α_1 (which is taken from Figure 18), is marked off the vertical through σ_1. Its sloped line intersects Mohr's circle at a point which, together with its coordinates σ_n and τ_n, yields the components of the wanted stress vector $\boldsymbol{\sigma}_n$. The procedure of the method is slightly altered if the angle α_1 is not marked off the vertical through σ_1, but the central angle $2\alpha_1$ is marked off from the σ_n-axis. Its sloped line does also meet the image of σ_n on Mohr's circle.

Let us note the following facts. If Figure 18 is termed as 'free-body diagram' and Figure 19 as 'stress plane', then we notice that to get along Mohr's circle from the image point $(\sigma_1, 0)$ to image point (σ_n, τ_n), the same orientation must be given to the central angle $2\alpha_1$ in the stress plane, as angle α_1 admits if in the free-body diagram we change from the direction x_1 (normal direction of the reference surface of σ_1) to the direction of $\boldsymbol{n}$ (normal direc-

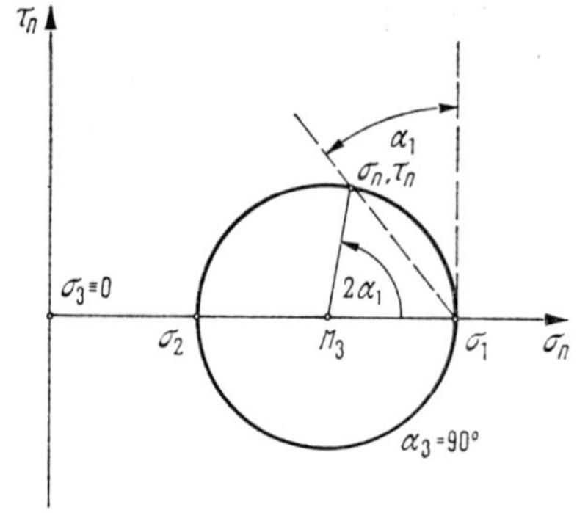

Figure 19

tion of the reference surface of σ_n). This *agreement of angle orientation* in the stress plane and the free-body diagram must be adhered to in all the problems to come. We also notice that τ_n, in Figure 19, is obtained as a positive quantity. Looking at Figure 18 shows that for a positive τ_n in the stress plane, the relative stress component vector runs in such a way in the free-body diagram that in its field of vision, as it were, that part of the body (the 'interior of the body'), which is affected by τ_n, lies on the right-hand side. To proceed with Mohr's method, *a special plus or minus convention* must be observed, which must be strictly distinguished from that applying to the stress-vector. It is: *positive values σ_n in the stress plane yield tensile stress component vectors in the free-body diagram; positive values τ_n in the stress plane lead to shearing stress component vectors, for which the interior of the body in the field of vision of the shearing stress-vectors lies to the right-hand side.*

Let us use Mohr's Method in the case of the plane state of stress, to solve three more basic problems:

a) Let the stress-tensor be given relative to the axes x_1, x_2 at a point P

$$\begin{pmatrix} \sigma_{11} & \sigma_{12} & 0 \\ \sigma_{21} & \sigma_{22} & 0 \\ 0 & 0 & 0 \end{pmatrix}. \tag{2.2.40}$$

Wanted are the principal directions 1, 2 and the principal strains σ_1, σ_2 in P.

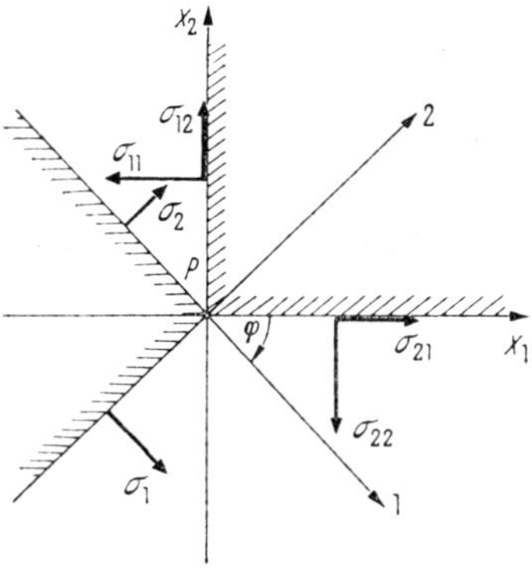

Figure 20

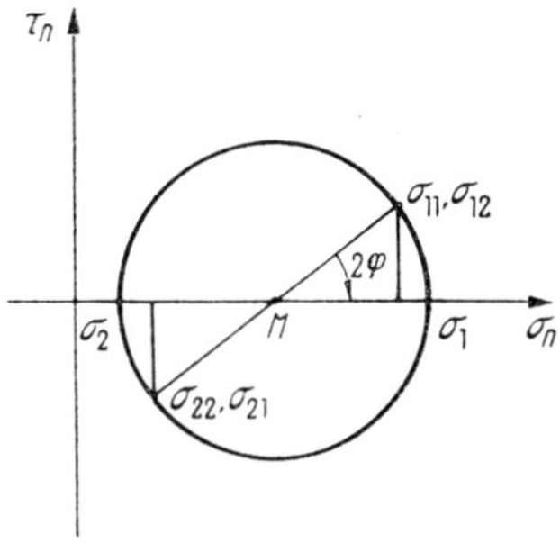

Figure 21

Looking at the free-body diagram (Figure 20), the diagonal components of the stress-tensor are positive for this example, since σ_{11} and σ_{22} are tensile stresses. The components $\sigma_{12} = \sigma_{21}$ are negative in comparison since their relative stress-vectors indicate on planes with normals, having the direction of negative axes, in the direction of positive axes. To work with Mohr's method, a different plus and minus sign agreement must be used. Accordingly, the following is true for this example: σ_{11} and σ_{22} remain positive as tensile stresses. The shearing stress σ_{12} is positive, since the interior of the body lies, as seen in direction of its stress-vector, on the right-hand side. Similarly, σ_{21} is negative. Considering these plus or minus signs and the known magnitudes, the image points $(\sigma_{11}, \sigma_{12})$ and $(\sigma_{22}, \sigma_{21})$ in the σ_n, τ_n plane in Figure 21 (stress plane) can be marked. The connecting line of these points cut off on the σ_n axis the centre M of Mohr's circle, which is drawn about M through the image points $(\sigma_{11}, \sigma_{12})$ and $(\sigma_{22}, \sigma_{21})$. The circle intersects with the σ_n-axis in the image points $(\sigma_1, 0)$ and $(\sigma_2, 0)$. These yield the principal stresses. They are both positive for our example, and are thus tensile stresses. In the stress plane we arrive at image point $(\sigma_1, 0)$ from image point $(\sigma_{11}, \sigma_{12})$ on Mohr's circle about angle 2φ in a clockwise-direction. Therefore, we must mark off in the free-body diagram angle φ with the same orientation of the x_1-direction in order to obtain principal direction 1. Principal direction 2 is vertical to it. The already known principal stresses σ_1 and σ_2 can be used to mark off in the free-body diagram their stress vectors in the order of magnitude and direction.

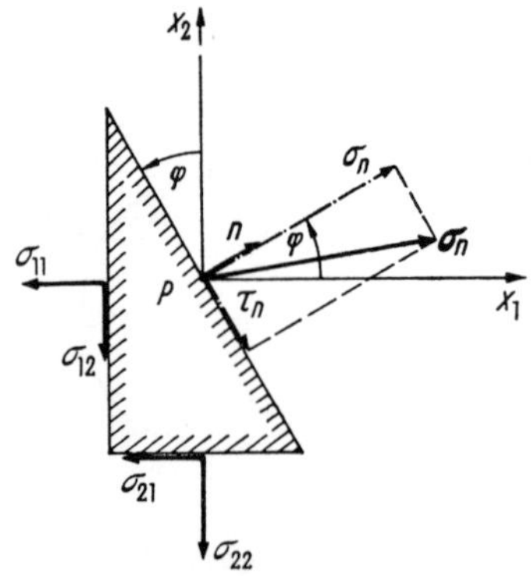

Figure 22

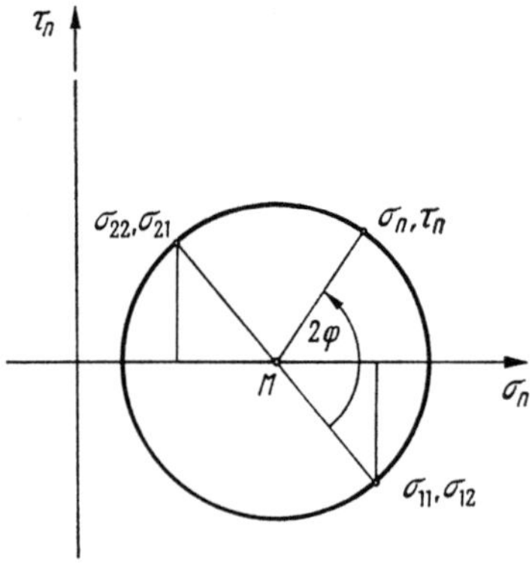

Figure 23

b) For a point P is given the stress-tensor (2.2.40), relative to axes x_1, x_2. To find stress vector $\boldsymbol{\sigma}_n$ for the section through P which forms angle φ with the x_2-axis.

It is evident from the free-body diagram (Figure 22), that this time not only the diagonal components of stress-tensor σ_{11}, σ_{22} are positive as tensile stresses, but that the shearing stress components $\sigma_{12} = \sigma_{21}$ have a positive sign also, since their stress-vectors point on the respective cuts (with normals having negative axial-direction) in the direction of negative axes. According to Mohr's plus or minus sign definition, which we will now use, the tensile stresses σ_{11}, σ_{12} are also positive, while the shearing stress σ_{21} is positive (body interior lies on the right-hand side) and the shearing stress σ_{12} is negative (body interior lies on the left-hand side). Using the corresponding magnitudes and plus or minus signs, according to Mohr's definition, the image points $(\sigma_{11}, \sigma_{12})$ and $(\sigma_{22}, \sigma_{21})$ are sought in the stress plane (Figure 23), which determine Mohr's circle. Therefore, it can be drawn. From the free-body diagram, angle φ is taken, having an orientation which is obtained by changing from the direction x_1 to the direction of normal $\boldsymbol{n}$ of the sloped cut. Using the same orientation, the angle 2φ is marked off from the radius of the circle through $(\sigma_{11}, \sigma_{12})$ as a central angle in the stress plane. Its free side intersects the circle in the wanted image point (σ_n, τ_n). The coordinates σ_n, τ_n are used as components in order to draw the stress-vector $\boldsymbol{\sigma}_n$, operating in the sloped cut in the free body diagram. To draw in the stress component vectors belonging to σ_n, τ_n, the plus or minus signs must be considered. σ_n, τ_n are positive in the stress plane, thus the σ_n and τ_n are drawn into the free-body diagram as tensile stress and shearing stress, respectively, so that the body interior lies to the right-hand side of the shearing stress.

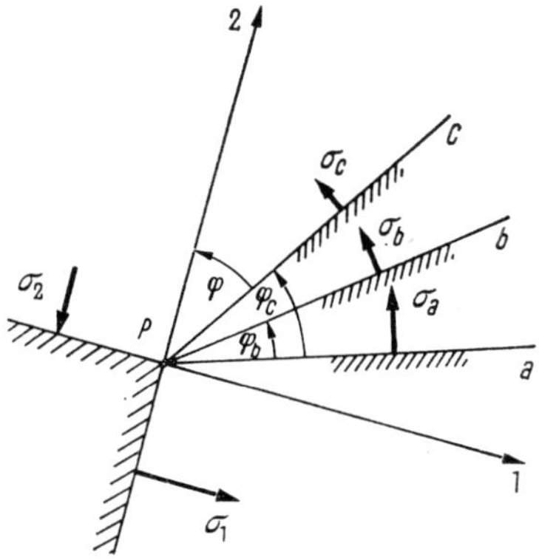

Figure 24

c) For a point P, three sections a, b, c are given together with the relative normal stresses σ_a, σ_b, σ_c. To find are the principal directions and principal stresses in P.

In the stress plane (Figure 25a), distances corresponding to the normal stresses σ_a, σ_b, σ_c are marked off vertically from the τ_n axis, with consideration of the plus and minus signs, and we draw three parallels to the τ_n-axis at these intervals. The position of the σ_n-axis is, as yet, unknown. Let us assume any point A on the parallel which is at an interval σ_a. At point A angles φ_b, φ_c, taken from the free-body diagram (Figure 24), are marked off from the parallel in their prescribed orientation. The free sides of the angles intersect on the other two parallels at points B and C. Thus, three points of Mohr's circles are being given by A, B, C, which can now be drawn. Since centre M is then known, the σ_n-axis is subsequently indicated. It is the straight line through M and vertical to the τ_n-axis. If the σ_n-axis has been found, then the image points are obtained, by which the wanted principal stresses σ_1 and σ_2 are determined. They are points of intersection of the σ_n-axis with Mohr's circle. In the stress plane (Figure 25b), we find the angle 2φ with its orientation.

It indicates how we arrive at the image point for σ_2 from the image point for σ_c on Mohr's circle. In this orientation, angle φ is marked off from

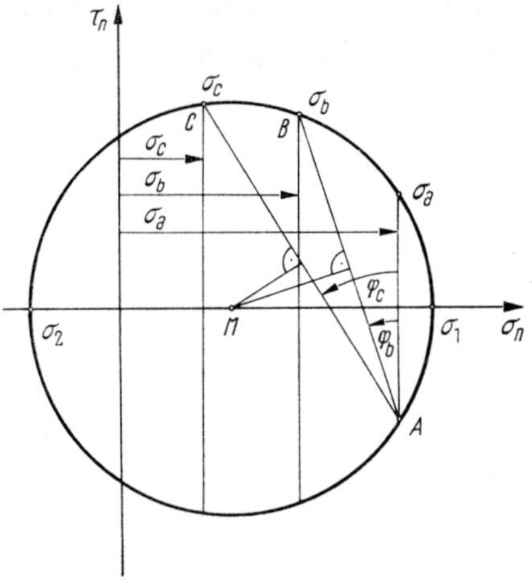

Figure 25a

the *c*-direction in the free-body diagram. Thus, the principal direction 2 is determined. Principal direction 1 is vertical to it. We read off the stress plane that σ_1 is positive and σ_2 is negative. Therefore, in the free-body diagram, we mark off, on the respective sections, σ_1 as a tensile stress and σ_2 as a compressive stress. The correctness of the described construction becomes evident from Figure 25b, when considering the relation consisting between central and peripheral angles.

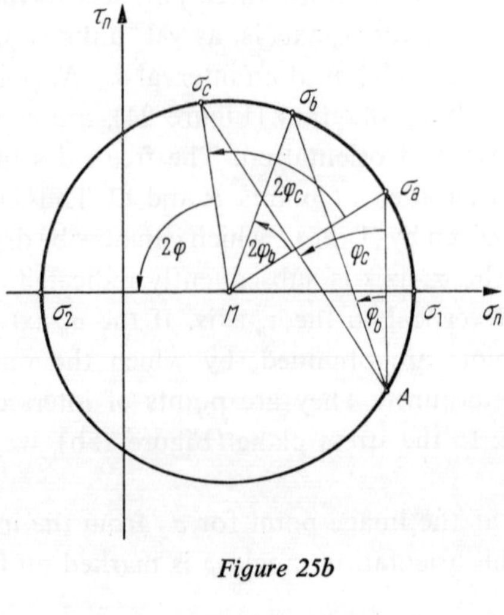

Figure 25b

As a supplement to 2.1.4, let us mention that Mohr's method can be used similarly for the strain tensor. Instead of working in an σ_n, τ_n-plane, we work in an ε_n, $\gamma_n/2$-plane, with the strain tensor. Let ε_n and $\gamma_2/2$ be the longitudinal strain and shearing strain of an intersection with normal $\boldsymbol{n}$ through reference point P. Instead of normal stresses and shearing stresses of the stress-tensor, longitudinal and shearing strain of the strain-tensor will take their place. Rational translation of the method is left to the reader.

2.3 The material law

2.3.1 *The anisotropic medium*

The material law represents a relation between the components of the stress and strain tensors or vice-versa. Up to now we are basically not able to indicate a relation of this kind, since we have discussed the strain tensor in 2.1, using Lagrange's coordinates and the stress tensor in 2.2, using Euler's coordinates. The existence of a relation between the components of both tensors pre-supposes that both are taken with respect to a common co-ordinate system. Let us overcome these difficulties by using the fact that the difference between Lagrange's and Euler's coordinates is so small for an infinitesimal state of deformation and strain, that we are permitted to neglect it. This will be shown later. We are permitted to assume in the following that both tensors to be combined are related to a common cartesian x_i-system.

Through tension tests with steel bars, and bars of other materials, it is known that for the relation between stress σ and strain ε, the following Hooke's law is valid in the elastic range,

$$\sigma = E \cdot \varepsilon. \tag{2.3.1}$$

This is a linear relation between stress and strain. Proportionality-constant E is the *modulus of elasticity* or *Young's modulus*. This originally phe-nomenological law has, in the meantime, been proven physically. See [10].

In accordance with (2.3.1), and in generalizing Hooke's law for multi-dimensional states of stress and strain, let us write

$$\sigma_{ij} = C_{ijkl}\varepsilon_{kl}, \quad i, j, k, l = 1, 2, 3. \tag{2.3.2}$$

Let the medium be homogeneous. Then the coefficients C_{ijkl} are independent from the position of the reference point in the medium, and are therefore 'elastic constants'. It can be shown that they are components of a fourth order tensor, which is termed *elasticity-tensor*. Considering the symmetry of the stress-tensor, then the elasticity-tensor is symmetric relative to the first two indices:

$$C_{ijkl} = C_{jikl}, \tag{2.3.3}$$

and considering the symmetry of the strain-tensor, it is also symmetric relative to the last two indices

$$C_{ijkl} = C_{ijlk}. \tag{2.3.4}$$

For this reason, and by introducing new notations instead of C_{ijkl}, we can write for (2.3.2)

$$\begin{aligned}
\sigma_{11} &= A_{11}\varepsilon_{11}+A_{12}\varepsilon_{22}+A_{13}\varepsilon_{33}+A_{14}\varepsilon_{23}+A_{15}\varepsilon_{31}+A_{16}\varepsilon_{12},\\
\sigma_{22} &= A_{21}\varepsilon_{11}+A_{22}\varepsilon_{22}+A_{23}\varepsilon_{33}+A_{24}\varepsilon_{23}+A_{25}\varepsilon_{31}+A_{26}\varepsilon_{12},\\
&\cdots\cdots\cdots\cdots\cdots\cdots\\
\sigma_{23} &= A_{41}\varepsilon_{11}+A_{42}\varepsilon_{22}+A_{43}\varepsilon_{33}+A_{44}\varepsilon_{23}+A_{45}\varepsilon_{31}+A_{46}\varepsilon_{12},\\
&\cdots\cdots\cdots\cdots\cdots\cdots\\
\sigma_{12} &= A_{61}\varepsilon_{11}+A_{62}\varepsilon_{22}+A_{63}\varepsilon_{33}+A_{64}\varepsilon_{23}+A_{65}\varepsilon_{31}+A_{66}\varepsilon_{12}.
\end{aligned} \tag{2.3.5}$$

Thus, the generalized Hooke's law for *anisotropic bodies* having 36 elasticity constants A_{ik}, $i, k = 1, 2, 3, \ldots, 6$ has been set up. If a volume element in a state of stress is subjected to a virtual strain $\delta\varepsilon_{ij}$, then the stress components σ_{ij} yield the work

$$\delta A_{(i)} = -\sigma_{ij}\delta\varepsilon_{ij}. \tag{2.3.6}$$

Introducing *specific strain energy, specific elastic potential*, respectively, through definition $U = -A_{(i)}$, then for the variation of the strain energy, because of $\delta U = -\delta A_{(i)}$, we have following (2.3.6)

$$\delta U = \sigma_{ij}\delta\varepsilon_{ij}. \tag{2.3.7}$$

It is assumed for U that it is a single-valued function

$$U = U(\varepsilon_{ij}) \tag{2.3.8}$$

of the strain components. Then

$$\delta U = \frac{\partial U}{\partial \varepsilon_{ij}} \delta \varepsilon_{ij} \tag{2.3.9}$$

is also valid. Comparing (2.3.9) and (2.3.7), yields

$$\sigma_{ij} = \frac{\partial U}{\partial \varepsilon_{ij}}. \tag{2.3.10}$$

Let us use the relation

$$\frac{\partial^2 U}{\partial \varepsilon_{ij} \partial \varepsilon_{kl}} = \frac{\partial^2 U}{\partial \varepsilon_{kl} \partial \varepsilon_{ij}}.$$

Because of (2.3.10), it yields

$$\frac{\partial \sigma_{ij}}{\partial \varepsilon_{kl}} = \frac{\partial \sigma_{kl}}{\partial \varepsilon_{ij}}. \tag{2.3.11}$$

Let us write, for example, $i = 1$, $j = 2$, $k = 2$, $l = 3$. Then we obtain from (2.3.11)

$$\frac{\partial \sigma_{12}}{\partial \varepsilon_{23}} = \frac{\partial \sigma_{23}}{\partial \varepsilon_{12}}.$$

According to (2.3.5), we obtain

$$A_{64} = A_{46}$$

from it, and generally $A_{ik} = A_{ki}$. Coefficients of (2.3.5) are thus symmetric. On the condition that a specific elastic potential does exist, which is a single-valued function of strain-tensors components, *the number of elastic components is thus reduced from 36 to 21.*

2.3.2 *The isotropic medium*

If an elastic continuum is the same in any direction, it is called *isotropic.* An isotropic medium requires that the elasticity-tensor is not influenced by any rotation of the system of axes.

Let the elasticity-tensor, with respect to a cartesian reference system x_i be represented by C_{ijkl} and with respect to a rotated system x_i' by C'_{ijkl}. Then, on one hand, because C_{ijkl} represents a 4th order tensor, the transformation law is

$$C_{pqrs} = a_{ip}\,a_{jp}\,a_{kr}\,a_{ls}\,C'_{ijkl}, \; a_{mn} = \cos(x'_m, x_n) \tag{2.3.12}$$

and, on the other hand, because of the immunity against the rotation of the reference system,

$$C'_{ijkl} = C_{ijkl}. \tag{2.3.13}$$

From (2.3.12) and (2.3.13) follows the condition

$$C_{pqrs} = a_{ip}\,a_{jq}\,a_{kr}\,a_{ls}\,C_{ijkl}, \tag{2.3.14}$$

which is only satisfied if C_{ijkl} is

$$C_{pqrs} = \lambda\delta_{pq}\,\delta_{rs} + \mu\delta_{pr}\,\delta_{qs} + \kappa\delta_{ps}\,\delta_{qr}. \tag{2.3.15}$$

λ, μ and κ are elastic constants in (2.3.15). The accuracy of (2.3.15) is checked by inserting this relation in (2.3.14). We observe that (2.3.14) is actually satisfied by the relation (2.3.15).

The symmetry of C_{ijkl} in relation to the two front and two back indices, mentioned in 2.3.1, must yield

$$C_{ijkl} = C_{jikl} = C_{ijlk} \tag{2.3.16}$$

in accordance with (2.3.3) and (2.3.4). Notating the condition (2.3.16) by using (2.3.15) it becomes evident, when considering the symmetry of the unit tensor δ_{ij}, that (2.3.16) is only possible if

$$(\kappa-\mu)(\delta_{ik}\delta_{jl}-\delta_{il}\delta_{jk}) = 0. \tag{2.3.17}$$

Let $i = k, j = l$ in (2.3.17). Then, $\delta_{ik}\delta_{jl} = 9$, $\delta_{il}\delta_{jk} = 3$ is valid, and (2.3.17) can only be accurate if $\kappa - \mu = 0$ is true. It follows that $\kappa = \mu$. Thus *the number of elastic constants for a homogeneous isotropic medium has been reduced to two*. They are *Lamé's constants* λ and μ. (2.3.15) is written as

$$C_{ijkl} = \lambda\delta_{ij}\delta_{kl} + \mu(\delta_{ik}\delta_{jl} + \delta_{jl}\delta_{jk}) \tag{2.3.18}$$

in its simpler form.

Let us notate the generalized version of Hooke's law (2.3.2) by using (2.3.18). It changes to

$$\sigma_{ij} = \lambda\delta_{ij}\delta_{kl}\varepsilon_{kl} + \mu(\delta_{ik}\delta_{jl} + \delta_{il}\delta_{jk})\varepsilon_{kl},$$

and after conversion, by using the characteristics of the unit tensor, it is

$$\sigma_{ij} = \lambda\delta_{ij}\varepsilon_{kk} + 2\mu\varepsilon_{ij}. \tag{2.3.19}$$

In (2.3.19) we have obtained Hooke's law in its special form, valid for homogeneous, isotropic, elastic bodies.

It is of interest to solve (2.3.19) for ε_{ij}. It follows from (2.3.19)

$$\varepsilon_{ij} = \frac{1}{2\mu}\sigma_{ij} - \frac{\lambda}{2\mu}\delta_{ij}\varepsilon_{kk}. \tag{2.3.20}$$

Through contraction ($i = j$, $\delta_{ii} = 3$) and subsequent renaming of dummy indices, we obtain from (2.3.20)

$$\varepsilon_{kk} = \frac{\sigma_{kk}}{3\lambda + 2\mu}. \tag{2.3.21}$$

Thus (2.3.20) is changed to

$$\varepsilon_{ij} = \frac{1}{2\mu}\sigma_{ij} - \frac{\lambda}{2\mu(3\lambda + 2\mu)}\delta_{ij}\sigma_{kk}. \tag{2.3.22}$$

We introduce new notations by using *Poisson's ratio* ν and Young's modulus E. These two other elastic constants are defined by

$$\frac{1}{2\mu} = \frac{1+\nu}{E}, \quad \frac{\lambda}{2\mu(3\lambda+2\mu)} = \frac{\nu}{E}, \tag{2.3.23}$$

which indicates that they are not independent from the initially introduced Lamé constants λ, μ. Through (2.3.23), (2.3.22) obtains the form

$$\varepsilon_{ij} = \frac{1+\nu}{E}\sigma_{ij} - \frac{\nu}{E}\delta_{ij}\sigma_{kk}, \tag{2.3.24}$$

which subsequently will often be used.

Another important fact for an isotropic medium is as follows. In this, the strain and stress tensors have the same principal direction, i.e., the same system of principal axes. They are termed as being *coaxial.*

To prove this fact let us assume that the strain tensor has been represented in its system of principal axes. The exceptional feature in this representation is that all components ε_{ij} containing $i \neq j$ vanish. Thus

$$\varepsilon_{ij} = 0 \text{ for } i \neq j \tag{2.3.25}$$

is valid for the principal axes system. To use Hooke's law in the form (2.3.19) we must assume that σ_{ij} is relative to the system of axes elected for ε_{ij}. Stress tensor σ_{ij} has thus been taken for the principal axes system of the strain tensor. Writing (2.3.19) for $i \neq j$, then because of $\delta_{ij} = 0$ for $i \neq j$, and because of (2.3.25)

$$\sigma_{ij} = 0 \text{ for } i \neq j$$

is yielded. It can be deduced that the used system of axes represents a principal axes system for the stress tensor also. Thus, the principal axes systems of strain and stress tensor coincide, and both tensors are actually coaxial.

2.3.3 *The relation to the basic laws of rheology*

Within the limits of rheology, separate descriptions are given for the cubical dilatation and the shear deformation of bodies. Thus, two rheological basic laws are set up. The first law indicates that the bodies reacts elastically to volume change. Thus it is

$$\sigma_M = K\Theta. \tag{2.3.26}$$

In (2.3.26) $\sigma_M = \sigma_{ii}/3$ represents the *normal mean stress*, $\Theta = \varepsilon_{ii}$ the relative *cubical dilatation*, and the proportionality factor K represents the *bulk modulus*. (2.3.26) can be represented as

$$\sigma_{ii} = 3K\varepsilon_{ii} \tag{2.3.27}$$

because of the significance of σ_M and Θ. To describe the shear deformation, the relation

$$a_0 + a_1\,\varepsilon_{ij}^{(0)} + a_2\,\dot{\varepsilon}_{ij}^{(0)} + a_3\,\sigma_{ij}^{(0)} + a_4\,\dot{\sigma}_{ij}^{(0)} = 0 \tag{2.3.28}$$

is generally set up, within the limits of a linear theory, as the second rheological basic law. In (2.3.28), the deviators of the strain tensor, stress tensor, respectively, as well as their derivatives with respect to time, $\dot{\varepsilon}_{ij}^{(0)}$, $\dot{\sigma}_{ij}{}^{(0)}$, come to hand. These quantities are related in (2.3.28) through constants a_0 to a_4. The special laws for the different rheological bodies are obtained by imposing special conditions on these constants. The elastic *Hooke's body* is described, for example, by writing in (2.3.28)

$$\begin{aligned} a_0 &= a_2 = a_4 = 0, \\ a_1/a_3 &= -2\mu. \end{aligned} \tag{2.3.29}$$

Therefore, (2.3.28) changes to

$$\sigma_{ij}^{(0)} = 2\mu\varepsilon_{ij}^{(0)}. \tag{2.3.30}$$

Let us show that the two rheological partial laws (2.3.27) and (2.3.30) of Hooke's body do in actual fact, lead to Hooke's law. Thus would be proven that Hooke's body being a subject of the theory of elasticity, does actually adapt itself within the limits of *rheology*, as a special case.

Let us recalculate (2.3.30). According to (2.1.79), we write

$$\sigma_{ij}^{(0)} = 2\mu(\varepsilon_{ij} - \tfrac{1}{3}\Theta\cdot\delta_{ij}). \tag{2.3.31}$$

Inserting (2.3.26) and (2.3.31) in (2.2.30),

$$\sigma_{ij} = K\Theta\delta_{ij} + 2\mu\varepsilon_{ij} - \frac{2\mu}{3}\,\Theta\delta_{ij}$$

is obtained. Transforming this by using $\Theta = \varepsilon_{kk}$, we obtain

$$\sigma_{ij} = (K - \tfrac{2}{3}\mu)\varepsilon_{kk}\delta_{ij} + 2\mu\varepsilon_{ij}. \tag{2.3.32}$$

From the relation

$$\lambda = K - \tfrac{2}{3}\mu \tag{2.3.33}$$

between bulk modulus K and Lamé's constants λ and μ, we obtain

$$\sigma_{ij} = \lambda\varepsilon_{kk}\delta_{ij} + 2\mu\varepsilon_{ij}$$

instead of (2.3.32).

Thus we have arrived at (2.3.19) i.e., Hooke's law, which had been set up in 2.3.2.

2.3.4 *The relation between the constants of an isotropic medium*

As is known, there are only two independent elastic constants for the isotropic continuum. A great number of different terms and definitions are used for these. Thus, *Lamé's constants* λ and μ, *Young's modulus* (i.e., *elasticity modulus*) E, *Poisson's ratio* ν and the *bulk modulus* K have been dealt with. Another constant, *shear modulus* G, is often used. Since only two of them are independent, certain relations must exist between them. Several of them we have met in (2.3.23) and (2.3.33).
Another exists in the observation $\mu \equiv G$. Various relations completely and systematically arranged are contained in the following table.

2.4 Elements of a theory of non-linear elasticity

2.4.1 *Different kinds of non-linearities*

Within the limit of elasticity theory, two kinds of non-linearities can occur simultaneously or independently from each other. Let us discuss each of them individually.

Let us assume in the first case that displacements u_i and their derivations, $\partial u_i/\partial \xi_i$ are not infinitesimal, but finite. Hence, the so-called *geometric*

	λ	$\mu \equiv G$	E	ν	K
λ, μ	λ	μ	$\frac{\mu(3\lambda+2\mu)}{\lambda+\mu}$	$\frac{\lambda}{2(\lambda+\mu)}$	$\lambda+\frac{2}{3}\mu$
λ, E	λ	$\frac{e-3\lambda+r^*}{4}$	E	$\frac{2\lambda}{E+\lambda+r}$	$\frac{E+3\lambda+r}{6}$
λ, ν	λ	$\frac{\lambda(1-2\nu)}{2\nu}$	$\frac{\lambda(1+\nu)(1-2\nu)}{\nu}$	ν	$\frac{\lambda(1+\nu)}{3\nu}$
λ, K	λ	$\frac{3}{2}(K-\lambda)$	$\frac{9K(K-\lambda)}{3K-\lambda}$	$\frac{\lambda}{3K-\lambda}$	K
μ, E	$\frac{\mu(E-2\mu)}{3\mu-E}$	μ	E	$\frac{E-2\mu}{2\mu}$	$\frac{\mu E}{3(3\mu-E)}$
μ, ν	$\frac{2\mu\nu}{1-2\nu}$	μ	$2\mu(1+\nu)$	ν	$\frac{2\mu(1+\nu)}{3(1-2\nu)}$
μ, K	$K-\frac{2}{3}\mu$	μ	$\frac{9K\mu}{6K+\mu}$	$\frac{3K-2\mu}{6K+2\mu}$	K
E, ν	$\frac{E\nu}{(1+\nu)(1-2\nu)}$	$\frac{E}{2(1+\nu)}$	E	ν	$\frac{E}{3(1-2\nu)}$
E, K	$\frac{3K(3K-E)}{9K-E}$	$\frac{3KE}{9K-E}$	E	$\frac{3K-E}{6K}$	K
ν, K	$\frac{3K\nu}{1+\nu}$	$\frac{3K(1-2\nu)}{2(1+\nu)}$	$3K(1-2\nu)$	ν	K

$^*r$ represents $\sqrt{(E^2+9\lambda^2+2E\lambda)}$ here, and next, and next to it.

non-linearity occurs. Since the condition of the elastic body at the occurrence of finite deformations is quite different before loading than it is afterwards, then the difference between Lagrange's and Euler's representation must be strongly differentiated when the strain condition is described. In 2.1.3 to 2.1.5, we have so far used coordinates ξ_i of the non-deformed body as independent variable, and have thus used Lagrange's description method for the state of strain. Therefore, let us make up for Euler's description method on the condition of finite deformations. In addition, non-linear terms will occur in Lagrange's as well as in Euler's presentation of components belonging to the strain tensors, since, different from 2.1.4, the products of the displacement derivatives are not to be neglected. The conditions of equilibrium are to be written for the deformed volume element as in 2.2.2, because of the comparative difference between the undeformed and deformed state of the bodies. Furthermore, it is necessary to go beyond the representation of 2.2.2, and to change the original Euler description of the equilibrium conditions to the Lagrange description, so that in this connection, two methods of description are at our disposal. In the process of change to Lagrange's notation method of the equilibrium conditions, again several non-linear terms originating from the finite deformation will be included in the equations.

The second case is at hand, if because of the extent of deformations, the proportionality limit is crossed, so that Hooke's law, which was introduced in Chapter 2, looses its validity. Then the linear connection determined in the material equations (2.3.27) and (2.3.30) between the respective variables, is no longer valid. Instead, non-linear relations must be introduced, and we talk about the occurrence of *physical non-linearity* of the continuum. That is, substituting linear by non-linear material equations, will be discussed below.

2.4.2 *Finite deformations*

Through the presence of finite deformations occurs, at least, the geometric non-linearity of the continuum. Let us consider these. As we have seen, these set up consequences for the representation of strain, as well as for the derivation of the equilibrium conditions. We shall discuss these problems in detail, carrying out the methods in one each of the following paragraphs.

2.4.2.1 *The state of strain in finite deformations.* In Lagrange's representation, (2.1.1*) is valid. Using cartesian coordinates, there is for the square of the line-elements, before deformation occurs,

$$ds_0^2 = d\xi_j \xi_j = \delta_{jk} d\xi_j d\xi_k . \tag{2.4.1}$$

Following deformation it is

$$ds^2 = dx_i dx_i = \frac{\partial x_i}{\partial \xi_j}\frac{\partial x_i}{\partial \xi_k} d\xi_j d\xi_k. \tag{2.4.2}$$

The components of the metric tensor

$$g_{jk} = \frac{\partial x_i}{\partial \xi_j}\frac{\partial x_i}{\partial \xi_k} \tag{2.4.3}$$

occur in (2.4.2), the latter determining the continuum's metric relative to Lagrange's coordinates. As a measure for the occurring strain, let us take $ds^2 - ds_0^2$. Using (2.4.1), (2.4.2) and (2.4.3), we find for it

$$ds^2 - ds_0^2 = \left(\frac{\partial x_i}{\partial \xi_j}\frac{\partial x_i}{\partial \xi_k} - \delta_{jk}\right) d\xi_j d\xi_k = (g_{jk} - \delta_{jk}) d\xi_j d\xi_k. \tag{2.4.4}$$

Usually we write $g_{jk} - \delta_{jk} = 2e_{jk}$, where e_{jk} is termed as *Lagrange's strain tensor*. For this tensor,

$$e_{jk} = \tfrac{1}{2}(g_{jk} - \delta_{jk}) \tag{2.4.5}$$

is valid. This relation is converted to

$$e_{jk} = \frac{1}{2}\left(\frac{\partial u_j}{\partial \xi_k} + \frac{\partial u_k}{\partial \xi_j} + \frac{\partial u_i}{\partial \xi_j}\frac{\partial u_i}{\partial \xi_k}\right) \tag{2.4.6}$$

by using (2.4.3) and (2.1.5). Comparison with the corresponding quantity ε_{jk}, which is obtained by assuming infinitesimal deformations and represented by (2.1.38), shows that e_{jk} is different from ε_{jk} by the occurrence of non-linear terms

$$\frac{1}{2}\frac{\partial u_i}{\partial \xi_j}\frac{\partial u_i}{\partial \xi_k} \tag{2.4.7}$$

in (2.4.6). Assuming that the partial derivatives $\partial u_i/\partial \xi_j$ of the displacements are very small, then we are permitted to neglect the sum (2.4.7) of the non-linear terms of e_{jk} and Lagrange's strain tensor e_{jk} changes to the simpler tensor ε_{jk}.

In Euler's representation, (2.1.6*) is valid. From this, the relations

$$\mathrm{d}s_0^2 = \mathrm{d}\xi_i \mathrm{d}\xi_i = \frac{\partial \xi_i}{\partial x_j}\frac{\partial \xi_i}{\partial x_k}\mathrm{d}x_j \mathrm{d}x_k \tag{2.4.8}$$

for cartesian coordinates are valid, as well as

$$\mathrm{d}s^2 = \mathrm{d}x_j \mathrm{d}x_j = \delta_{jk}\mathrm{d}x_j \mathrm{d}x_k. \tag{2.4.9}$$

The measure of strain is now indicated by

$$\mathrm{d}s^2 - \mathrm{d}s_0^2 = \left(\delta_{jk} - \frac{\partial \xi_i}{\partial x_j}\frac{\partial \xi_i}{\partial x_k}\right)\mathrm{d}x_j \mathrm{d}x_k. \tag{2.4.10}$$

Also

$$h_{jk} = \frac{\partial \xi_i}{\partial x_j}\frac{\partial \xi_i}{\partial x_k} \tag{2.4.11}$$

is introduced as a metric tensor, determining the metric of the continuum relative to Euler's coordinates. Then (2.4.10) changes to

$$\mathrm{d}s^2 - \mathrm{d}s_0^2 = (\delta_{jk} - h_{jk})\mathrm{d}x_j \mathrm{d}x_k. \tag{2.4.12}$$

Quite similarly as previously, we use *Euler's strain tensor*

$$\eta_{jk} = \tfrac{1}{2}(\delta_{jk} - h_{jk}). \tag{2.4.13}$$

Through (2.4.11) and (2.1.10), (2.4.13) is changed to

$$\eta_{jk} = \frac{1}{2}\left(\frac{\partial u_j}{\partial x_k} + \frac{\partial u_k}{\partial x_j} - \frac{\partial u_i}{\partial x_j}\frac{\partial u_i}{\partial x_k}\right). \tag{2.4.14}$$

It becomes evident that in the components of Euler's strain tensor sums containing non-linear terms occur, i.e., the expressions

$$\frac{1}{2}\frac{\partial u_i}{\partial x_j}\frac{\partial u_i}{\partial x_k}. \tag{2.4.15}$$

Let us mention that other notations are used for the strain tensor. Thus, *Lagrange's* is sometimes called *Green's*, and *Euler's* is also known as *Almansian strain-tensor* [11].

As we have seen in finite deformations, whose displacements u_i and derivations $\partial u_i/\partial \xi_j$, $\partial u_i/\partial x_j$ are not small, we must strictly differentiate between the representations of strain in Lagrange's or Euler's coordinates, and use either Lagrange's strain tensor e_{jk} or Euler's η_{jk}. However using as a basis infinitesimal deformations, in this case, everything becomes simpler. Because of the smallness of the derivatives, their products can be neglected. In (2.4.6) and (2.4.14), sums (2.4.7), (2.4.15) disappear, and to begin with we obtain

$$e_{jk} \approx \frac{1}{2}\left(\frac{\partial u_j}{\partial \xi_k} + \frac{\partial u_k}{\partial \xi_j}\right) = \varepsilon_{jk},$$

$$\eta_{jk} \approx \frac{1}{2}\left(\frac{\partial u_j}{\partial x_k} + \frac{\partial u_k}{\partial x_j}\right),$$

respectively.

The relation (2.1.4*) and (2.1.9*) exists between coordinates ξ_i and x_i. In infinitesimal deformations the displacements u_i will be very small, so that they can be made equal to nought. Thus $x_i \approx \xi_i$ is obtained, i.e., there is no difference between the coordinates before and after deformation. Therefore, as mentioned in 2.1.2, no difference need be made in the description of the strain tensor, and it is at the same time

$$e_{jk} \approx \varepsilon_{jk}, \quad \eta_{jk} \approx \varepsilon_{jk}.$$

Whenever infinitesimal deformations are assumed, we shall use the *classic strain tensor* in our calculation, which is determined by

$$\varepsilon_{jk} = \frac{1}{2}\left(\frac{\partial u_j}{\partial x_k} + \frac{\partial u_k}{\partial x_j}\right). \tag{2.4.16}$$

Since we no longer differentiate between the coordinates ξ_i and x_i, all relations can be written in x_i, and we introduce for the partial derivative of any dependent variable f, the abbreviation

$$\frac{\partial f}{\partial x_k} = f_{,k}, \tag{2.4.17}$$

since a confusion is no longer possible. Using (2.4.17), the classic strain tensor is written as

$$\varepsilon_{jk} = \tfrac{1}{2}(u_{j,k} + u_{k,j}). \tag{2.4.18}$$

Let us give another interpretation of the components for Lagrange's strain tensor. At first, let us observe the undeformed line element $ds_0 = d\xi_1$ for which $d\xi_2 = d\xi_3 = 0$ is valid. Following deformation, it changes to

$$ds = (1 + E_1)ds_0, \tag{2.4.19}$$

where E_1 represents extension. Because of (2.4.19),

$$ds^2 - ds_0^2 = [(1 + E_1)^2 - 1]ds_0^2. \tag{2.4.20}$$

According to (2.4.4) and (2.4.5), together with $d\xi_1 = ds_0$, $d\xi_2 = d\xi_3 = 0$,

$$ds^2 - ds_0^2 = 2e_{11}\, ds_0^2. \tag{2.4.21}$$

Comparison of (2.4.20) and (2.4.21) yields

$$E_1 = \sqrt{(1 + 2e_{11})} - 1. \tag{2.4.22}$$

Because of (2.4.6), (2.1.38) and $\partial u_i/\partial \xi_j = \varepsilon_{ij} + \omega_{ij}$ (cf. 2.1.4, page 79), (2.4.22) changes to

$$E_1 = \sqrt{[1 + 2\varepsilon_{11} + \varepsilon_{11}^2 + (\varepsilon_{21} + \omega_{21})^2 + (\varepsilon_{31} + \omega_{31})^2]} - 1. \tag{2.4.23}$$

This almost corresponds to the given expression (2.1.57a), for extension E_1 in the discussion of affine infinitesimal transformation. The difference between (2.4.23) and (2.1.57a) only consists of the fact that we have been more accurate and have considered the influence of rotations given by ω_{21}, ω_{31}.

For infinitesimal quantities e_{11}, ε_{11}, ε_{21}, ε_{31}, ω_{21} and ω_{31}, respectively, we obtain from (2.4.22) and (2.4.23), in agreement with what has been said in 2.1.4,

$$E_1 \approx e_{11} \approx \varepsilon_{11},$$

which clears up the significance of e_{11}, and similarly, of the other diagonal terms e_{22}, e_{33}.

Because of (2.4.6), instead of (2.4.22), for E_1 and quite similarly for E_2 and E_3, we can write

$$\begin{aligned} E_1 &= \sqrt{\left[\left(1+\frac{\partial u_1}{\partial \xi_1}\right)^2 + \left(\frac{\partial u_2}{\partial \xi_1}\right)^2 + \left(\frac{\partial u_3}{\partial \xi_1}\right)^2\right]} - 1, \\ E_2 &= \sqrt{\left[\left(1+\frac{\partial u_2}{\partial \xi_2}\right)^2 + \left(\frac{\partial u_1}{\partial \xi_2}\right)^2 + \left(\frac{\partial u_3}{\partial \xi_2}\right)^2\right]} - 1, \\ E_3 &= \sqrt{\left[\left(1+\frac{\partial u_3}{\partial \xi_3}\right)^2 + \left(\frac{\partial u_1}{\partial \xi_3}\right)^2 + \left(\frac{\partial u_2}{\partial \xi_3}\right)^2\right]} - 1, \end{aligned} \tag{2.4.24}$$

which we shall need in the following. Let us look at the undeformed line elements

$$\mathrm{d}s_0^* = \mathrm{d}\xi_2, \ \mathrm{d}\xi_1 = \mathrm{d}\xi_3 = 0$$

and

$$\mathrm{d}s_0 = \mathrm{d}\xi_3^*, \ \mathrm{d}\xi_1^* = \mathrm{d}\xi_2^* = 0$$

forming angle $\pi/2$. Following deformation, they represent the line elements $\mathrm{d}s$, $\mathrm{d}s^*$, forming angle ϑ, which is calculated from

$$\mathrm{d}s\,\mathrm{d}s^* \cos\vartheta = \mathrm{d}x_i \mathrm{d}x_i^* = \frac{\partial x_i}{\partial \xi_j}\frac{\partial x_i}{\partial \xi_k}\mathrm{d}\xi_j \mathrm{d}\xi_k^* = \frac{\partial x_i}{\partial \xi_2}\frac{\partial x_i}{\partial \xi_3}\mathrm{d}\xi_2 \mathrm{d}\xi_3^*.$$

Because of (2.4.3) and (2.4.5),

$$\frac{\partial x_i}{\partial \xi_2}\frac{\partial x_i}{\partial \xi_3} = 2e_{23},$$

thus yielding

$$\cos\vartheta = 2e_{23}\frac{\mathrm{d}\xi_2 \mathrm{d}\xi_3^*}{\mathrm{d}s\,\mathrm{d}s^*}. \tag{2.4.25}$$

Furthermore, $\mathrm{d}s = (1+E_2)\mathrm{d}s_0 = \sqrt{(1+2e_{22})}\mathrm{d}\xi_2$ and $\mathrm{d}s^* = (1+E_3)\mathrm{d}s_0^* =$ $= \sqrt{(1+2e_{33})}\mathrm{d}\xi^*$, are valid by which follows

$$\cos \vartheta = \frac{2e_{23}}{\sqrt{1+2e_{22}}\sqrt{1+2e_{33}}} \tag{2.4.26}$$

from (2.4.25). Because of

$$\cos \vartheta = \cos\left(\frac{\pi}{2} - \gamma_{23}\right) = \sin \gamma_{23},$$

and for small deformations, in agreement with (2.1.65) from (2.4.26), we obtain

$$\gamma_{23} \approx 2e_{23} \approx 2\varepsilon_{23},$$

by which is given a clear interpretation for e_{23} by the connection with angle γ_{23}, and similarly for the other components of Lagrange's strain-tensor outside the diagonal.

2.4.2.2 *The state of stress in finite deformations.* It is at hand to examine the state of stress, and especially the setting up of equilibrium conditions in Euler's coordinates, which has already been done in 2.2.2. Only after loading and deformation does the stress state and state of equilibrium occur, which is of interest to us. But then, as we have said, the description of the deformed body and its conditions is most directly possible in x_i as independent coordinates. Thus, we have derived in 2.2.2 equilibrium conditions

$$\frac{\partial \sigma_{ij}}{\partial x_j} + K_i = 0 \tag{2.2.9}$$

and boundary conditions

$$\sigma_{ij} n_j = p_i. \tag{2.2.10}$$

There are important reasons to convert (2.2.9) and (2.2.10) to Lagrange coordinates. It is not only for the sake of formal completion that we wish to notate the state of strain and stress and the equilibrium conditions in terms of Lagrange coordinates, but, based on fundamental consideration, we would prefer to use Lagrange's method of representation. In (2.2.9) and (2.2.10) we are working with coordinates x_i, which themselves contain unknown displacements u_i. Determining the u_i is one of the aims of the calculations which are carried out within the limits of the theory of elasticity. Therefore, it will be useful to change over to Lagrange's method of representation, *in which all unknowns, as well as the* u_i, occur explicitly.

To begin with, let us deal with the conversion of the equilibrium conditions (2.2.9) and with *Euler's stress tensor* σ_{ij}, which they contain. Let us write

$$\frac{\partial \sigma_{ij}}{\partial \xi_k} \frac{\partial \xi_k}{\partial x_j} + K_i = 0. \tag{2.4.27}$$

Let us carry out an intermediate calculation: From $x_k = \xi_k + u_k$ follows

$$\mathrm{d}x_k = \frac{\partial x_k}{\partial \xi_j} \mathrm{d}\xi_j = \left(\delta_{kj} + \frac{\partial u_k}{\partial \xi_j}\right) \mathrm{d}\xi_j = \alpha_{kj} \mathrm{d}\xi_j, \tag{2.4.28}$$

i.e., we can write

$$\frac{\partial x_k}{\partial \xi_j} = \delta_{kj} + \frac{\partial u_k}{\partial \xi_j} = \alpha_{kj}. \tag{2.4.29}$$

Similarly

$$\mathrm{d}\xi_j = \frac{\partial \xi_j}{\partial x_k} \mathrm{d}x_k. \tag{2.4.30}$$

Interpreting (2.4.30) as a solution of the linear system of equations, (2.4.28), with respect to $\mathrm{d}\xi_j$, then we write for (2.4.30)

$$\mathrm{d}\xi_j = \alpha_{kj}^{-1} \mathrm{d}x_k, \tag{2.4.31}$$

and find that

$$\frac{\partial \xi_j}{\partial x_k} = \alpha_{kj}^{-1} = \frac{\mathrm{co}\,(\alpha_{jk})}{|\alpha_{kj}|} \tag{2.4.32}$$

is valid. Furthermore

$$\mathrm{d}V_x = \mathrm{d}x_1\,\mathrm{d}x_2\,\mathrm{d}x_3\,,\ \mathrm{d}V_\xi = \mathrm{d}\xi_1\,\mathrm{d}\xi_2\,\mathrm{d}\xi_3$$

is true, and hence

$$\mathrm{d}V_x = \mathrm{d}x_1\,\mathrm{d}x_2\,\mathrm{d}x_3 = \det\left(\frac{\partial x_k}{\partial \xi_j}\right)\mathrm{d}\xi_1\,\mathrm{d}\xi_2\,\mathrm{d}\xi_3 = \det\left(\frac{\partial x_k}{\partial \xi_j}\right)\mathrm{d}V_\xi\,.$$

From this follows that

$$\det\left(\frac{\partial x_k}{\partial \xi_j}\right) = \frac{\mathrm{d}V_x}{\mathrm{d}V_\xi},$$

and because of (2.4.29) also

$$|\alpha_{kj}| = \frac{\mathrm{d}V_x}{\mathrm{d}V_\xi}\,. \tag{2.4.33}$$

Using the result of this intermediate calculation, i.e., (2.4.32) and (2.4.33), then (2.4.27) is changed into

$$\frac{\partial \sigma_{ij}}{\partial \xi_k}\,\mathrm{co}\,(\alpha_{jk}) + \frac{\mathrm{d}V_x}{\mathrm{d}V_\xi}\,K_i = 0. \tag{2.4.34}$$

There follows another intermediate calculation. Let

$$(\alpha_{jk}) = \begin{pmatrix} 1+\dfrac{\partial u_1}{\partial \xi_1} & \dfrac{\partial u_1}{\partial \xi_2} & \dfrac{\partial u_1}{\partial \xi_3} \\[2ex] \dfrac{\partial u_2}{\partial \xi_1} & 1+\dfrac{\partial u_2}{\partial \xi_2} & \dfrac{\partial u_2}{\partial \xi_3} \\[2ex] \dfrac{\partial u_3}{\partial \xi_1} & \dfrac{\partial u_3}{\partial \xi_2} & 1+\dfrac{\partial u_3}{\partial \xi_3} \end{pmatrix}, \tag{2.4.35}$$

and hence

$$
\begin{aligned}
\mathrm{co}\,(\alpha_{11}) &= \left(1+\frac{\partial u_2}{\partial \xi_2}\right)\left(+ \frac{\partial u_3}{\partial \xi_3}\right) - \frac{\partial u_2}{\partial \xi_3}\,\frac{\partial u_3}{\partial \xi_2} \\
&= 1+\frac{\partial u_2}{\partial \xi_2}+\frac{\partial u_3}{\partial \xi_3}+\frac{\partial u_2}{\partial \xi_2}\,\frac{\partial u_3}{\partial \xi_3}-\frac{\partial u_2}{\partial \xi_3}\,\frac{\partial u_3}{\partial \xi_2}, \\
\mathrm{co}\,(\alpha_{12}) &= -\left(1+\frac{\partial u_3}{\partial \xi_3}\right)\frac{\partial u_2}{\partial \xi_1}+\frac{\partial u_2}{\partial \xi_3}\,\frac{\partial u_3}{\partial \xi_1} \\
&= -\frac{\partial u_2}{\partial \xi_1}-\frac{\partial u_3}{\partial \xi_3}\,\frac{\partial u_2}{\partial \xi_1}+\frac{\partial u_2}{\partial \xi_3}\,\frac{\partial u_3}{\partial \xi_1}, \\
\mathrm{co}\,(\alpha_{13}) &= \frac{\partial u_2}{\partial \xi_1}\,\frac{\partial u_3}{\partial \xi_2}-\left(1+\frac{\partial u_2}{\partial \xi_2}\right)\frac{\partial u_3}{\partial \xi_1} \\
&= \frac{\partial u_2}{\partial \xi_1}\,\frac{\partial u_3}{\partial \xi_2}-\frac{\partial u_3}{\partial \xi_1}-\frac{\partial u_2}{\partial \xi_2}\,\frac{\partial u_3}{\partial \xi_1}.
\end{aligned}
\tag{2.4.36}
$$

Let us write

$$\frac{\partial}{\partial \xi_k}\,\mathrm{co}\,(\alpha_{jk}) = \frac{\partial}{\partial \xi_1}\,\mathrm{co}\,(\alpha_{j1})+\frac{\partial}{\partial \xi_2}\,\mathrm{co}\,(\alpha_{j2})+\frac{\partial}{\partial \xi_3}\,\mathrm{co}\,(\alpha_{j3}). \tag{2.4.37}$$

Then (2.4.37), because of (2.4.36), yields for $j = 1$

$$\frac{\partial}{\partial \xi_k}\cdot \mathrm{co}\,(\alpha_{1k}) = 0.$$

This result can be generalized, yielding

$$\frac{\partial}{\partial \xi_k}\cdot \mathrm{co}\,(\alpha_{jk}) = 0, \quad j = 1, 2, 3. \tag{2.4.38}$$

The relation (2.4.38) has been the aim of the second intermediate calculation. Because of (2.4.38) we are permitted to write

$$\frac{\partial}{\partial \xi_k}\,[\sigma_{ij}\cdot \mathrm{co}\,(\alpha_{jk})]+\frac{\mathrm{d}V_x}{\mathrm{d}V_\xi}\,K_i = 0 \tag{2.4.39}$$

in place of (2.4.34). Using $\sigma_{ij}\cdot \mathrm{co}\,(\alpha_{jk})$, we have converted in (2.4.39) to *Lagrange's stress tensor*. But since it is not symmetric, we shall not be content with the version (2.4.39) of the equilibrium condition, but shall make further conversions.

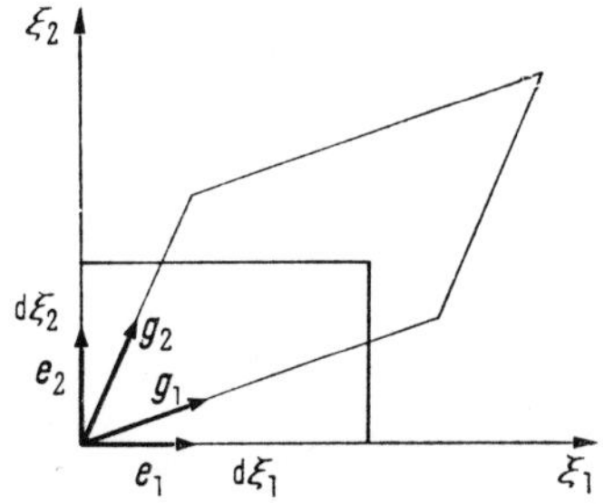

Figure 26

Let us consider a volume element, whose edges $d\xi_i$ coincide with the coordinate axes before the deformation (Figure 26). The unit vectors $\boldsymbol{e}_i$ are in the coordinate axes. Following deformation, the volume element changes into another one, in whose edges lie the *grid vectors* $\boldsymbol{g}_i$, which are images of the unit vectors $\boldsymbol{e}_i$. To calculate the $\boldsymbol{g}_i$ from the $\boldsymbol{e}_i$, let us go back to (2.1.35), ignoring the translation, i.e., we leave out term u_{i0} in (2.1.35). The mapping x_i of a vector ξ_i is given by

$$x_i = \left(\delta_{ij} + \frac{\partial u_i}{\partial \xi_j}\right) \xi_j . \tag{2.4.40}$$

Using (2.4.40) on vector $\boldsymbol{e}_1 = (1, 0, 0)$, we obtain for $\boldsymbol{g}_1$ the components

$$g_{1i} = \left(\delta_{1i} + \frac{\partial u_i}{d\xi_i}\right), \tag{2.4.41}$$

and similarly for the grid vectors $\boldsymbol{g}_2$ and $\boldsymbol{g}_3$

$$\begin{aligned} g_{2i} &= \left(\delta_{2i} + \frac{\partial u_i}{\partial \xi_2}\right), \\ g_{3i} &= \left(\delta_{3i} + \frac{\partial u_i}{\partial \xi_3}\right). \end{aligned} \tag{2.4.42}$$

Let us calculate the cosines of angles between vectors $\boldsymbol{e}_i$ and $\boldsymbol{g}_i$. It is for example,

$$\boldsymbol{e}_1 = (1, 0, 0),\ |\boldsymbol{e}_1| = 1,$$
$$\boldsymbol{g}_1 = \left(1+\frac{\partial u_1}{\partial \xi_1}, \frac{\partial u_2}{\partial \xi_1}, \frac{\partial u_3}{\partial \xi_1}\right),\ |\boldsymbol{g}_1| = \tag{2.4.43}$$
$$= \sqrt{\left(1+\frac{\partial u_1}{\partial \xi_1}\right)^2+\left(\frac{\partial u_2}{\partial \xi_1}\right)^2+\left(\frac{\partial u_3}{\partial \xi_1}\right)^2}.$$

A glance at (2.4.24) indicates that $|\boldsymbol{g}_1| = 1+E_1$ is valid. From the scalar product of $\boldsymbol{e}_1$ with $\boldsymbol{g}_1$ follows because of (2.4.43)

$$\cos(\boldsymbol{e}_1, \boldsymbol{g}_1) = \frac{1+\dfrac{\partial u_1}{\partial \xi_1}}{1+E_1}.$$

Generally,

$$\boldsymbol{e}_j = (\delta_{ji}),\ \boldsymbol{g}_k = \left(\delta_{ki}+\frac{\partial u_i}{\partial \xi_k}\right),\ |\boldsymbol{e}_j| = 1,\ |\boldsymbol{g}_k| = 1+E_k,$$

and hence

$$\cos(e_j, g_k) = \frac{\delta_{kj}+\dfrac{\partial u_j}{\partial \xi_k}}{1+E_k}. \tag{2.4.44}$$

Equation (2.4.44) is used to convert co (α_{11}). To do this we need to convert (2.4.44) into

$$\delta_{kj}+\frac{\partial u_j}{\partial \xi_k} = [\cos(e_j, g_k)](1+E_k) \tag{2.4.45}$$

and to insert it in the first line of (2.4.36). Thus we obtain

$$\text{co}(\alpha_{11}) = [\cos(e_2, g_2)\cos(e_3, g_3)-\cos(e_3, g_2)\cos(e_2, g_3)]$$
$$(1+E_2)(1+E_3). \tag{2.4.46}$$

Let us consider the area spanned by grid vectors $\boldsymbol{g}_2$ and $\boldsymbol{g}_3$ of the deformed volume element. It is $\mathrm{d}F_{x1} = \mathrm{d}x_2 \mathrm{d}x_3 \sin(\boldsymbol{g}_2, \boldsymbol{g}_3)$, and the normal vector of this area is

$$\boldsymbol{N}_1 = \boldsymbol{g}_2 \times \boldsymbol{g}_3, \tag{2.4.47}$$

which because of $|\boldsymbol{g}_k| = \mathrm{d}x_k/\mathrm{d}\xi_k$, k constant, has the magnitude

$$|N_1| = \frac{\mathrm{d}x_2}{\mathrm{d}\xi_2} \frac{\mathrm{d}x_3}{\mathrm{d}\xi_3} \sin(\boldsymbol{g}_2, \boldsymbol{g}_3) = \frac{\mathrm{d}F_{x1}}{\mathrm{d}F_{\xi 1}}. \tag{2.4.48}$$

Using (2.4.45), the following representation is obtained for vectors $\boldsymbol{g}_2$ and $\boldsymbol{g}_3$ from (2.4.42):

$$\boldsymbol{g}_2 = (1+E_2)\begin{pmatrix}\cos(\boldsymbol{e}_1, \boldsymbol{g}_2)\\ \cos(\boldsymbol{e}_2, \boldsymbol{g}_2)\\ \cos(\boldsymbol{e}_3, \boldsymbol{g}_2)\end{pmatrix}, \quad \boldsymbol{g}_3 = (1+E_3)\begin{pmatrix}\cos(\boldsymbol{e}_1, \boldsymbol{g}_3)\\ \cos(\boldsymbol{e}_2, \boldsymbol{g}_3)\\ \cos(\boldsymbol{e}_3, \boldsymbol{g}_3)\end{pmatrix}. \tag{2.4.49}$$

With (2.4.49) we obtain from (2.4.47)

$$\boldsymbol{N}_1 = (1+E_2)(1+E_3)\begin{vmatrix}\boldsymbol{e}_1 & \boldsymbol{e}_2 & \boldsymbol{e}_3\\ \cos(\boldsymbol{e}_1, \boldsymbol{g}_2) & \cos(\boldsymbol{e}_2, \boldsymbol{g}_2) & \cos(\boldsymbol{e}_3, \boldsymbol{g}_2)\\ \cos(\boldsymbol{e}_1, \boldsymbol{g}_3) & \cos(\boldsymbol{e}_2, \boldsymbol{g}_3) & \cos(\boldsymbol{e}_3, \boldsymbol{g}_3)\end{vmatrix} = \begin{pmatrix}N_{11}\\ N_{12}\\ N_{13}\end{pmatrix}. \tag{2.4.50}$$

This enables us to calculate the component N_{11} of the normal vector $\boldsymbol{N}_1$. Evaluating (2.4.50) yields

$$\begin{aligned} N_{11} &= \boldsymbol{N}_1 \boldsymbol{e}_1 = |N_1| \cdot \cos(\boldsymbol{N}_1, \boldsymbol{e}_1)\\ &= [\cos(\boldsymbol{e}_2, \boldsymbol{g}_2)\cos(\boldsymbol{e}_3, \boldsymbol{g}_3) - \cos(\boldsymbol{e}_3, \boldsymbol{g}_2)\cos(\boldsymbol{e}_2, \boldsymbol{g}_3)] \cdot (1+E_2)\\ &\quad (1+E_3). \end{aligned} \tag{2.4.51}$$

Comparison of (2.4.46), (2.4.48) with (2.4.51) yields

$$\mathrm{co}(\alpha_{11}) = N_{11} = \frac{\mathrm{d}F_{x1}}{\mathrm{d}F_{\xi 1}} \cos(\boldsymbol{N}_1, \boldsymbol{e}_1). \tag{2.4.52}$$

Through similar calculation we find generally

$$\text{co}\,(\alpha_{jk}) = \frac{\mathrm{d}F_{x(k)}}{\mathrm{d}F_{\xi(k)}} \cos\,(\boldsymbol{N}_k, \boldsymbol{e}_j) = \frac{\mathrm{d}F_{x(k)}}{\mathrm{d}F_{\xi(k)}} n_{jk}, \tag{2.4.53}$$

since $\cos\,(\boldsymbol{N}_k, \boldsymbol{e}_j)$ is the component $\boldsymbol{n}_{jk}$ of unit normal vector $\boldsymbol{n}_k$. Through (2.4.53) we can change (2.4.39) into

$$\frac{\partial}{\partial \xi_k}\left[\frac{\mathrm{d}F_{x(k)}}{\mathrm{d}F_{\xi(k)}} \sigma_{ij} n_{jk}\right] + \frac{\mathrm{d}V_x}{\mathrm{d}V_\xi} K_i = 0. \tag{2.4.54}$$

According to (2.2.5), $\sigma_{ij}n_{jk} = s_{ik}$, where s_{ik} are the components of the stress vector related to the deformed area element $\mathrm{d}F_{xk}$.

Hence

$$\frac{\mathrm{d}F_{x(k)}}{\mathrm{d}F_{\xi(k)}} \sigma_{ij} n_{jk} = \frac{\mathrm{d}F_{x(k)}}{\mathrm{d}F_{\xi(k)}} s_{ik} = s_{ik}^0, \tag{2.4.55}$$

and now s_{ik}^0 is the stress vector related to the *undeformed* area element $\mathrm{d}F_{\xi k}$. Since K_i is the volume force related to the deformed volume element $\mathrm{d}V_x$, we can write similarly

$$\frac{\mathrm{d}V_x}{\mathrm{d}V_\xi} K_i = K_i^0, \tag{2.4.56}$$

and thus introduce the volume force K_i^0 related to the *undeformed* volume element $\mathrm{d}V_\xi$. Then (2.4.54), because of (2.4.55) and (2.4.56), changes to

$$\frac{\partial}{\partial \xi_k} s_{ik}^0 + K_i^0 = 0. \tag{2.4.57}$$

This system of equations can also be represented in vector notation by

$$\frac{\partial}{\partial \xi_k} \boldsymbol{\sigma}_k^0 + \boldsymbol{K}^0 = 0. \tag{2.4.58}$$

Vectors $\boldsymbol{\sigma}_k^0$ and $\boldsymbol{K}^0$ are, as mentioned, related to the *undeformed* area elements, to the *undeformed* volume element, respectively.

Let stress tensor $\boldsymbol{\sigma}_k^0$ have the components σ_{kj}^{0g} in the direction of the grid vectors. Then, in the direction of the ξ_i-axes of the undeformed body it has the components $\sigma_{kj}^{0g} \cos(\boldsymbol{e}_i, \boldsymbol{g}_j)$. Hence, (2.4.58) can be given in component notation as

$$\frac{\partial}{\partial \xi_k} [\sigma_{kj}^{0g} \cos(\boldsymbol{e}_i \cdot \boldsymbol{g}_j)] + K_i^0 = 0 \tag{2.4.59}$$

in relation to the ξ_i axes. Other notation methods are also used. Because of (2.4.44) we obtain from (2.4.59)

$$\frac{\partial}{\partial \xi_k} \left[\frac{\sigma_{kj}^{0g}}{1+E_j} \left(\delta_{ji} + \frac{\partial u_i}{\partial \xi_j} \right) \right] + K_i^0 = 0. \tag{2.4.60}$$

If the symmetric tensor

$$\frac{\sigma_{kj}^{0g}}{1+E_j} = k_{kj} \tag{2.4.61}$$

is termed *Kirchhoff 'stress' tensor*, and we use the relation (2.1.5) leading to

$$\delta_{ji} + \frac{\partial u_i}{\partial \xi_j} = \frac{\partial x_i}{\partial \xi_j}, \tag{2.4.29}$$

then (2.4.60) changes to

$$\frac{\partial}{\partial \xi_k} \left(k_{kj} \frac{\partial x_i}{\partial \xi_j} \right) + K_i^0 = 0. \tag{2.4.62}$$

This form of equilibrium conditions is also used ([11], page 184).

Considering the requirement that all unknowns are to occur explicitly, we prefer to use the system of equations

$$\frac{\partial}{\partial \xi_k} \left[k_{kj} \left(\delta_{ji} + \frac{\partial u_i}{\partial \xi_j} \right) \right] + K_i^0 = 0. \tag{2.4.63}$$

In (2.4.63) we have given the required conversion of (2.2.9) to Lagrange's method of representation. Similarly, we arrive at

$$k_{ij} n_j = p_i^0 \tag{2.4.64}$$

from (2.2.10), by which the conversion of the boundary conditions to Lagrange's method of representation is given.

Finally, let us set the case that small deformations are present, then the difference between the deformed and undeformed quantities, as well as between the coordinates x_i and ξ_i will be slight. Therefore, in (2.4.63) and (2.4.64), k_{ij} can be substituted by σ_{ij}, K_i^0 by K_i, p_i^0 by p_i. In addition, $\partial u_i/\partial \xi_j$, as being small in relation to '1', can be neglected. Furthermore, using abbreviation (2.4.17), Euler's representation (2.2.9), (2.2.10), as well as Lagrange's representation (2.4.63), (2.4.64) changes to 'the classic' representation

$$\begin{aligned} \sigma_{ji,j} + K_i &= 0, \\ \sigma_{ji} n_j &= p_i. \end{aligned} \tag{2.4.65}$$

We shall always use this set of equations in the following chapters, because if no special reference is made, then the presence of infinitesimal deformations is assumed.

2.4.3 *Physical non-linearity*

Independent from permitting or not permitting a geometric linearization, let us consider whether the proportionality limits valid for the material equations (2.3.26) and (2.3.30) are being exceeded. This exceeding of limits can actually occur in strains which, on one hand are so small that geometric linearization remains intact, but on the other hand are so large that Hooke's law which is based on the theorem of proportionality, does not remain valid. Then the linear equations (2.3.26) and (2.3.30) are to be replaced by non-linear equations, thus arriving at *physical non-linearity* of the medium. Let us continue considering the case dealing with the occurrence of small strain.

We can still work with the components of the classic strain tensor ε_{ij} and the classic stress tensor σ_{ij}, because of the permitted geometric linearization. But in consideration of the already existing physical non-linearity, these tensors are to be linked by a non-linear law. H. Kauderer [12] has made a proposition dealing with this particular aspect. Hence, instead of (2.3.27),

$$\sigma_{ii} = 3K\kappa(\varepsilon_{ii})\varepsilon_{ii}, \tag{2.4.66}$$

and instead of (2.3.30),

$$\sigma_{ij}^{(0)} = 2\mu\gamma(\psi_0^2)\varepsilon_{ij}^{(0)}. \tag{2.4.67}$$

Non-linearity of these two material laws is given by the presence of the non-linear *tension function*

$$\kappa(\varepsilon_{ii}) = 1+\kappa_1\varepsilon_{ii}+\kappa_2\varepsilon_{ii}^2+\ldots \tag{2.4.68}$$

and of the *non-linear shearing function*

$$\begin{aligned}\gamma(\psi_0^2) &= 1+\gamma_2\psi_0^2+\gamma_4\psi_0^4+\ldots,\\ \psi_0 &= \frac{2}{\sqrt{3}}[(\varepsilon_{11}^2+\varepsilon_{22}^2+\varepsilon_{33}^2-\varepsilon_{11}\varepsilon_{22}-\varepsilon_{22}\varepsilon_{33}-\varepsilon_{33}\varepsilon_{11})\\ &\quad+2(\varepsilon_{12}^2+\varepsilon_{23}^2+\varepsilon_{31}^2)]^{\frac{1}{2}}.\end{aligned} \tag{2.4.69}$$

For strain which is below the proportionality limits, the powers of ε_{ii} and of ψ_0 respectively, can be neglected, so that the equations (2.4.66) and (2.4.67) without difficulty, go back to (2.3.27) and (2.3.30), i.e., to equations of Hooke's law.

Other non-linear material laws are used, e.g., by referring to the common principal-axes – system of strain – and stress-tensor, a general, non-linear law

$$\sigma_i-\tfrac{1}{3}J_{1s} = f_i(\varepsilon_i-\tfrac{1}{3}J_{1V}),\; i = 1, 2, 3, \tag{2.4.70}$$

is given, in which σ_i and ε_i are the principal values, J_{1s} and J_{1V} are the first invariants of the tensors and f_i are suitably chosen non-linear functions. By (2.4.70), a generalization of (2.3.30) has been given.

2.4.4 *Combination of non-linearities*

In ending this chapter, let us discuss the manner of combination of non-linearities.

The first case that *physical non-linearity* is linked with *geometric linearity*, has been discussed in 2.4.3.

Let the second case be that *geometric non-linearity* occurs at the same time as *physical linearity*. Then, the linear Hooke's law is written for the components e_{ij} of Lagrange's strain tensor and the components k_{ij} of Kirchhoff's strain tensor. This yields the material law

$$k_{ij} = \lambda\delta_{ij}e_{kk} + 2\mu e_{ij} \tag{2.4.71}$$

instead of (2.3.19).

The third case contains the *geometric non-linearity* combined with *physical non-linearity*. Then the components e_{ij} of Lagrange's strain tensor and the components k_{ij} of Kirchhoff's stress tensor are to be linked by a non-linear material law, similarly to (2.4.70).

3

Problems and methods of solution

3.1 Problems

3.1.1 *Boundary value problems*

The most general problem of the mathematical theory of elasticity consists of estimating the *distribution of stress and strain as well as the displacements* at all points of a body when certain boundary conditions are given. Thus, we are faced with boundary value problems. If not stated otherwise, we shall assume the occurrence of infinitesimal deformations. Without differentiating between the theories of Lagrange and Euler, we may then calculate (as shown in Chapter 2) by simply using the coordinates x_i which, in addition, we assume to be cartesian coordinates. We are also permitted to use classic strain tensor ε_{ij}, the classic stress tensor σ_{ij} (compare 2.2.2), the classic equations of equilibrium and boundary conditions (2.4.65), Hooke's law (2.3.19) and (2.3.14) respectively, as well as the abbreviation (2.4.17). We then have for our estimation of the fifteen unknowns (*six stress components* σ_{ij}, *six strain components* ε_{ij}, and *three displacement components* u_i) exactly the necessary fifteen equations, i.e., the *three conditions of equilibrium*

$$\sigma_{ji,j} + K_i = 0, \tag{3.1.1}$$

the *six geometric equations*

$$\varepsilon_{ij} = \tfrac{1}{2}(u_{i,j} + u_{j,i}), \tag{3.1.2}$$

and the *six physical equations*

$$\sigma_{ij} = \lambda\delta_{ij}\varepsilon_{kk}+2\mu\varepsilon_{ij}, \quad \varepsilon_{ij} = \frac{1+\nu}{E}\,\sigma_{ij}-\frac{\nu}{E}\,\delta_{ij}\sigma_{kk}, \tag{3.1.3a, b}$$

respectively.

In addition, there are also the boundary conditions – for *specified surface forces*

$$\sigma_{ij}n_j = p_i, \tag{3.1.4}$$

and for *specified surface displacements*

$$u_i(0) = u_{i0}. \tag{3.1.5}$$

It is, in fact, quite possible to solve all problems, using the system of equations and boundary conditions at hand. We shall do this only in the very rarest of cases. The set problems are usually of a more specific nature, and according to their corresponding peculiarities, special ways of solving them have been developed. These, however, shall be discussed later.

To begin with there may be restrictions, as far as the formulation of the problems is concerned. For example, we may be asked to find the *distribution of stress and strain* only, ignoring the displacements. Hence, we are dealing with only twelve unknowns: *six stresses* σ_{ij}, and *six strain components* ε_{ij}. In order to estimate these, we shall use twelve equations, and these are the *three conditions of equilibrium*, the *six physical equations*, and the *equations of compatibility*

$$\varepsilon_{ij,kl}+\varepsilon_{kl,ij}-\varepsilon_{ik,jl}-\varepsilon_{jl,ik} = 0. \tag{3.1.6}$$

As we already know from 2.1.5, only three of the equations (3.1.6) are independent of each other. Thus, a total of twelve independent equations are available. The introduction of (3.1.6) into the calculation vouches for those strains as a solution to the problem, which satisfy the demand for unique displacements.

Another restriction of the problem may consist of the fact that only the calculation of the *stress distribution* and of *displacements* is asked for. Then nine unknowns will occur, these being the *six stresses* σ_{ij} and the *three displacements* u_i. We shall need nine equations, which will actually be available to the *three conditions of equilibrium*, and to the *six physical equations* expressed in terms of the u_i.

We shall now show how, on the basis of these general statements, specific systems of equations have been developed for the diverse boundary value problems.

We differentiate between three kinds of boundary value problems. In the *first boundary value problem* the displacements u_i, on the surface O, are given. We, therefore, have to use the boundary conditions (3.1.5). To begin with we are looking for the displacements, and as we shall find out later, the stress and strain components can be deduced from these. Therefore, nine unknowns (σ_{ij}, u_i) and nine equations, (3.1.1) and (3.1.3), the latter expressed by u_i, will be used.

At first we shall convert the physical equations into

$$\sigma_{ij} = \lambda\delta_{ij}\varepsilon_{kk} + 2\mu\varepsilon_{ij}, \tag{3.1.3a}$$

by way of

$$\varepsilon_{kk} = u_{k,k} \tag{3.1.7}$$

and (3.1.2) introducing the displacements u_i. This yields

$$\sigma_{ij} = \lambda\delta_{ij}u_{k,k} + \mu(u_{i,j} + u_{j,i}). \tag{3.1.8}$$

With the aid of (3.1.8), we can change the conditions of equilibrium (3.1.1) into

$$(\lambda+\mu)u_{k,kj} + \mu u_{j,ii} + K_j = 0 \tag{3.1.9}$$

and have obtained, in this way, the *Lamé-Navier equations*. Using a symbolic notation, these equations read as follows

$$(\lambda+\mu) \operatorname{grad} \operatorname{div} \boldsymbol{u} + \mu\nabla^2\boldsymbol{u} + \boldsymbol{K} = 0. \tag{3.1.9*}$$

The system of equations (3.1.9), together with the boundary conditions (3.1.5), is adequate for calculating the needed displacements u_i. Furthermore, once these have been found by calculation, then the stress components may be found by (3.1.8), and the strain components by (3.1.2).

Let us emphasize the important case when no volume forces occur. Then, instead of (3.1.9)

$$(\lambda+\mu)u_{k,kj}+\mu u_{j,ii} = 0. \tag{3.1.10}$$

Differentiating with respect to x_j and by addition we obtain from (3.1.10)

$$(\lambda+\mu)(u_{k,k})_{,jj}+\mu(u_{j,ii})_{,j} = 0. \tag{3.1.11}$$

If we exchange the sequence of differentiation in the second term of (3.1.11) we have

$$(\lambda+\mu)(u_{k,k})_{,jj}+\mu(u_{j,j})_{,ii} = 0. \tag{3.1.12}$$

In the second term of (3.1.12) we shall rename the dummy indices j into k and i into j. We may then put

$$(\lambda+2\mu)(u_{k,k})_{,jj} = 0$$

instead of (3.1.12). From this we get in turn

$$(u_{k,k})_{,jj} = 0, \;\; \nabla^2(u_{k,k}) = 0, \tag{3.1.13}$$

which because of $u_{k,k} = \varepsilon_{kk} = J_{1v}$ is the same as

$$\nabla^2 J_{1v} = 0. \tag{3.1.14}$$

Integrating (3.1.11) after division by μ,

$$\left(\frac{\lambda}{\mu}+1\right)(u_{k,k})_{,j}+u_{j,ii} = 0. \tag{3.1.15}$$

In a symbolic notation (3.1.15) equals

$$\left(\frac{\lambda}{\mu}+1\right)\frac{\partial J_{1v}}{\partial x_j}+\nabla^2 u_j = 0. \tag{3.1.16}$$

When operator ∇^2 is applied to (3.1.16), then we get, by interchanging the sequence of the differentiation in the first term,

$$\nabla^2\nabla^2 u_j+\left(\frac{\lambda}{\mu}+1\right)\frac{\partial}{\partial x_j}\nabla^2 J_{1v} = 0. \tag{3.1.17}$$

Because of (3.1.14), (3.1.17) changes into the biharmonic differential equation

$$\nabla^2\nabla^2 u_j = 0. \tag{3.1.18}$$

(3.1.18), together with (3.1.5), suffice for the evaluation of the displacements u_i. We can finally say that in the case of missing volume forces, the first boundary value problem in elasticity theory proves to be the general boundary value problem of the biharmonic differential equation. It is a peculiarity of the mathematical theory of elasticity that this differential equation occurs frequently elsewhere.

The *second boundary value problem* is characterized by the fact that the external forces p_i are prescribed on the surface O of the body concerned. Therefore, boundary condition (3.1.4) has to be applied. In this case, with reference to this boundary condition, the determination of the stress components is of primary importance, although the strain components may also be found. Thus, we have twelve unknowns $(\sigma_{ij}, \varepsilon_{ij})$ and we shall, therefore, use the twelve (independent) equations (3.1.1), (3.1.3b) and (3.1.6).

To begin with, we shall introduce Hooke's physical equations (3.1.3b) to the equations of compatibility (3.1.6), which yields

$$\sigma_{ij,kl}+\sigma_{kl,ij}-\sigma_{ik,jl}-\sigma_{jl,ik} = \frac{\nu}{1+\nu}(\delta_{ij}\sigma_{mm,kl}+\delta_{kl}\sigma_{mm,ij} - \delta_{ik}\sigma_{mm,jl}-\delta_{jl}\sigma_{mm,ik}). \tag{3.1.19}$$

Through contraction [inserting $l = k$ in (3.1.19)], we get from (3.1.19)

$$\sigma_{ij,kk}+\sigma_{kk,ij}-\sigma_{ik,jk}-\sigma_{jk,ik} = \frac{\nu}{1+\nu}(\delta_{ij}\sigma_{mm,kk}+3\sigma_{mm,ij} - \sigma_{mm,ji}-\sigma_{mm,ij}).$$

When $\sigma_{ij,kk} = \nabla^2\sigma_{ij}$ is used, and the dummy indices on the right side are changed from m to k and from k to m, then we obtain

$$\nabla^2\sigma_{ij}+\sigma_{kk,ij}-\sigma_{ik,jk}-\sigma_{jk,ik} = \frac{\nu}{1+\nu}(\delta_{ij}\nabla^2\sigma_{kk}+\sigma_{kk,ij}). \tag{3.1.20}$$

From (3.1.1) we obtain by differentiating

$$\sigma_{ik,\,jk} = -K_{i,\,j}, \tag{3.1.21}$$

and using this, (3.1.20) changes into

$$\nabla^2\sigma_{ij} + \frac{1}{1+\nu}\,\sigma_{kk,\,ij} - \frac{\nu}{1+\nu}\,\delta_{ij}\nabla^2\sigma_{kk} = -(K_{i,\,j}+K_{j,\,i}). \tag{3.1.22}$$

Let us carry out two contractions in (3.1.19) by writing down $k = i$ and $l = j$. In this way we obtain

$$\sigma_{ij,\,ij}+\sigma_{ij,\,ij}-\sigma_{ii,\,jj}-\sigma_{jj,\,ii} = \frac{\nu}{1+\nu}\,(\sigma_{mm,\,jj}+\sigma_{mm,\,jj}-$$

$$-3\sigma_{mm,\,jj}-3\sigma_{mm,\,ii}).$$

Introducing operator ∇^2 and renaming the dummy indices, we obtain

$$2\sigma_{ij,\,ij}-2\nabla^2\sigma_{kk} = \frac{4\nu}{1+\nu}\,\nabla^2\sigma_{kk},$$

which is recalculated into

$$\sigma_{ij,\,ij} = \frac{1-\nu}{1+\nu}\,\nabla^2\sigma_{kk}.$$

Because of (3.1.21) we obtain

$$\nabla^2\sigma_{kk} = -\frac{1+\nu}{1-\nu}\,K_{i,\,i}. \tag{3.1.23}$$

When the volume forces are either missing or constant, we write

$$\nabla^2\sigma_{kk} = 0. \tag{3.1.24}$$

Substituting (3.1.23) for (3.1.22) we get

$$\nabla^2\sigma_{ij} + \frac{1}{1+\nu}\,\sigma_{kk,\,ij} = -\frac{\nu}{1-\nu}\,\delta_{ij}\,K_{k,\,k}-(K_{i,\,j}+K_{j,\,i}). \tag{3.1.25}$$

These are *Michell's equations*, which together with the boundary conditions (3.1.4) let us find the unknown stresses.

For missing or constant volume forces, Michell's equations change into *Beltrami's* simpler *equations*

$$\nabla^2 \sigma_{ij} + \frac{1}{1+\nu} \sigma_{kk,\,ij} = 0. \tag{3.1.26}$$

By applying operator ∇^2 to (3.1.26), and by interchanging the sequence of differentiation in the second term, (3.1.26) changes into

$$\nabla^2 \nabla^2 \sigma_{ij} + \frac{1}{1+\nu} (\nabla^2 \sigma_{kk})_{,\,ij} = 0.$$

Because of (3.1.24) this, in turn, yields the biharmonic differential equation

$$\nabla^2 \nabla^2 \sigma_{ij} = 0. \tag{3.1.27}$$

It is plain that this particular differential equation plays an excellent part in the theory of elasticity under the previously mentioned assumptions for the volume forces.

When the stresses σ_{ij} have been determined from (3.1.25) or (3.1.27) and from the corresponding boundary conditions, then the strain components ε_{ij} are also determined by applying

$$\varepsilon_{ij} = \frac{1+\nu}{E} \sigma_{ij} - \frac{\nu}{E} \delta_{ij} \sigma_{kk} . \tag{3.1.3b}$$

That these are quite meaningful is obvious, since they lead to well-defined displacements u_i through integration, as in the previous calculation we have taken into account (3.1.6), which are the conditions of compatibility.

The *third boundary value problem* is on hand, when mixed boundary conditions are to be satisfied, i.e., if on one part of the surface the forces are specified and, on the other, the displacements. This means that partly (3.1.4) and partly (3.1.5) is valid. This may give rise to complicated problems, which will not be discussed at this point. They can be dealt with by either going back to the originally drawn up equations (3.1.1) to (3.1.3) of stress, strain and displacement, considering the boundary conditions as they occur; or we can try to transform back to the first or second boundary value problem, eliminating and transforming quantities involved in the interfering parts of the mixed boundary conditions such that uniform boundary conditions of the first or second type are obtained.

3.1.2 *Methods of solution*

Problems of the mathematical theory of elasticity are usually so complicated that they cannot be solved in general in a closed form. We shall, therefore, discuss some fundamental methods, which will either facilitate or render possible an eventual approach.

Since the boundary value problems in the theory of elasticity are linear, the *superposition principle* can be used on the assumption that the given loads are independent from the displacements. This principle represents an important expedient for the solution of problems. It is possible, for instance, to find firstly the general solution to the equation (3.1.9), and (3.1.25) respectively, which have been made homogeneous by neglecting mass forces. Furthermore, it may be possible to indicate a particular solution for the actual equations (3.1.9) and (3.1.25) respectively, which are inhomogeneous. The general solution to the actual inhomogeneous problem is obtained by superposition of the two previously found solutions.

A further possibility of finding a solution to the given problem is constituted by setting up separate problems suitably dividing the given external loads. It will probably be easy supplying solutions for these separate problems. The superposition of these separate solutions will then lead to the solution of the original problem as a whole.

Difficulties in a problem will often develop from the boundary conditions. In such a case, the *principle of Saint-Vénant* can be of assistance. It indicates that stresses and deformations will vary only slightly if the external surface forces are altered on a small part of a body's surface in such a way that the supplemental forces, which caused the change, are in equilibrium. The principle is valid when stresses and deformations are only considered in a suitable distance of the region where the external surface forces have been changed. With the assistance of this principle, complicated general boundary value problems are substituted by possibly simpler ones for which solutions can be given. The solution of a substitute problem of this kind represents, strictly speaking, only an approximate solution of the original problem, but is usually quite accurate.

Finally, there are the *inverse* and the *semi-inverse methods*. In the first case, a known solution of the displacements is assumed with the aid of which strain and stress states are determined. Finally, using the boundary conditions, the body itself and its load and reactions are determined. In the second case of the semi-inverse method, part of the unknowns is given, and the missing quantities are determined in such a way that the differential equations and boundary conditions are being satisfied. At the same time, these solutions should correspond to problems of a practical nature.

It is, of course, possible to combine the different methods sensibly. But it is especially important that *unique solutions* do exist when the inverse and semi-inverse methods are to be successfully applied. Therefore, the following deals with the question of uniqueness of solutions of problems in the mathematical theory of elasticity.

The *theorem of Clapeyron* states: if a body which is affected by a system of volume forces and surface forces is in a state of equilibrium, then the deformation energy is half that of the work done by the external forces in the displacements u_i. This is the case when u_i are the displacements leading from the stress free state to the state of equilibrium. In order to prove this theorem, we set out with the identity

$$\int_O \sigma_{ij} n_j u_i \mathrm{d}O = \int_V (\sigma_{ij} u_i)_{,j} \mathrm{d}V. \tag{3.1.28}$$

This is derived from the *Gaussian integral theorem*. Through differentiation on the right side of (3.1.28) and application of

$$u_{i,j} = \varepsilon_{ij} + \omega_{ij}, \tag{3.1.29}$$

and because of $\sigma_{ij}\omega_{ij} = 0$, we obtain for (3.1.28)

$$\int_O \sigma_{ij} n_j u_i \mathrm{d} = \int_V (\sigma_{ij,j} u_i + \sigma_{ij} \varepsilon_{ij}) \mathrm{d}V. \tag{3.1.30}$$

Because of (2.3.2) and (2.3.10),

$$\frac{\partial U}{\partial \varepsilon_{ij}} = \sigma_{ij} = C_{ijkl} \varepsilon_{kl}$$

is valid. Therefore, the specific deformation energy U must be given by

$$U = \tfrac{1}{2}\sigma_{ij}\varepsilon_{ij}. \tag{3.1.31}$$

In addition, the symmetry of σ_{ij} and the relation (3.1.1) are of importance, hence $\sigma_{ij,j} = -K_i$ is true, and (3.1.30) changes into

$$\int_O \sigma_{ij} n_j u_i \mathrm{d}O = -\int_V K_i u_i \mathrm{d}V + 2\int_V U \mathrm{d}V. \tag{3.1.32}$$

Again this equation is changed into

$$\int_V K_i u_i \mathrm{d}V + \int_O p_i u_i \mathrm{d}O = 2\int_V U \mathrm{d}V \tag{3.1.33}$$

using (3.1.4).

In this way the theorem of Clapeyron has been expressed mathematically, since on the left side of the equation we have the work of the external forces and on the right side twice the deformation energy.

Let there be two solutions ($\sigma_{ij}^{(1)}$, $\sigma_{ij}^{(2)}$ for the stresses and $u_i^{(1)}$, $u_i^{(2)}$ for the displacements). Then

$$\sigma_{ij,j}^{(1)} = \sigma_{ij,j}^{(2)} = -K_i, \tag{3.1.34}$$

$$\sigma_{ij}^{(1)} n_j = \sigma_{ij}^{(2)} n_j = p_i \tag{3.1.35}$$

are both valid for these. Let us write $\sigma_{ij}^* = \sigma_{ij}^{(1)} - \sigma_{ij}^{(2)}$. Hence, according to (3.1.34) $\sigma_{ij,j}^* = 0$ is valid, and it is evident that

$$\sigma_{ij}^* = \sigma_{ij}^{(1)} - \sigma_{ij}^{(2)},\ u_i^* = u_i^{(1)} - u_i^{(2)} \tag{3.1.36}$$

is a special solution according to the superposition principle, thus fulfilling the equilibrium conditions in the special case of $K_i \equiv 0$. If in (3.1.36) we were to use the theorem of Clapeyron, then allowance for the special condition $K_i = 0$ should be made. Therefore, (3.1.32) would yield specially

$$\int_O \sigma_{ij}^* n_j u_i^* \mathrm{d}O = 2\int_V U^* \mathrm{d}V,\ U^* = \tfrac{1}{4}\sigma_{ij}^*(u_{i,j}^* + u_{j,i}^*). \tag{3.1.37}$$

But since (3.1.35) holds, $\sigma_{ij}^{*}n_j = \sigma_{ij}^{(1)}n_j - \sigma_{ij}^{(2)}n_j = 0$ is true. Therefore we get from (3.1.37)

$$\int_V U^* \mathrm{d}V = 0, \tag{3.1.38}$$

i.e., the deformation energy disappears for the solution (3.1.36). As the deformation energy is a positive definite quantity, then (3.1.38) can only remain true if

$$U^* = \tfrac{1}{4}\sigma_{ij}^{*}(u_{i,j}^{*} + u_{j,i}^{*}) = 0$$

is true. This is possible for $\sigma_{ij}^{*} = 0$, from which $\sigma_{ij}^{(1)} = \sigma_{ij}^{(2)}$ is derived, i.e., the *uniqueness of solutions of stresses* is evident. Once this fact has been established, then using Hooke's Law (3.1.3b), and because of (3.1.2), it is followed by the *uniqueness of solutions of displacements.* This proof of uniqueness does not only presuppose infinitesimal deformations, but also the fact that the load is not affected by displacement. Thus, we can disregard the displacement of the points of application of the loads resulting from deformation. Once we are dealing with finite deformations, e.g., when dealing with problems of stability, then the theorem of uniqueness does not apply. Branching of solutions is possible, which constitutes a mathematical evidence for the existence of nontrivial positions of equilibrium and the possibility of buckling.

The classic boundary value problems of the theory of elasticity deal with partial differential equations as we have seen in (3.1.9). Hence, we can also refer to the statements of the mathematical theory dealing with partial differential equations in order to find the uniqueness of the solutions. For example, reference can be made to Morgenstern-Szabo [13], who deal with this problem.

At the beginning of this paragraph, we have named two principles and derived certain methods of solution from them. In all, we have shown three fundamental methods – superposition of solutions, introduction of substitute problems, and reversal of problems. It remains to be seen how the specific solutions we are looking for are arrived at using the various mathematical methods along the different approaches. We shall discuss the most important analytical procedures which deal with two-dimensional and three-dimensional problems in the following. We also shall see that the general principles of analytical mechanics in particular give rise to methods of the calculus of variations which are very profitable, and can be used as the basis of various approximation methods. Several other approximation methods are based on the finite difference method. The various aspects of the mathematical theory of elasticity generally result from the fact that a diversity of analytical and numerical, as well as precise and approximate mathematical methods have been used in dealing with basic problems. Reference may be made to a selection of books by Collatz [14], Kantorowitsch-Krylow [15], Michlin [16] and Muskhelishvili [17].

3.1.3 *Curvilinear coordinates*

The chance of successfully solving a boundary value problem depends on whether a suitable set of coordinates for the particular problem has been chosen. Until now, we have only used cartesian coordinates, which will, of course, not always be advisable. In certain cases, and in specific problems, special curvilinear coordinates should be used. For this reason, it becomes important to rewrite the fundamental equations of the theory of elasticity in terms of curvilinear coordinates. However, we shall restrict ourselves even now to *orthogonal* systems of coordinates only. Once these various fundamental equations have been rewritten, then problems in any, and all forms, can be tackled.

Suppose we are changing from the cartesian coordinates x_i to the curvilinear coordinates q_i, when relations

$$x_i = x_i(q_1, q_2, q_3), \ i = 1, 2, 3, \tag{3.1.39}$$

are known. Then the following is true for the square of the line element,

$$\mathrm{d}s^2 = \frac{\partial x_i}{\partial q_j}\frac{\partial x_i}{\partial q_k}\,\mathrm{d}q_j\,\mathrm{d}q_k. \tag{3.1.40}$$

Let

$$g_{jk} = \frac{\partial x_i}{\partial q_j}\frac{\partial x_i}{\partial q_k} \tag{3.1.41}$$

be the *covariant metric tensor*. Since q_i are *orthogonal* coordinates, then

$$g_{jk} = 0 \text{ for } j \neq k, \tag{3.1.42}$$

so that (3.1.41) represents a diagonal tensor having diagonal components

$$g_{(i)i} = \frac{\partial x_k}{\partial q_{(i)}} \frac{\partial x_k}{\partial q_i}, \quad i = 1, 2, 3, \text{ do not sum over } i, \tag{3.1.43}$$

which we abbreviate to

$$h_i = \sqrt{g_{(i)i}}. \tag{3.1.44}$$

The fundamental equations of the theory of elasticity, i.e., conditions of equilibrium, geometric and physical equations, can be used only in the given versions of (3.1.1), (3.1.2) and (3.1.3) when dealing with cartesian coordinates. Should we want to change to curvilinear coordinates, then these formulae require substitution of the usual partial derivatives by covariant ones, e.g., substitute $u_{i,j}$ by $u_{i|j}$. However, this change does not suffice. To use the formulae in their new notation, all vectors and tensors are to be converted into *physical components*. As covariant derivatives are also vectors and tensors, then physical components must be used for them. Let us shorten the term 'physical components of the covariant derivatives' to *physical derivatives*. Whenever quantities are notated in component representation, then the *base vectors* $\boldsymbol{b}_i$, $i = 1, 2, 3$, should not be used in this connection, but the appropriate unit vectors

$$\boldsymbol{e}_i = \frac{\boldsymbol{b}_i}{h_{(i)}}, \quad i = 1, 2, 3, \tag{3.1.45}$$

should be used. We shall name the latter *physical base vectors*.

Let us summarize. When formulae of the theory of elasticity – which are valid for cartesian coordinates – are to be changed to formulae which are valid for curvilinear coordinates, then in the first case all vector components and tensor components should be substituted by physical components, all partial derivatives by physical derivatives and all base vectors by physical base vectors. In short, the existing tensorial quantities should be changed into their corresponding physical quantities.

3.1.3.1 *Physical derivative of a vector with respect to curvilinear orthogonal coordinates.* Let there be a vector $\boldsymbol{A}$ with the *covariant components* A_i.

In the chapter dealing with the introduction to tensor calculus it has been shown that for its physical components A_i^*, the relation

$$A_i^* = \frac{A_i}{\sqrt{g_{(ii)}}} = \frac{A_i}{h_{(i)}} \tag{3.1.46}$$

is true. According to (1.2.75), the *covariant derivative* of A_i is

$$A_{i|j} = \frac{\partial A_i}{\partial q_j} - \begin{Bmatrix} k \\ ij \end{Bmatrix} A_k, \tag{3.1.47}$$

which because of (1.2.55) can be transcribed into

$$A_{i|j} = \frac{\partial A_i}{\partial q_j} - g^{kl}[ij, l]A_k. \tag{3.1.48}$$

If the Christoffel symbol of the first kind is rewritten according to (1.2.54) in (3.1.48), then the result is

$$A_{i|j} = \frac{\partial A_i}{\partial q_j} - g^{kl} \cdot \frac{1}{2}\left(\frac{\partial g_{il}}{\partial q_j} + \frac{\partial g_{jl}}{\partial q_i} - \frac{\partial q_{ij}}{\partial q_l}\right) A_k. \tag{3.1.49}$$

In calculating (3.1.49), notice should be taken of the relation (3.1.42), because orthogonal coordinates are required, and hence we arrive at

$$A_{i|j} = \frac{\partial A_i}{\partial q_j} - \frac{1}{2}\left(g^{(ii)} \frac{\partial g_{(i)i}}{\partial q_j} A_{(i)} + g^{(jj)} \frac{\partial q_{(j)j}}{\partial q_i} A_{(j)} - g^{kl} \frac{\partial q_{ij}}{\partial q_l} A_k\right). \tag{3.1.50}$$

In the relation (3.1.50), i and j are not to be taken into account for the summation.

All orthogonal coordinates have in common

$$g^{(i)i} = \frac{1}{g_{(i)i}} \tag{3.1.51}$$

as a connection between the contra-variant and the covariant components of the metric tensor. Through (3.1.51) and (3.1.44) we can transcribe (3.1.50) into

$$A_{i|j} = \frac{\partial A_i}{\partial q_j} - \frac{1}{2}\left(\frac{1}{h_i^2}\frac{\partial h_i^2}{\partial q_j}A_i + \frac{1}{h_j^2}\frac{\partial h_j^2}{\partial q_i}A_j - \delta_{ij}\sum_l \frac{\partial h_j^2}{\partial q_l}\frac{A_l}{h_l^2}\right), \tag{3.1.52}$$

whereas the summation convention for (3.1.52) is cancelled. Vector A_i has changed into tensor $A_{i|j}$ through covariant derivation. We have already seen in (1.2.121) that a tensor $A_{i|j}$ of this kind has the physical components

$$A^*_{i,j} = \frac{A_{i|j}}{h_i h_j} \tag{3.1.53}$$

when the *orthogonal* curvilinear coordinates are used. As we have mentioned previously, we shall name $A^*_{i,j}$, i.e., the physical component of the covariant derivative of the vector, the *physical derivative of this vector with respect to curvilinear orthogonal coordinates.* It is obtained in accordance with (3.1.53) from (3.1.52), dividing the latter equation by $h_i h_j$. It yields

$$A^*_{i,j} = \frac{1}{h_i h_j}\frac{\partial A_i}{\partial q_j} - \frac{1}{2}\left(\frac{1}{h_i^2}\frac{\partial h_i^2}{\partial q_j}\frac{A_i}{h_i h_j} + \frac{1}{h_j^2}\frac{\partial h_j^2}{\partial q_i}\frac{A_j}{h_i h_j} - \right.$$
$$\left. -\delta_{ij}\sum_l \frac{\partial h_j^2}{\partial q_l}\frac{A_l}{h_l^2 h_i h_j}\right). \tag{3.1.54}$$

Moreover, (3.1.54) is easily converted into

$$A^*_{ij} = \frac{1}{h_j}\left[\frac{\partial}{\partial q_j}\left(\frac{A_i}{h_i}\right) - \frac{1}{h_i}\left(\frac{A_j}{h_j}\right)\frac{\partial h_j}{\partial q_i} + \delta_{ij}\sum_l \frac{1}{h_l}\left(\frac{A_l}{h_l}\right)\frac{\partial h_j}{\partial q_l}\right]. \tag{3.1.55}$$

If (3.1.46) is used in (3.1.55), the result is

$$A^*_{i,j} = \frac{1}{h_j}\left(\frac{\partial A_i^*}{\partial q_j} - \frac{A_j^*}{h_i}\frac{\partial h_j}{\partial q_i} + \delta_{ij}\sum_l \frac{A_l^*}{h_l}\frac{\partial h_j}{\partial q_l}\right). \tag{3.1.56}$$

We note that the summation convention from (3.1.54) to (3.1.56) is no more effective.

3.1.3.2 *Physical derivation of a scalar with respect to curvilinear, orthogonal coordinates.* Following (2.4.17), the usual partial derivative of f is

$$f_{,j} = \frac{\partial f}{\partial q_j}. \tag{3.1.57}$$

This, since it is a scalar, is at the same time its covariant derivation. Through differentiation, scalar f has become vector $f_{,j}$. Its physical components have to be used when calculating further, using curvilinear coordinates. After what has been said, this is the same as using the physical components

$$f_{,j}^{*} = \frac{1}{h_j}\frac{\partial f}{\partial q_j} \text{ (do not sum with respect to } j!) \tag{3.1.58}$$

of the covariant derivative (3.1.57) of scalar f in the calculation. Analogous to the given definition in 3.1.3, (3.1.58) represents the physical derivative of the scalar f.

3.1.3.3 *Derivation of the physical base vectors (unit vectors) with respect to curvilinear, orthogonal coordinates.* According to (1.2.105) the equation

$$\frac{\partial \boldsymbol{b}_i}{\partial q_j} = \begin{Bmatrix} k \\ ij \end{Bmatrix} \boldsymbol{b}_k$$

is true. Because of (1.2.55) and (1.2.54) it can be converted into

$$\frac{\partial \boldsymbol{b}_i}{\partial q_j} = g^{kn}[ij, n]\boldsymbol{b}_k = \tfrac{1}{2}g^{kn}\left(\frac{\partial g_{in}}{\partial q_j} + \frac{\partial g_{jn}}{\partial q_i} - \frac{\partial g_{ij}}{\partial q_n}\right)\boldsymbol{b}_k . \tag{3.1.59}$$

Because of (3.1.42), and together with (3.1.44), we get

$$\frac{\partial \boldsymbol{b}_i}{\partial q_j} = \frac{1}{2}\left(\frac{1}{h_i^2}\frac{\partial h_i^2}{\partial q_j}\boldsymbol{b}_i + \frac{1}{h_j^2}\frac{\partial h_j^2}{\partial q_i}\boldsymbol{b}_j - \delta_{ij}\sum_n \frac{1}{h_n^2}\frac{\partial h_j^2}{\partial q_n}\boldsymbol{b}_n\right). \tag{3.1.60}$$

In the following calculations and in (3.1.60), the summation convention shall be invalid. (3.1.60) is easily converted into

$$\frac{\partial \boldsymbol{b}_i}{\partial q_j} = \frac{1}{h_i}\frac{\partial h_i}{\partial q_j}\boldsymbol{b}_i + \frac{1}{h_j}\frac{\partial h_i}{\partial q_i}\boldsymbol{b}_j - \delta_{ij}\sum_n \frac{h_j}{h_n^2}\frac{\partial h_j}{\partial q_n}\boldsymbol{b}_n ,$$

and dividing by h_i it is again changed into

$$\frac{1}{h_i}\frac{\partial \boldsymbol{b}_i}{\partial q_j} = \frac{\boldsymbol{b}_i}{h_i^2}\frac{\partial h_i}{\partial q_j} + \frac{1}{h_i}\left(\frac{\boldsymbol{b}_j}{h_j}\right)\frac{\partial h_j}{\partial q_i} - \delta_{ij}\sum_n \frac{\partial h_j}{\partial q_n}\frac{1}{h_n}\left(\frac{\boldsymbol{b}_n}{h_n}\right).$$

This relation is the same as

$$\frac{\partial}{\partial q_j}\left(\frac{\boldsymbol{b}_i}{h_i}\right) = \frac{1}{h_i}\left(\frac{\boldsymbol{b}_j}{h_j}\right)\frac{\partial h_j}{\partial q_i} - \delta_{ij}\sum_n \frac{1}{h_n}\left(\frac{\boldsymbol{b}_n}{h_n}\right)\frac{\partial h_j}{\partial q_n},$$

which changes into

$$\frac{\partial \boldsymbol{e}_i}{\partial q_j} = \frac{\boldsymbol{e}_j}{h_i}\frac{\partial h_j}{\partial q_i} - \delta_{ij}\sum_n \frac{\boldsymbol{e}_n}{h_n}\frac{\partial h_j}{\partial q_n} \tag{3.1.61}$$

because of (3.1.45). (3.1.61) is the wanted derivative of the physical base vectors. It is also, although with slight variation, what we have termed the physical derivative of a vector in accordance with (3.1.56). The slight variation is more obvious, when the following substitutions in (3.1.56) are made: for A_i^* (i.e., A_j^*, i.e., A_l^*), formally $\boldsymbol{e}_i$ (i.e., $\boldsymbol{e}_j$, i.e., $\boldsymbol{e}_l$) and for $\boldsymbol{e}_{i,j}$ a nought. When further reduction is made after $\partial \boldsymbol{e}_i/\partial q_j$, then (3.1.61) results from (3.1.56).

3.1.3.4 *Conversion of the basic equations of the theory of elasticity to curvilinear orthogonal coordinates.* Some difficulties occur if the conditions of equilibrium (3.1.1), the geometric equations (3.1.2) and the physical equations (3.1.3) are to be used in the calculation with curvilinear, orthogonal coordinates. The same is true for the Lamé-Navier and Beltrami-Michell equations (3.1.9) and (3.1.25), which will be discussed later. All these relations are applicable to cartesian coordinates only. To change them into a formula applicable to curvilinear orthogonal coordinates, the following substitutions should be made. Vector and tensor components are substituted by physical components, partial and covariant derivatives of eligible quantities are substituted by physical derivatives, and base vectors are substituted by unit vectors. These substitutions are possible because conversion was anticipated in 3.1.3.1 to 3.1.3.3, and the relations (3.1.56), (3.1.58) and (3.1.61) were derived as supplements.

To begin with, the geometric equations

$$\varepsilon_{ij} = \tfrac{1}{2}(u_{i,j}+u_{j,i}) \tag{3.1.2}$$

are converted, and changed into

$$\varepsilon_{ij}^* = \tfrac{1}{2}(u_{i,j}^* + u_{j,i}^*). \tag{3.1.62}$$

Let u_i^* be the physical components of the displacement vector. Using these, we obtain from (3.1.56)

$$u_{i,j}^* = \frac{1}{h_j}\left(\frac{\partial u_i^*}{\partial q_j} - \frac{u_j^*}{h_i}\frac{\partial h_j}{\partial q_i} + \delta_{ij}\sum_l \frac{u_l^*}{h_l}\frac{\partial h_j}{\partial q_l}\right). \tag{3.1.63}$$

When (3.1.63) is substituted in (3.1.62), we obtain by simple calculation

$$\varepsilon_{ij}^* = \frac{1}{2}\left[\frac{h_i}{h_j}\frac{\partial}{\partial q_j}\left(\frac{u_i^*}{h_i}\right) + \frac{h_j}{h_i}\frac{\partial}{\partial q_i}\left(\frac{u_j^*}{h_j}\right) + 2\delta_{ij}\sum_l \frac{u_l^*}{h_i h_l}\frac{\partial h_i}{\partial q_l}\right]. \tag{3.1.64}$$

These are the geometric equations in a new form applicable to curvilinear coordinates, or differently, they are the physical components of the strain tensor.

Let σ_{ij}^* be the physical components of the stress tensor. These, together with the physical components ε_{ij}^* of the strain tensor, can be used to write down Hooke's Law, i.e., the physical equations (3.1.9) are to be converted using quantities σ_{ij}^*, ε_{ij}^* which are marked with an asterisk. It, therefore, yields

$$\sigma_{ij}^* = \lambda\delta_{ij}\varepsilon_{kk}^* + 2\mu\varepsilon_{ij}^*, \; \varepsilon_{ij}^* = \frac{1+\nu}{E}\sigma_{ij}^* - \frac{\nu}{E}\delta_{ij}\sigma_{kk}^*, \tag{3.1.65a, b}$$

respectively.

The physical equations apply to curvilinear, orthogonal coordinates when they are expressed in this formula.

When the conditions of equilibrium are to be converted, they first have to be represented differently from (3.1.1). Let $\boldsymbol{S}$ be the symbol for the stress tensor, then

$$\boldsymbol{S} = \begin{pmatrix} \sigma_{11} & \sigma_{12} & \sigma_{13} \\ \sigma_{21} & \sigma_{22} & \sigma_{23} \\ \sigma_{31} & \sigma_{32} & \sigma_{33} \end{pmatrix} \tag{3.1.66}$$

is true. Hence (3.1.1), using $\boldsymbol{S}$ and the vector $\boldsymbol{K}$ in symbolic notation, reads

$$\operatorname{div} \boldsymbol{S} + \boldsymbol{K} = 0. \tag{3.1.67}$$

Notated by physical components, we have

$$S = \sigma_{ij}^{*} e_i e_j, \tag{3.1.68}$$

when $e_i e_j$ is the tensorial product (dyad) of the unit vectors. We can state for a fact that

$$s_i = \sigma_{ij}^{*} e_j \tag{3.1.69}$$

represents the stress vector on the surface element with norm n_i. Therefore, instead of (3.1.68) there is

$$S = s_i e_i, \tag{3.1.70}$$

where again a tensorial product (dyad) is on the right side. At this point, an intermediate calculation becomes necessary. We know that div $A = A_{i,i} = A_{1,1} + A_{2,2} + A_{3,3}$. When curvilinear, orthogonal coordinates are used, then physical derivatives and components must be assumed at all times, and hence the equation $A = A_{1,1}^{*} + A_{2,2}^{*} + A_{3,3}^{*}$ will be used for calculation. According to (3.1.56) and following a few simple modifications, we should, therefore, arrive at

$$\begin{aligned} \operatorname{div} A &= \frac{1}{h}\left[\frac{\partial}{\partial q_1}(h_2 h_3 A_1^{*}) + \frac{\partial}{\partial q_2}(h_1 h_3 A_2^{*}) + \frac{\partial}{\partial q_3}(h_1 h_2 A_3^{*})\right], \\ h &= h_1 h_2 h_3 . \end{aligned} \tag{3.1.71}$$

This result, applicable to a vector, is now formally used for tensor S. According to (3.1.70), and analogous to the component representation of a vector, the tensor is taken to be marked by the 'components' s_i. We shall, therefore, substitute in (3.1.71) S for A and s_i for A_i^{*} ($i = 1, 2, 3$). This yields

$$\operatorname{div} S = \frac{1}{h}\left[\frac{\partial}{\partial q_1}(h_2 h_3 s_1) + \frac{\partial}{\partial q_2}(h_1 h_3 s_2) + \frac{\partial}{\partial q_3}(h_1 h_2 s_3)\right]. \tag{3.1.72}$$

Let us continue the calculation by using the expression (3.1.69) in (3.1.72) for s_i, $i = 1, 2, 3$. We shall then differentiate what is contained in the round brackets, in accordance with the product rule. For reasons of (3.1.69), these brackets contain the $e_j, j = 1, 2, 3$, thus leading to derivatives of the physical base vectors (unit vectors), which are carried out in accordance with (3.1.61). After lengthy calculations, an extensive expression is obtained from (3.1.72), containing terms connected with the base vectors e_i, $i = 1, 2, 3$. If all terms multiplied by e_i were collected, then the first component $(\operatorname{div} S)_i^*$ from div S with respect to curvilinear coordinates would be obtained, i.e.,

$$(\operatorname{div} S)_1^* = \frac{1}{h}\left\{\frac{\partial}{\partial q_1}(h_2 h_3 \sigma_{11}^*) + \frac{1}{h_1}\left[\frac{\partial(h_1^2 h_3 \sigma_{12}^*}{\partial q_2} + \frac{\partial(h_1^2 h_2 \sigma_{13}^*)}{\partial q_3}\right]\right. \\ \left. - h_3 \frac{\partial h_2}{\partial q_1}\sigma_{22}^* - h_2 \frac{\partial h_3}{\partial q_1}\sigma_{33}^*\right\}. \tag{3.1.73}$$

The remainder of the components from div S are derived from (3.1.73) by cyclic exchange of indices. This enables us to give the conditions of equilibrium in terms of curvilinear coordinates. For this purpose, the physical components for (3.1.67), i.e.,

$$(\operatorname{div} S)_i^* + K_i^* = 0, \quad i = 1, 2, 3, \tag{3.1.74}$$

have to be used. For $i = 1$, this yields, through (3.1.73), as a first condition of equilibrium

$$\frac{\partial}{\partial q_1}(h_2 h_3 \sigma_{11}^*) + \frac{1}{h_1}\left[\frac{\partial(h_1^2 h_3 \sigma_{12}^*)}{\partial q_2} + \frac{\partial(h_1^2 h_2 \sigma_{13}^*)}{\partial q_3}\right] \\ - h_3 \frac{\partial h_2}{\partial q_1}\sigma_{22}^* - h_2 \frac{\partial h_3}{\partial q_1}\sigma_{33}^* + hK_1^* = 0. \tag{3.1.75}$$

By cyclic exchange of indices, two remaining conditions of equilibrium are obtained. It can be seen that the conditions of equilibrium can be summed up by

$$\sum_{j=1}^{3} \frac{\partial}{\partial q_j}\left(\frac{hh_i^2 \sigma_{ij}^*}{h_i h_j}\right) - \tfrac{1}{2}\sum_{j=1}^{3} \frac{h\sigma_{jj}^*}{h_j^2}\frac{\partial h_j^2}{\partial q_i} + hh_i K_i^* = 0, \\ h = h_1 h_2 h_3, \quad i = 1, 2, 3. \tag{3.1.76}$$

We should note that the summation convention does not apply to this formula.

(3.1.64), (3.1.65) and (3.1.76) is the fundamental set of equations, corresponding with relations from (3.1.1) to (3.1.3). This fundamental set of equations is obtained for solving problems of the theory of elasticity, dealing with curvilinear orthogonal coordinates. Let us demonstrate this in an example for cylindrical coordinates. The following is true for the coordinates

$$q_1 = r,\ q_2 = \varphi,\ q_3 = z, \tag{3.1.77}$$

and for (3.1.39) we write

$$x_1 = r\cos\varphi,\ x_2 = r\sin\varphi,\ x_3 = z. \tag{3.1.78}$$

From (3.1.41) and (3.1.78) follows

$$\begin{aligned}
g_{11} &= \left(\frac{\partial x_1}{\partial r}\right)^2 + \left(\frac{\partial x_2}{\partial r}\right)^2 + \left(\frac{\partial x_3}{\partial r}\right)^2 = \cos^2\varphi + \sin^2\varphi = 1,\\
g_{22} &= \left(\frac{\partial x_1}{\partial \varphi}\right)^2 + \left(\frac{\partial x_2}{\partial \varphi}\right)^2 + \left(\frac{\partial x_3}{\partial \varphi}\right)^2 = (-r\sin\varphi)^2 + (r\cos\varphi)^2 = r^2,\\
g_{33} &= \left(\frac{\partial x_1}{\partial z}\right)^2 + \left(\frac{\partial x_2}{\partial z}\right)^2 + \left(\frac{\partial x_3}{\partial z}\right)^2 = 1.
\end{aligned} \tag{3.1.79}$$

From (3.1.79) we obtain

$$h_1 = 1,\ h_2 = r,\ h_3 = 1 \tag{3.1.80}$$

because of (3.1.44). For the geometric equations (3.1.64), i.e., for the physical components of the strain tensor, because of (3.1.77) and (3.1.80), and by using the terms

$$\begin{aligned}
&u_1^* = u_r,\ u_2^* = u_\varphi,\ u_3^* = u_z,\\
&\varepsilon_{11}^* = \varepsilon_{rr},\ \varepsilon_{12}^* = \varepsilon_{r\varphi},\ \varepsilon_{13}^* = \varepsilon_{rz},\\
&\varepsilon_{22}^* = \varepsilon_{\varphi\varphi},\ \varepsilon_{23}^* = \varepsilon_{\varphi z},\ \varepsilon_{33}^* = \varepsilon_{zz},
\end{aligned} \tag{3.1.81}$$

we are able to determine by simple calculation the system of equations

$$\varepsilon_{rr} = \frac{\partial u_r}{\partial r}, \; \varepsilon_{r\varphi} = \frac{1}{2}\left(\frac{1}{r}\frac{\partial u_r}{\partial \varphi} + \frac{\partial u_\varphi}{\partial r} - \frac{u_\varphi}{r}\right),$$
$$\varepsilon_{rz} = \frac{1}{2}\left(\frac{\partial u_r}{\partial z} + \frac{\partial u_z}{\partial r}\right), \; \varepsilon_{\varphi\varphi} = \frac{1}{r}\frac{\partial u_\varphi}{\partial \varphi} + \frac{u_r}{r}, \tag{3.1.82}$$
$$\varepsilon_{\varphi z} = \frac{1}{2}\left(\frac{\partial u_\varphi}{\partial z} + \frac{1}{r}\frac{\partial u_z}{\partial \varphi}\right), \; \varepsilon_{zz} = \frac{\partial u_z}{\partial z}.$$

Using for the physical components σ^*_{ij} of the stress tensor the terms

$$\begin{aligned} \sigma^*_{11} &= \sigma_{rr}, \; \sigma^*_{12} = \sigma_{r\varphi}, \; \sigma^*_{13} = \sigma_{rz}, \\ \sigma^*_{22} &= \sigma_{\varphi\varphi}, \; \sigma^*_{23} = \sigma_{\varphi z}, \; \sigma^*_{33} = \sigma_{zz}, \end{aligned} \tag{3.1.83}$$

and for the physical components of the strain tensor (3.1.81), then from (3.1.65a), Hooke's Law is obtained in the form

$$\begin{aligned} \sigma_{rr} &= \lambda(\varepsilon_{rr}+\varepsilon_{\varphi\varphi}+\varepsilon_{zz})+2\mu\varepsilon_{rr}, \; \sigma_{r\varphi} = 2\mu\varepsilon_{r\varphi}, \\ \sigma_{\varphi\varphi} &= \lambda(\varepsilon_{rr}+\varepsilon_{\varphi\varphi}+\varepsilon_{zz})+2\mu\varepsilon_{\varphi\varphi}, \; \sigma_{rz} = 2\mu\varepsilon_{rz}, \\ \sigma_{zz} &= \lambda(\varepsilon_{rr}+\varepsilon_{\varphi\varphi}+\varepsilon_{zz})+2\mu\varepsilon_{zz}, \; \sigma_{\varphi z} = 2\mu\varepsilon_{\varphi z}. \end{aligned} \tag{3.1.84}$$

Accordingly, the version (3.1.56) of Hooke's Law leads to

$$\begin{aligned} \varepsilon_{rr} &= \frac{1+\nu}{E}\sigma_{rr} - \frac{\nu}{E}(\sigma_{rr}+\sigma_{\varphi\varphi}+\sigma_{zz}), \; \varepsilon_{r\varphi} = \frac{1+\nu}{E}\sigma_{r\varphi}, \\ \varepsilon_{\varphi\varphi} &= \frac{1+\nu}{E}\sigma_{\varphi\varphi} - \frac{\nu}{E}(\sigma_{rr}+\sigma_{\varphi\varphi}+\sigma_{zz}), \; \varepsilon_{rz} = \frac{1+\nu}{E}\sigma_{rz}, \\ \varepsilon_{zz} &= \frac{1+\nu}{E}\sigma_{zz} - \frac{\nu}{E}(\sigma_{rr}+\sigma_{\varphi\varphi}+\sigma_{zz}), \; \varepsilon_{\varphi z} = \frac{1+\nu}{E}\sigma_{\varphi z}. \end{aligned} \tag{3.1.85}$$

When we write down

$$K^*_1 = K_r, \; K^*_2 = K_\varphi, \; K^*_3 = K_z, \tag{3.1.86}$$

then because of (3.1.77), (3.1.80), (3.1.83) and (3.1.86), the conditions of equilibrium in cylindrical coordinates follow from (3.1.76), and are expressed in the formula

$$\frac{\partial\sigma_{rr}}{\partial r}+\frac{1}{r}\frac{\partial\sigma_{r\varphi}}{\partial\varphi}+\frac{\partial\sigma_{rz}}{\partial z}+\frac{\sigma_{rr}-\sigma_{\varphi\varphi}}{r}+K_r=0,$$
$$\frac{\partial\sigma_{r\varphi}}{\partial r}+\frac{1}{r}\frac{\partial\sigma_{\varphi\varphi}}{\partial\varphi}+\frac{\partial\sigma_{\varphi z}}{\partial z}+\frac{2\sigma_{r\varphi}}{r}+K_\varphi=0, \tag{3.1.87}$$
$$\frac{\partial\sigma_{rz}}{\partial r}+\frac{1}{r}\frac{\partial\sigma_{\varphi z}}{\partial\varphi}+\frac{\partial\sigma_{zz}}{\partial z}+\frac{\sigma r_z}{r}+K_z=0.$$

This result can quite easily be checked.

3.1.3.5 *Conversion of the Lamé-Navier equations to curvilinear coordinates.* Using the formula (1.1.34)

$$\nabla^2\boldsymbol{u}=\text{grad div }\boldsymbol{u}-\text{curl curl }\boldsymbol{u}$$

which is known to us through vector analysis, (3.1.9*) is converted into

$$(\lambda+2\mu)\text{ grad div }\boldsymbol{u}-\mu\text{ curl curl }\boldsymbol{u}+\boldsymbol{K}=0. \tag{3.1.88}$$

We shall use this expression of the Lamé-Navier equations for the conversion to curvilinear coordinates.

To begin with, let us calculate curl $\boldsymbol{u}$, e.g., we shall write for the first component of this vector

$$(\text{curl }\boldsymbol{u})_1=u^*_{3,2}-u^*_{2,3}. \tag{3.1.89}$$

Following the method, we substitute the usual derivatives $u_{3,2}$ and $u_{2,3}$ in the particular formula for cartesian coordinates by the physical derivatives $u^*_{3,2}$ and $u^*_{2,3}$. By applying (3.1.56) we arrive at

$$u^*_{3,2}-u^*_{2,3}=\frac{1}{h_2}\left(\frac{\partial u^*_3}{\partial q_2}-\frac{u^*_2}{h_3}\frac{\partial h_2}{\partial q_3}\right)-\frac{1}{h_3}\left(\frac{\partial u^*_2}{\partial q_3}-\frac{u^*_3}{h_2}\frac{\partial h_3}{\partial q_2}\right). \tag{3.1.90}$$

This expression can be summarized differently by obtaining

$$(\text{curl }\boldsymbol{u})_1=\left[\frac{1}{h_2h_3}\frac{\partial}{\partial q_2}(h_3u^*_3)-\frac{\partial}{\partial q_3}(h_2u^*_2)\right] \tag{3.1.91}$$

for (3.1.89) because of (3.1.90). By cyclic exchanges of indices further components of curl $\boldsymbol{u}$ are obtained. They are

$$\begin{aligned}(\text{curl}\,\boldsymbol{u})_2 &= \left[\frac{1}{h_1 h_3}\,\frac{\partial}{\partial q_3}(h_1 u_1^*) - \frac{\partial}{\partial q_1}(h_3 u_3^*)\right],\\ (\text{curl}\,\boldsymbol{u})_3 &= \left[\frac{1}{h_1 h_2}\,\frac{\partial}{\partial q_1}(h_2 u_2^*) - \frac{\partial}{\partial q_2}(h_1 u_1^*)\right].\end{aligned} \tag{3.1.92}$$

When formula (3.1.91) is used for vector curl $\boldsymbol{u}$ instead of for vector $\boldsymbol{u}$, then the first component of the vector curl curl $\boldsymbol{u}$ is obtained,

$$(\text{curl curl}\,\boldsymbol{u})_1 = \frac{1}{h_2 h_3}\left\{\frac{\partial}{\partial q_2}\,[h_3(\text{curl}\,\boldsymbol{u})_3] - \frac{\partial}{\partial q_3}\,[h_2(\text{curl}\,\boldsymbol{u})_2]\right\}. \tag{3.1.93}$$

We insert the equations (3.1.92) in (3.1.93) which yields

$$\begin{aligned}(\text{curl curl}\,\boldsymbol{u})_1 = \frac{1}{h_2 h_3}&\left\{\frac{\partial}{\partial q_2}\left[\frac{h_3}{h_1 h_2}\left(\frac{\partial}{\partial q_1}(h_2 u_2^*) - \frac{\partial}{\partial q_2}(h_1 u_1^*)\right)\right]\right.\\ &\left.- \frac{\partial}{\partial q_3}\left[\frac{h_2}{h_1 h_3}\left(\frac{\partial}{\partial q_3}(h_1 u_1^*) - \frac{\partial}{\partial q_1}(h_3 u_3^*)\right)\right]\right\}.\end{aligned} \tag{3.1.94}$$

The remaining components of curl curl $\boldsymbol{u}$ result from (3.1.94) by cyclic exchange of indices.

Let us calculate grad div $\boldsymbol{u}$. Let us call div $\boldsymbol{u}$ temporarily f, and we have to write down grad f. We adhere to the rule which applies for a change of coordinates into curvilinear coordinates and write $f_{,j}^*$ instead of $f_{,j}$ (valid for cartesian coordinates). Thus, because of (3.1.58), grad f in indicial notation, as valid for curvilinear coordinates, is defined as follows:

$$\frac{1}{h_j}\,\frac{\partial f}{\partial q_j}, \quad j = 1, 2, 3, \text{ not to sum with respect to } j.$$

Therefore,

$$(\text{grad div}\,\boldsymbol{u})_1 = (\text{grad}\,f)_1 = \frac{1}{h_1}\,\frac{\partial f}{\partial q_1}, \quad f = \text{div}\,\boldsymbol{u}, \tag{3.1.95}$$

is valid for the first component of div $\boldsymbol{u}$. In addition, from (3.1.71) we read off for $f = \text{div}\,\boldsymbol{u}$

$$f = \operatorname{div} \boldsymbol{u} = \frac{1}{h}\left[\frac{\partial}{\partial q_1}(h_2 h_3 u_1^*) + \frac{\partial}{\partial q_2}(h_1 h_3 u_2^*) + \frac{\partial}{\partial q_3}(h_1 h_2 u_3^*)\right],$$

$$h = h_1 h_2 h_3. \tag{3.1.96}$$

Substituting (3.1.96) for (3.1.95) yields

$$(\operatorname{grad} \operatorname{div} \boldsymbol{u})_1 = \frac{1}{h_1}\frac{\partial}{\partial q_1}\left\{\frac{1}{h}\left[\frac{\partial}{\partial q_1}(h_2 h_3 u_1^*) + \frac{\partial}{\partial q_2}(h_1 h_3 u_2^*) + \frac{\partial}{\partial q_3}(h_1 h_2 u_3^*)\right]\right\}, \quad h = h_1 h_2 h_3. \tag{3.1.97}$$

The remaining components of grad div $\boldsymbol{u}$ can be established from (3.1.97) by cyclic exchange of indices. Through (3.1.94) and (3.1.97), the first scalar equation of the vector equation (3.1.88) for curvilinear coordinates is obtained,

$$(\lambda+2\mu)\frac{1}{h_1}\frac{\partial}{\partial q_1}\left\{\frac{1}{h_1 h_2 h_3}\left[\frac{\partial}{\partial q_1}(h_2 h_3 u_1^*) + \frac{\partial}{\partial q_2}(h_1 h_3 u_2^*) + \frac{\partial}{\partial q_3}(h_1 h_2 u_3^*)\right]\right\} - \frac{\mu}{h_2 h_3}\left\{\frac{\partial}{\partial q_2}\left[\frac{h_3}{h_1 h_2}\left(\frac{\partial}{\partial q_1}(h_2 u_2^*) - \frac{\partial}{\partial q_2}(h_1 u_1^*)\right)\right] - \frac{\partial}{\partial q_3}\left[\frac{h_2}{h_1 h_3}\left(\frac{\partial}{\partial q_3}(h_1 u_1^*) - \frac{\partial}{\partial q_1}(h_3 u_3^*)\right)\right]\right\} + K_1^* = 0. \tag{3.1.98}$$

The remaining two equations by Lamé-Navier are derived from (3.1.98) by the cyclic exchange of indices. Thus, the conversion of Lamé-Navier equations to curvilinear coordinates has been given.

When for example cylindrical coordinates are required, then the equation

$$(\lambda+2\mu)\frac{\partial}{\partial r}\left\{\frac{1}{r}\left[\frac{\partial}{\partial r}(r u_r) + \frac{\partial u_\varphi}{\partial \varphi} + \frac{\partial}{\partial z}(r u_z)\right]\right\} - \frac{\mu}{r}\left\{\frac{\partial}{\partial \varphi}\left[\frac{1}{r}\left(\frac{\partial}{\partial r}(r u_\varphi) - \frac{\partial u_r}{\partial \varphi}\right)\right] - \frac{\partial}{\partial z}\left[r\left(\frac{\partial u_r}{\partial z} - \frac{\partial u_z}{\partial r}\right)\right]\right\} + K_r = 0 \tag{3.1.99}$$

is obtained from (3.1.98) because of (3.1.77), (3.1.80), the first line of (3.1.81) and (3.1.86). The remaining two equations for cylindrical coordinates by Lamé-Navier, are obtained correspondingly, and their calculation is easy. We shall then refrain from citing them at this point.

3.1.3.6 *Conversion of Beltrami equations to curvilinear coordinates.* Beltrami's equations are known to us from 3.1.1. They are as follows:

$$\nabla^2\sigma_{ij} + \frac{1}{1+\nu}\sigma_{kk,ij} = 0. \tag{3.1.26}$$

To convert to curvilinear coordinates, (3.1.26) is given in symbolic notation, and we change over to the physical components of the stress tensor. Thus, the tensor equation

$$\text{div grad}\, \boldsymbol{S} + \frac{1}{1+\nu}(\text{grad grad})\, J_{1s} = 0 \tag{3.1.100}$$

is obtained. The equation contains

$$\boldsymbol{S} = \sigma^*_{ij}\boldsymbol{e}_i\boldsymbol{e}_j, \tag{3.1.68}$$

$$J_{1s} = \sigma^*_{11} + \sigma^*_{22} + \sigma^*_{33}, \tag{3.1.101}$$

and (grad grad) represents a tensor which yields from the tensorial product (dyad) of the Nabla-vector

$$\nabla = \boldsymbol{e}_i \frac{1}{h_{(i)}} \frac{\partial}{\partial q_i} \tag{3.1.102}$$

by itself. In the usual manner we obtain the gradient by multiplying the Nabla-vector by a scalar and the divergence as the scalar product of ∇ and a vector.

For the following we need several more results which are obtained by the following calculation. Let us go back to

$$\frac{\partial \boldsymbol{e}_i}{\partial q_j} = \frac{\boldsymbol{e}_j}{h_i}\frac{\partial h_j}{\partial q_i} - \delta_{ij}\sum_n \frac{\boldsymbol{e}_n}{h_n}\frac{\partial h_j}{\partial q_n}. \tag{3.1.61}$$

This is represented in a shortened version as

$$\frac{\partial \boldsymbol{e}_i}{\partial q_j} = L^n_{ij}\,\boldsymbol{e}_n. \tag{3.1.103}$$

The indices i, j are fixed, thus the summation is to be carried out with respect to n only. The three indices symbol L^n_{ij} has the definition

$$L^n_{ij} = \frac{\delta_{jn}}{h_{(i)}}\frac{\partial h_{(j)}}{\partial q_i} - \frac{\delta_{ji}}{h_{(n)}}\frac{\partial h_{(j)}}{\partial q_n}. \tag{3.1.104}$$

Hence, (3.1.103) and (3.1.61) are equivalent, a fact which is easily checked. The following properties are important for L^n_{ij}:

$$\begin{aligned} &L^n_{ij} = 0 \text{ for } n \neq i \neq j,\\ &L^n_{(n)j} = 0,\ L^n_{(nn)} = 0,\\ &L^n_{(ii)} = -\frac{1}{h_{(n)}}\frac{\partial h_i}{\partial q_n},\ L^n_{i(n)} = \frac{1}{h_{(i)}}\frac{\partial h_{(n)}}{\partial q_i} \text{ for } n \neq i. \end{aligned} \tag{3.1.105}$$

Since $\nabla^2 f = \text{div grad} f$ will be used quite often, let us derivate this expression. To begin with

$$\text{div grad } f = \frac{\boldsymbol{e}_0}{h_{(0)}}\cdot\frac{\partial}{\partial q_0}\frac{\boldsymbol{e}_i}{h_{(i)}}\frac{\partial f}{\partial q_i}$$

is expressed in component notation, and the multiplication involved is carried out. The result is

$$\text{div grad } f = \frac{\boldsymbol{e}_0\cdot\boldsymbol{e}_i}{h_{(0)}}\frac{\partial}{\partial q_0}\left(\frac{1}{h_{(i)}}\frac{\partial f}{\partial q_i}\right) + \frac{\boldsymbol{e}_0}{h_{(0)}}\cdot\frac{1}{h_{(i)}}\frac{\partial f}{\partial q_i}\frac{\partial \boldsymbol{e}_i}{\partial q_0}.$$

We now use $\boldsymbol{e}_0\cdot\boldsymbol{e}_1 = \delta_{0i}$, etc. and (3.1.103), which yields

$$\begin{aligned} \text{div grad } f &= \frac{\delta_{0i}}{h_{(0)}}\frac{\partial}{\partial q_0}\left(\frac{1}{h_{(i)}}\frac{\partial f}{\partial q_i}\right) + \frac{\boldsymbol{e}_0}{h_{(0)}}\frac{1}{h_{(i)}}\frac{\partial f}{\partial q_i}L^n_{i0}\,\boldsymbol{e}_n\\ &= \frac{\delta_{0i}}{h_{(0)}}\frac{\partial}{\partial q_0}\left(\frac{1}{h_{(i)}}\frac{\partial f}{\partial q_i}\right) + \frac{\delta_{0n}}{h_{(0)}}\frac{1}{h_{(i)}}\frac{\partial f}{\partial q_i}L^n_{i0}. \end{aligned}$$

We finally arrive at

$$\nabla^2 f \equiv \operatorname{div} \operatorname{grad} f = \frac{1}{h_{(i)}} \left[\frac{\partial}{\partial q_i} \left(\frac{1}{h_{(i)}} \frac{\partial f}{\partial q_i} \right) + \frac{1}{h_{(0)}} \frac{\partial f}{\partial q_i} L_{i0}^{0} \right], \tag{3.1.106}$$

when Kronecker's symbols are considered and through conversion. Another useful representation of $\nabla^2 f$ is derived simply by using operation (3.1.71) on vector

$$\operatorname{grad} f = \begin{pmatrix} \dfrac{1}{h_1} \dfrac{\partial f}{\partial q_1} \\ \dfrac{1}{h_2} \dfrac{\partial f}{\partial q_2} \\ \dfrac{1}{h_3} \dfrac{\partial f}{\partial q_3} \end{pmatrix}.$$

It leads to

$$\nabla^2 f \equiv \operatorname{div} \operatorname{grad} f = \frac{1}{h} \left[\frac{\partial}{\partial q_1} \left(\frac{h_2 h_3}{h_1} \frac{\partial f}{\partial q_1} \right) + \frac{\partial}{\partial q_2} \left(\frac{h_1 h_3}{h_2} \frac{\partial f}{\partial q_2} \right) + \frac{\partial}{\partial q_3} \left(\frac{h_1 h_2}{h_3} \frac{\partial f}{\partial q_3} \right) \right], \quad h = h_1 h_2 h_3 . \tag{3.1.107}$$

We can now convert (3.1.100). First we calculate

$$(\operatorname{grad} \operatorname{grad}) J_{1s} = \frac{\boldsymbol{e}_i}{h_{(i)}} \frac{\partial}{\partial q_i} \frac{\boldsymbol{e}_j}{h_{(j)}} \frac{\partial J}{\partial q_j},$$

and write for the time being J which is short for J_{1s}. Differentiating according to the product rule, using (3.1.103), and forming dyads with the unit vectors as factors, we obtain

$$(\operatorname{grad} \operatorname{grad}) J = \frac{\boldsymbol{e}_i \boldsymbol{e}_j}{h_{(i)} h_{(j)}} \frac{\partial^2 J}{\partial q_i \partial q_j} + \frac{\boldsymbol{e}_i \boldsymbol{e}_j}{h_{(i)}} \frac{\partial J}{\partial q_j} \cdot \left(- \frac{1}{h_{(j)}^2} \frac{\partial h_{(j)}}{\partial q_i} \right) + \frac{\boldsymbol{e}_i}{h_{(i)} h_{(j)}} \frac{\partial J}{\partial q_j} L_{ji}^{n} \boldsymbol{e}_n ,$$

which is changed to

$$(\text{grad grad})J = \boldsymbol{e}_i \boldsymbol{e}_j \left[\frac{1}{h_{(i)} h_{(j)}} \frac{\partial^2 J}{\partial q_i \partial q_j} - \frac{1}{h_{(i)} h_{(j)}^2} \frac{\partial J}{\partial q_i} \frac{\partial h_{(j)}}{\partial q_i} + \frac{1}{h_{(i)} h_{(n)}} \frac{\partial J}{\partial q_n} L_{ni}^j \right]. \tag{3.1.108}$$

It is evident that we are dealing with a second order tensor. To find tensor components, let us choose i and j as fixed numbers. Thus, we obtain as an expression for the diagonal components $[(\text{grad grad})J]_{(ii)}$, $i = j$ (fixed),

$$[(\text{grad grad})J]_{(ii)} = \frac{1}{h_{(i)}^2} \frac{\partial^2 J}{\partial q_i^2} - \frac{1}{h_{(i)}^3} \frac{\partial J}{\partial q_i} \frac{\partial h_{(i)}}{\partial q_{(i)}} + \frac{1}{h_{(i)} h_{(n)}} \frac{\partial J}{\partial q_n} L_{n(i)}^i .$$

According to (3.1.105),

$$L_{n(i)}^i = \frac{1}{h_{(n)}} \frac{\partial h_{(i)}}{\partial q_n} \quad n \neq i,$$

so that

$$[(\text{grad grad})J]_{(ii)} = \frac{1}{h_{(i)}^2} \frac{\partial^2 J}{\partial q_i^2} - \frac{1}{h_{(i)}^3} \frac{\partial J}{\partial q_{(i)}} \frac{\partial h_{(i)}}{\partial q_{(i)}} + \frac{1}{h_{(i)} h_{(n)}^2} \frac{\partial J}{\partial q_n} \frac{\partial h_{(i)}}{\partial q_n}, \tag{3.1.109}$$

(summation with respect to n for $n \neq i$!),

is the final result. Hence we have for example for $i = 1$,

$$[(\text{grad grad})J]_{11} = \frac{1}{h_1^2} \frac{\partial^2 J}{\partial q_1^2} - \frac{1}{h_1^3} \frac{\partial J}{\partial q_1} \frac{\partial h_1}{\partial q_1} + \frac{1}{h_1 h_2^2} \frac{\partial J}{\partial q_2} \frac{\partial h_1}{\partial q_2} + \frac{1}{h_1 h_3^2} \frac{\partial J}{\partial q_3} \frac{\partial h_1}{\partial q_3}. \tag{3.1.110}$$

$i \neq j$ is fixed for components outside the diagonal. Accordingly, let us take L_{ni}^j for fixed $j \neq i$, and let us sum with respect to n in (3.1.108). But according to (3.1.105), L_{ni}^j, $i \neq j$, differs from nought only if $n = i$. Hence, only one term remains from the sum with respect to n, which is multiplied by

$$L^j_{(ii)} = -\frac{1}{h_{(j)}}\frac{\partial h_i}{\partial q_j}.$$

Therefore, we obtain for the components outside the diagonal

$$\begin{aligned}[(\text{grad grad})J]_{(ij)} &= \frac{1}{h_{(i)}h_{(j)}}\frac{\partial^2 J}{\partial q_i \partial q_j} - \frac{1}{h_{(i)}h^2_{(j)}}\frac{\partial J}{\partial q_i}\frac{\partial h_{(j)}}{\partial q_i} \\ &\quad - \frac{1}{h^2_{(i)}h_{(j)}}\frac{\partial J}{\partial q_i}\frac{\partial h_{(i)}}{\partial q_j}, \quad i \neq j, \text{ fixed.}\end{aligned} \tag{3.1.111}$$

Next, let us calculate grad $\boldsymbol{S}$. In component notation we express it as

$$\text{grad } \boldsymbol{S} = \frac{\boldsymbol{e}_i}{h_{(i)}}\frac{\partial}{\partial q_i}\sigma^*_{jk}\,\boldsymbol{e}_j\boldsymbol{e}_k.$$

This expression has to be differentiated using the product rule and taking (3.1.103) into account. Moreover, unit vectors are to be linked tensorially forming tryads

$$\text{grad } \boldsymbol{S} = \frac{\boldsymbol{e}_i\boldsymbol{e}_j\boldsymbol{e}_k}{h_{(i)}}\frac{\partial \sigma^*_{jk}}{\partial q_i} + \frac{\sigma^*_{jk}}{h_{(i)}}\boldsymbol{e}_i\boldsymbol{e}_j L^n_{ki}\boldsymbol{e}_n + \frac{\sigma^*_{jk}}{h_{(i)}}\boldsymbol{e}_i L^n_{ji}\boldsymbol{e}_n\boldsymbol{e}_k.$$

By renaming the dummy indices we finally arrive at

$$\text{grad } \boldsymbol{S} = \frac{\boldsymbol{e}_i\boldsymbol{e}_j\boldsymbol{e}_k}{h_{(i)}}\left(\frac{\partial \sigma^*_{jk}}{\partial q_i} + \sigma^*_{jn}L^k_{ni} + \sigma^*_{nk}L^j_{ni}\right). \tag{3.1.112}$$

We notice grad $\boldsymbol{S}$ to be a third order tensor. Let us pass on to the construction of div grad $\boldsymbol{S}$. By way of (3.1.112) we find

$$\text{div grad } \boldsymbol{S} = \frac{\boldsymbol{e}_0}{h_{(0)}}\frac{\partial}{\partial q_0}\left[\frac{\boldsymbol{e}_i\boldsymbol{e}_j\boldsymbol{e}_k}{h_{(i)}}\left(\frac{\partial \sigma^*_{jk}}{\partial q_i} + \sigma^*_{jn}L^k_{ni} + \sigma^*_{nk}L^j_{ni}\right)\right]. \tag{3.1.113}$$

Care should be taken to multiply with $\boldsymbol{e}_0$ scalarly. Hence, we arrive back at a second order tensor for (3.1.113), since for example, $\boldsymbol{e}_0 \cdot \boldsymbol{e}_i = \delta_{0i}$ etc. This will be made use of in the following calculation. In addition, we shall differentiate in accordance with the product rule and using (3.1.103). To begin with we obtain

$$\text{div grad}\,\boldsymbol{S} = \frac{\boldsymbol{e}_j\boldsymbol{e}_k}{h_{(i)}}\frac{\partial}{\partial q_i}\left[\frac{1}{h_{(i)}}(\ldots)\right] + \frac{1}{h_{(i)}}(\ldots)\frac{\boldsymbol{e}_0}{h_{(0)}}\frac{\partial}{\partial q_0}[\boldsymbol{e}_i\boldsymbol{e}_j\boldsymbol{e}_k],$$

$$(\ldots) = \left(\frac{\partial\sigma^*_{jk}}{\partial q_i} + \sigma^*_{jn}L^k_{ni} + \sigma^*_{nk}L^j_{ni}\right). \tag{3.1.114}$$

Further calculation determines

$$\frac{\boldsymbol{e}_0}{h_{(0)}}\frac{\partial}{\partial q_0}[\boldsymbol{e}_i\boldsymbol{e}_j\boldsymbol{e}_k]$$

and finds

$$\begin{aligned}\frac{\boldsymbol{e}_0}{h_{(0)}}\cdot\frac{\partial}{\partial q_0}[\boldsymbol{e}_i\boldsymbol{e}_j\boldsymbol{e}_k] &= \frac{\boldsymbol{e}_0}{h_{(0)}}\cdot\frac{\partial \boldsymbol{e}_i}{\partial q_0}\boldsymbol{e}_j\boldsymbol{e}_k + \frac{\boldsymbol{e}_0}{h_{(0)}}\cdot\boldsymbol{e}_i\frac{\partial \boldsymbol{e}_j}{\partial q_0}\boldsymbol{e}_k + \frac{\boldsymbol{e}_0}{k_0}\boldsymbol{e}_i\boldsymbol{e}_j\frac{\partial \boldsymbol{e}_k}{\partial q_0}\\
&= \frac{\boldsymbol{e}_0}{h_{(0)}}\cdot L^m_{i0}\boldsymbol{e}_m\boldsymbol{e}_j\boldsymbol{e}_k + \frac{\boldsymbol{e}_0}{h_{(0)}}\cdot\boldsymbol{e}_i L^m_{j0}\boldsymbol{e}_m\boldsymbol{e}_k + \frac{\boldsymbol{e}_0}{h_{(0)}}\cdot\boldsymbol{e}_i\boldsymbol{e}_j L^m_{k0}\boldsymbol{e}_m\\
&= \boldsymbol{e}_j\boldsymbol{e}_k\frac{\delta_{0m}}{h_{(0)}}L^m_{i0} + \boldsymbol{e}_m\boldsymbol{e}_k\frac{\delta_{0i}}{h_{(0)}}L^m_{j0} + \boldsymbol{e}_j\boldsymbol{e}_m\frac{\delta_{0i}}{h_{(0)}}L^m_{k0}\\
&= \frac{1}{h_{(0)}}L^0_{i0}\boldsymbol{e}_j\boldsymbol{e}_k + \frac{1}{h_{(i)}}L^m_{ji}\boldsymbol{e}_m\boldsymbol{e}_k + \frac{1}{h_{(i)}}L^m_{ki}\boldsymbol{e}_j\boldsymbol{e}_m.\end{aligned} \tag{3.1.115}$$

Substitution in (3.1.114) yields

$$\begin{aligned}\text{div grad}\,\boldsymbol{S} = {}&\frac{\boldsymbol{e}_j\boldsymbol{e}_j}{h_{(i)}}\frac{\partial}{\partial q_i}\left[\frac{1}{h_{(i)}}(\ldots)\right] + \frac{1}{h_{(i)}}(\ldots)\left(\frac{1}{h_{(0)}}L^0_{i0}\boldsymbol{e}_j\boldsymbol{e}_k\right.\\
&\left.+\frac{1}{h_{(i)}}L^m_{ji}\boldsymbol{e}_m\boldsymbol{e}_k + \frac{1}{h_{(i)}}L^m_{ki}\boldsymbol{e}_j\boldsymbol{e}_m\right).\end{aligned} \tag{3.1.116}$$

The bracket (...) is written out in full in accordance with (3.1.114), also we multiply and rename the dummy indices, putting them into different order. This leads to:

$$\begin{aligned}\text{div grad}\,\boldsymbol{S} = {}&\frac{\boldsymbol{e}_j\boldsymbol{e}_k}{h_{(i)}}\left[\frac{\partial}{\partial q_i}\left(\frac{1}{h_{(i)}}\frac{\partial\sigma^*_{jk}}{\partial q_i}\right) + \frac{1}{h_{(0)}}L^0_{i0}\frac{\partial\sigma^*_{jk}}{\partial q_i}\right]\\
&+\boldsymbol{e}_j\boldsymbol{e}_k\left[\frac{1}{h_{(i)}}\frac{\partial}{\partial q_i}\frac{1}{h_{(i)}}(\sigma^*_{jn}L^k_{ni} + \sigma^*_{nk}L^j_{ni})\right.\end{aligned}$$

$$
\begin{aligned}
&+\frac{1}{h_{(i)}^2}\left(L_{mi}^j\frac{\partial\sigma_{mk}^*}{\partial q_i}+L_{mi}^k\frac{\partial\sigma_{jm}^*}{\partial q_i}\right)\\
&+\frac{1}{h_{(i)}h_{(0)}}L_{i0}^0(\sigma_{jn}^*L_{ni}^k+\sigma_{nk}^*L_{ni}^j)\\
&+\frac{1}{h_{(i)}^2}L_{mi}^j(\sigma_{mn}^*L_{ni}^k+\sigma_{nk}^*L_{ni}^m)\\
&+\frac{1}{h_{(i)}^2}L_{mi}^k(\sigma_{jn}^*L_{ni}^m+\sigma_{nm}^*L_{ni}^j)\Bigg],
\end{aligned}
\tag{3.1.117}
$$

which is a second order tensor. Comparing this result with (3.1.106), we notice that the first term of (3.1.117) is no more than $\boldsymbol{e}_j\boldsymbol{e}_k\nabla^2\sigma_{jk}^*$. This form is to be used as abbreviation in the following.

Now we are able to express the tensor equation (3.1.100) in component notation, because of (3.1.117) and (3.1.108), which is the same as expressing Beltrami's equations with respect to curvilinear coordinates. We obtain

$$
\begin{aligned}
\nabla^2\sigma_{jk}^* &+ \frac{1}{h_{(i)}}\frac{\partial}{\partial q_i}\frac{1}{h_{(i)}}(\sigma_{jn}^*L_{ni}^k+\sigma_{nk}^*L_{ni}^j)\\
&+\frac{1}{h_{(i)}^2}\left(L_{mi}^j\frac{\partial\sigma_{mk}^*}{\partial q_i}+L_{mi}^k\frac{\partial\sigma_{jm}^*}{\partial q_i}\right)+\frac{1}{h_{(i)}h_{(0)}}L_{i0}^0(\sigma_{jn}^*L_{ni}^k+\sigma_{nk}^*L_{ni}^j)\\
&+\frac{1}{h_{(i)}^2}L_{mi}^j(\sigma_{mn}^*L_{ni}^k+\sigma_{nk}^*L_{ni}^m)+\frac{1}{h_{(i)}^2}L_{mi}^k(\sigma_{jn}^*L_{ni}^m+\sigma_{nm}^*L_{ni}^j)\\
&+\frac{1}{1+\nu}\left(\frac{1}{h_{(j)}h_{(k)}}\frac{\partial^2 J}{\partial q_j\partial q_k}\cdot\frac{1}{h_{(j)}h_{(k)}^2}\frac{\partial J}{\partial q_k}\frac{\partial h_{(k)}}{\partial q_j}\right.\\
&\qquad\left.+\frac{1}{h_{(i)}h_{(j)}}\frac{\partial J}{\partial q_i}L_{ij}^k\right)=0.
\end{aligned}
\tag{3.1.118}
$$

To illustrate we evaluate (3.1.118) for cylindrical coordinates. Then (3.1.77), (3.1.80) and (3.1.83) are all valid, and all three index symbols L_{ij}^n are equal to nought with the exception of

$$
L_{(\varphi\varphi)}^r = -1,\ L_{r(\varphi)}^{\varphi} = 1. \tag{3.1.119}
$$

We write for example $j = k = r$, and by observing (3.1.105) and (3.1.119) we obtain

$$\nabla^2\sigma_{rr}+\frac{2}{r^2}(\sigma_{\varphi\varphi}-\sigma_{rr})-\frac{4}{r^2}\frac{\partial\sigma_{r\varphi}}{\partial\varphi}+\frac{1}{1+\nu}\frac{\partial^2 J}{\partial r^2}=0,$$

$$J=\sigma_{rr}+\sigma_{\varphi\varphi}+\sigma_{zz}. \tag{3.1.120}$$

The remaining equations are

$$\nabla^2\sigma_{\varphi\varphi}-\frac{2(\sigma_{\varphi\varphi}-\sigma_{rr})}{r^2}+\frac{4}{r^2}\frac{\partial\sigma r_{\varphi}}{\partial\varphi}+\frac{1}{1+\nu}\left(\frac{1}{r}\frac{\partial J}{\partial r}+\frac{\partial^2 J}{r^2\partial\varphi^2}\right)=0,$$

$$\nabla^2\sigma_{zz}+\frac{1}{1+\nu}\frac{\partial^2 J}{\partial z^2}=0,$$

$$\nabla^2\sigma_{r\varphi}-\frac{2}{r^2}\frac{\partial}{\partial\varphi}(\sigma_{\varphi\varphi}-\sigma_{rr})-\frac{4}{r^2}\sigma_{r\varphi}+\frac{1}{1+\nu}\frac{\partial}{\partial r}\left(\frac{1}{r}\frac{\partial J}{\partial\varphi}\right)=0, \tag{3.1.121}$$

$$\nabla^2\sigma_{\varphi z}-\frac{\sigma_{\varphi z}}{r^2}+\frac{2}{r^2}\frac{\partial\sigma_{rz}}{\partial\varphi}+\frac{1}{1+\nu}\frac{\partial^2 J}{r\partial\varphi\,\partial z}=0,$$

$$\nabla^2\sigma_{zr}-\frac{\sigma_{rz}}{r^2}-\frac{2}{r^2}\frac{\partial\sigma_{\varphi z}}{\partial\varphi}+\frac{1}{1+\nu}\frac{\partial^2 J}{\partial r\,\partial z}=0.$$

In (3.1.120) and (3.1.121), $\nabla^2\sigma_{rr}$, etc., must be calculated according to (3.1.106) or (3.1.107).

Finally it should be mentioned that conversion of Michell's equations (3.1.25) may be carried out following the pattern of Beltrami's equations, since they differ by only a few additional terms (3.1.26). The result of such a calculation has been given by C. E. Pearson [18].

3.1.4 *Simple examples*

This paragraph demonstrates the application of the basic equations (3.1.1), (3.1.2) and (3.1.3) and of the semi-inverse method to simple problems of the theory of elasticity.

1st Example. Let a homogeneous beam of edge lengths $2A$, $2B$ and L (Figure 27) and specific weight γ be fastened by its upper top area $x_3 = L$ in such a way that its weight is evenly distributed over this area. We shall find the displacements of all points of the beam, thus obtaining a statement which indicates the form which its surface will assume after having been deformed by gravity forces acting in the x_3-direction.

Let us begin with the assumption

$$\sigma_{11} = \sigma_{22} = \sigma_{12} = \sigma_{23} = \sigma_{31} = 0, \ \sigma_{33} = \gamma x_3 \qquad (3.1.122)$$

for the stress components according to the semi-inverse method. According to the boundary conditions, all surface stresses p on all lateral surface areas as well as on the lower top surface of the beam are equal to nought. Only on the beam's upper top surface $x_3 = L$ are they $p_1 = p_2 = 0$, $p_3 = \gamma L$. Since $\boldsymbol{n} = (0, 0, 1)$ is valid on the upper top surface, the boundary condition (3.1.4) is satisfied by the assumption (3.1.122). All other boundary conditions are satisfied in trivial manner by (3.1.122) also.

Let us pass on to the conditions of equilibrium (3.1.1). Let $\boldsymbol{K} = (0, 0, -\gamma)$ be valid. Then the set of equilibrium equations is satisfied when $\boldsymbol{K} = (0, 0, -\gamma)$ and the assumption (3.1.122) for the stresses are used in (3.1.1). Hence, the assumption for the stresses does not violate neither the boundary conditions nor the conditions of equilibrium and is, therefore, valid.

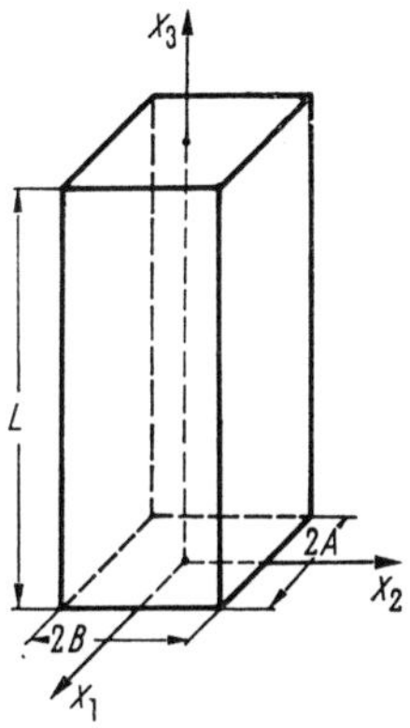

Figure 27

Let us calculate the components of the strain tensor using physical equations (3.1.3b) and (3.1.122):

$$\varepsilon_{11} = \varepsilon_{22} = -\frac{\nu}{E}\gamma x_3, \ \varepsilon_{33} = \frac{1}{E}\gamma x_3, \ \varepsilon_{12} = \varepsilon_{23} = \varepsilon_{31} = 0. \qquad (3.1.123)$$

Because of (3.1.123) and the geometric equation (3.1.2), we obtain for the displacements

$$u_{1,1} = -\frac{\nu}{E}\gamma x_3 = u_{2,2},\, u_{3,3} = \frac{\gamma}{E}x_3, \qquad (3.1.124)$$
$$u_{1,2}+u_{2,1} = u_{1,3}+u_{3,1} = u_{2,3}+u_{3,2} = 0.$$

Through integration, we obtain from $u_{3,3} = \gamma x_3/E$ the expression

$$u_3 = \frac{1}{2}\frac{\gamma}{E}x_3^2+u_3^0(x_1, x_2), \qquad (3.1.125)$$

u_3^0 being a yet undetermined integration function. From $u_{1,3}+u_{3,1} = 0$ and (3.1.125) follows $u_{1,3} = -u_{3,1} = -u_{3,1}^0$ which yields through integration

$$u_1 = -u_{3,1}^0 x_3+u_1^0(x_1, x_2). \qquad (3.1.126)$$

Similarly, we obtain from $u_{2,3}+u_{3,2} = 0$ and (3.1.125)

$$u_2 = -u_{3,2}^0 x_3+u_2^0(x_1, x_2). \qquad (3.1.127)$$

In these cases, u_1^0 and u_2^0 are again integration functions. At the same time

$$u_{1,1} = -\frac{\nu\gamma}{E}x_3 \text{ and } u_{1,1} = -u_{3,11}^0 x_3+u_{1,1}^0$$

are valid because of (3.1.124) and (3.1.126). But this can only be true if

$$u_{3,11}^0 = \frac{\nu\gamma}{E} \qquad (3.1.128)$$

and

$$u_{1,1}^0 = 0. \qquad (3.1.129)$$

In the same manner

$$u_{2,2} = -\frac{\nu\gamma}{E}x_3 \text{ and } u_{2,2} = -u_{3,22}^0 x_3+u_{2,2}^0$$

must be satisfied because of (3.1.124) and (3.1.127). This demands

$$u^0_{3,22} = \frac{\nu\gamma}{E} \tag{3.1.130}$$

and

$$u^0_{2,2} = 0. \tag{3.1.131}$$

Finally, from (3.1.124), (3.1.126) and (3.1.127) follows

$$u_{1,2}+u_{2,1} = -u^0_{3,12}x_3+u^0_{3,21}x_3+u^0_{2,1} = 0,$$

which leads to

$$u^0_{3,12} = u^0_{3,21} = 0 \tag{3.1.132}$$

and

$$u^0_{2,1}+u^0_{1,2} = 0. \tag{3.1.133}$$

From (3.1.129) and (3.1.131), we determine through integration

$$u^0_1 = F(x_2),\ u^0_2 = G(x_1).$$

Hence, because of (3.1.133), which is $G_1+F_{,2} = 0$,

$$F_{,2} = -G_{,1} = a = \text{const.}$$

But then yet again through integration

$$u^0_1 = ax_2+b,\ u^0_2 = -ax_1+c, \tag{3.1.134}$$

when b and c are integration constants. From (3.1.132) we obtain

$$u^0_{3,1} = \psi(x_1),\ u^0_{3,2} = \varphi(x_2), \tag{3.1.135}$$

from (3.1.128)

$$u^0_{3,1} = \frac{\nu\gamma}{E}x_1+\varphi^*(x_2) \tag{3.1.136}$$

and from (3.1.130)

$$u^0_{3,2} = \frac{\nu\gamma}{E}x_2 + \psi^*(x_1). \tag{3.1.137}$$

When (3.1.135) is compared with (3.1.136) and (3.1.137), it becomes evident that contradictions are avoidable only if the arbitrary functions $\psi(x_1)$, $\psi^*(x_1)$, $\varphi(x_2)$ and $\varphi^*(x_2)$ are determined in such a way that

$$u^0_{3,1} = \frac{\nu\gamma}{E}x_1 + d \text{ and } u^0_{3,2} = \frac{\nu\gamma}{E}x_2 + e. \tag{3.1.138}$$

The terms d and e are constants. From (3.1.138) we obtain through integration both

$$u^0_3 = \frac{\nu\gamma}{2E}x_1^2 + dx_1 + \alpha(x_2) \text{ and } u^0_3 = \frac{\nu\gamma}{2E}x_2^2 + ex_2 + \beta(x_1).$$

These are in agreement only when we write for the integration functions

$$\alpha(x_2) = \frac{\nu\gamma}{2E}x_2^2 + ex_2 + f, \quad \beta(x_1) = \frac{\nu\gamma}{2E}x_1^2 + dx_1 + f$$

which contain the new constant f. Eventually this leads to

$$u^0_3 = \frac{\nu\gamma}{2E}(x_1^2 + x_2^2) + dx_1 + ex_2 + f. \tag{3.1.139}$$

Through (3.1.134) and (3.1.139) we obtain from (3.1.125), (3.1.126) and (3.1.127) the wanted displacements,

$$\begin{aligned} u_1 &= -\frac{\nu\gamma}{E}x_1x_3 - dx_3 + ax_2 + b, \\ u_2 &= -\frac{\nu\gamma}{E}x_2x_3 - ex_3 - ax_1 + c, \\ u_3 &= \frac{\gamma}{2E}[x_3^2 + \nu(x_1^2 + x_2^2)] + dx_1 + ex_2 + f. \end{aligned} \tag{3.1.140}$$

In (3.1.140), the constants a to f are to be determined by boundary conditions. We require that since the beam is fastened in the upper top surface area

$$u_1 = u_2 = u_3 = u_{1,3} = u_{2,1} = u_{2,3} = 0 \text{ for } x_1 = x_2 = 0,\ x_3 = L \tag{3.1.141}$$

be valid.

This results in

$$a = b = c = d = e = 0,\ f = -\frac{\gamma}{2E}L^2.$$

Hence (3.1.140) finally becomes

$$\begin{aligned} u_1 &= -\frac{\nu\gamma}{E}x_1 x_3,\ u_2 = -\frac{\nu\gamma}{E}x_2 x_3, \\ u_3 &= \frac{\gamma}{2E}[x_3^2+\nu(x_1^2+x_2^2)-L^2]. \end{aligned} \tag{3.1.142}$$

Thus, all displacements have been calculated. To determine the new shape of the beam following deformation, let us remember that adhering to the condition of infinitesimal deformation, we have hitherto made no distinction between the coordinates ξ_i before, and x_i after deformation. We shall now take this difference into account. To begin with, let us substitute x_i by ξ_i in (3.1.142). It yields

$$u_1 = -\frac{\nu\gamma}{E}\xi_1 \xi_3,\ u_2 = -\frac{\nu\gamma}{E}\xi_2 \xi_3,$$

$$u_3 = \frac{\gamma}{2E}[\xi_3^2+\nu(\xi_1^2+\xi_2^2)-L^2]. \tag{3.1.142*}$$

Then, through $x_i = \xi_i+u_i$ and (3.1.142*), we calculate x_i, thus indicating the new shape of the beam, following deformation.

The original ground surface $\xi_3 = 0$ changes into

$$x_1 = \xi_1,\ x_2 = \xi_2,\ x_3 = \frac{\gamma}{2E}[\nu(\xi_1^2+\xi_2^2)-L^2] = \frac{\gamma}{2E}[\nu(x_1^2+x_2^2)-L^2],$$

which is a paraboloid of revolution. Its lowest point is on the x_3 axis having ordinate $x_3 = -(\gamma/2E)L^2$, which, in terms of value, represents the extensions of the beam's axis. The top surface area $\xi_3 = L$ changes into

$$x_1 = \xi_1\left(1-\frac{\nu\gamma}{E}L\right),\quad x_2 = \xi_2\left(1-\frac{\nu\gamma}{E}L\right),$$

$$x_3 = L+\frac{\nu\gamma}{2E}(\xi_1^2+\xi_2^2) = L+\frac{\nu\gamma}{2E}\,\frac{x_1^2+x_2^2}{\left(1-\frac{\nu\gamma}{E}L\right)^2}.$$

This is again a paraboloid of revolution, which, except for a very small factor $(1-\nu\gamma/E\,L)^2$, is congruent with the lower paraboloid. The anterior surface $\xi_1 = A$ changes into

$$x_1 = A\left(1-\frac{\nu\gamma}{E}\xi_3\right).$$

It means that, if in a first approximation ξ_3 is substituted by x_3, this is a plane, which approximates the axis when ξ_3 increases. The same can be said for the remaining lateral surfaces so that the beam changes into a truncated pyramid, which is bounded from below and from above by two almost congruent paraboloids of revolution (Figure 28).

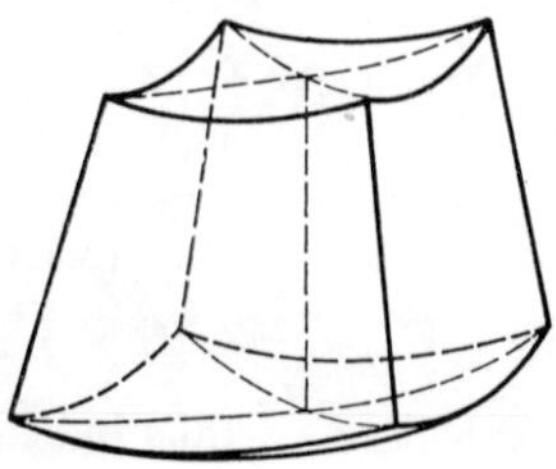

Figure 28

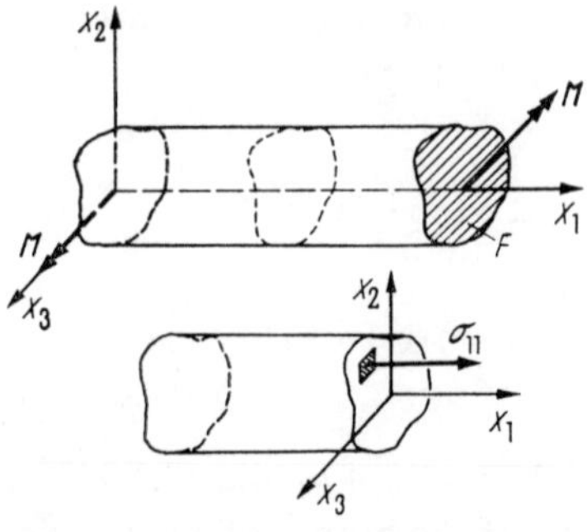

Figure 29

2nd Example. Let a cylindrical beam be exposed by its bases to bending moments (Figure 29). We shall calculate its stress distribution and its deformation. We assume the volume forces to be $\boldsymbol{K} \equiv 0$, and for the stresses we set up the trial solution

$$\sigma_{11} = cx_2, \; \sigma_{22} = \sigma_{33} = \sigma_{12} = \sigma_{23} = \sigma_{31} = 0. \tag{3.1.143}$$

It satisfies the conditions of equilibrium (3.1.1) because of $\boldsymbol{K} \equiv 0$, and by way of physical equations (3.1.3b) leads to the components

$$\varepsilon_{11} = \frac{c}{E} x_2, \; \varepsilon_{22} = \varepsilon_{33} = -\frac{\nu c}{E} x_2, \; \varepsilon_{12} = \varepsilon_{23} = \varepsilon_{31} = 0 \tag{3.1.144}$$

of the strain tensor. These in addition satisfy the compatibility conditions (2.1.88) and guarantee unique displacements. Let us deal with the boundary conditions. On the beam's lateral surface, $\boldsymbol{p} \equiv 0$. Therefore, the condition (3.1.4) states $\sigma_{ij} n_j = 0$ to be true. But since on the lateral surface $\boldsymbol{n} = (0, n_2, n_3)$ together with

$$-1 \leqq n_2, n_3 \leqq +1$$

is true, then $\sigma_{ij} n_j = 0$ is satisfied by (3.1.143).

On the bases, only moments occur about the x_3-axis but no normal forces are present. Therefore,

$$\int_F \sigma_{11} \, dF = c \int_F x_2 \, dF = 0.$$

Hence, also

$$\int_F x_2 \, dF = 0,$$

which means that the x_3-axis is the centroidal axis of the beam. Since no moments occur about the x_2-axis, then the following must be true

$$\int_F x_3 \sigma_{11} \, dF = c \int_F x_2 x_3 \, dF = 0.$$

Therefore

$$\int_F x_2 x_3 \, dF = 0$$

is true, which means that the x_2, x_3-axes are principle axes in the beam cross-sections. For the bases of the beam, the moment $|\boldsymbol{M}| \equiv M_3$ is specified. We must, therefore, demand

$$\int_F x_2 \sigma_{11} \, dF = c \int_F x_2^2 \, dF = M_3 .$$

If we write

$$\int_F x_2^2 \, dF = J_{33} ,$$

then we have

$$cJ_{33} = M_3 ,$$

which is followed by

$$c = \frac{M_3}{J_{33}} . \tag{3.1.145}$$

In addition to the known M_3, in (3.1.145) the moment of inertia J_{33} occurs, which is specified by the particular kind of beam cross-section. Therefore, on the right side of (3.1.145) only known quantities are found. Hence, the constant c of the trial solution (3.1.143) has been determined. It follows that

$$\sigma_{11} = \frac{M_3}{J_{33}} x_2 , \tag{3.1.146}$$

which shows that we are dealing with a linear stress distribution in the beam cross-section. The geometric equations (3.1.2) are used to determine the displacements. Considering (3.1.144), these equations yield

$$u_{1,1} = \frac{c}{E} x_2 , \tag{3.1.147}$$

$$u_{2,2} = u_{3,3} = -\frac{\nu c}{E} x_2, \tag{3.1.148}$$

$$u_{1,2}+u_{2,1} = u_{1,3}+u_{3,1} = u_{2,3}+u_{3,2} = 0. \tag{3.1.149}$$

The integration of (3.1.147) yields

$$u_1 = \frac{c}{E} x_1 x_2 + u_1^0(x_2, x_3). \tag{3.1.150}$$

From (3.1.149) and (3.1.150) follows

$$u_{2,1} = -u_{1,2} = -\frac{c}{E} x_1 - u_{1,2}^0, \tag{3.1.151}$$

and

$$u_{3,1} = -u_{1,3} = -u_{1,3}^0, \tag{3.1.152}$$

from which we obtain through integration

$$u_2 = -\frac{c}{2E} x_1^2 - u_{1,2}^0 x_1 + u_2^0(x_2, x_3), \tag{3.1.153}$$

and

$$u_3 = -u_{1,3}^0 x_1 + u_3^0(x_2, x_3). \tag{3.1.154}$$

If we substitute (3.1.153) and (3.1.154) in (3.1.148), we obtain

$$u_{1,22}^0 x_1 = u_{2,2}^0 + \frac{\nu c}{E} x_2 \text{ and } u_{1,33}^0 x_1 = u_{3,3}^0 + \frac{\nu c}{E} x_2.$$

This can only be true for

$$u_{1,22}^0 = 0, \; u_{1,33}^0 = 0. \tag{3.1.155}$$

Hence we obtain

$$u_{2,2}^0 = -\frac{\nu c}{E} x_2, \; u_{3,3}^0 = -\frac{\nu c}{E} x_2,$$

and from this, through integration,

$$u_2^0 = -\frac{\nu c}{2E} x_2^2 + h_2(x_3), \; u_3^0 = -\frac{\nu c}{E} x_2 x_3 + h_3(x_2). \tag{3.1.156}$$

If we substitute (3.1.156) in (3.1.153) and (3.1.154), we obtain

$$u_2 = -\frac{cx_1^2}{2E} - u_{1,2}^0 x_1 - \frac{\nu c}{2E} x_2^2 + h_2(x_3), \tag{3.1.157}$$

$$u_3 = -u_{1,3}^0 x_1 - \frac{\nu c}{E} x_2 x_3 + h_3(x_2). \tag{3.1.158}$$

In this way we can calculate from $u_{2,3} + u_{3,2} = 0$

$$h_{2,3} + h_{3,2} - \frac{\nu c}{E} x_3 - 2x_1 u_{1,23}^0 = 0.$$

The first three terms of this expression are independent from x_1. Hence, the expression is true only if

$$u_{1,23}^0 = 0, \tag{3.1.159}$$

and

$$h_{2,3} - \frac{\nu c}{E} x_3 = -h_{3,2}.$$

This in turn is valid only for

$$-h_{3,2} = a = \text{const},$$

$$h_{2,3} - \frac{\nu c}{E} x_3 = a. \tag{3.1.160}$$

From (3.1.155) and (3.1.159) we get

$$u_1^0 = bx_2 + dx_3 + e, \tag{3.1.161}$$

and from (3.1.160), by integration,

$$h_2 = \frac{\nu c}{2E} x_3^2 + a x_3 + f, \; h_3 = -a x_2 + g. \tag{3.1.162}$$

Through (3.1.161) and (3.1.162) we get from (3.1.150), (3.1.157) and (3.1.158) for the displacements

$$\begin{aligned} u_1 &= \frac{c}{E} x_1 x_2 + b x_2 + d x_3 + e, \\ u_2 &= -\frac{c}{2E} x_1^2 - b x_1 - \frac{\nu c}{2E} x_2^2 + \frac{\nu c}{2E} x_3^2 + a x_3 + f, \\ u_3 &= -d x_1 - \frac{\nu c}{E} x_2 x_3 - a x_2 + g. \end{aligned} \tag{3.1.163}$$

To determine the constants of integration a, b, d, e, f and g, let us adhere to the condition that at $x_1 = 0$, $x_2 = 0$, $x_3 = 0$, the displacements and all rotations are equal to nought, so that

$$u_1 = u_2 = u_3 = u_{1,2} = u_{2,1} = u_{2,3} = u_{3,2} = u_{3,1} = u_{1,3} = 0$$

is true. Hence, all integration constants are equal to nought, and for (3.1.163)

$$\begin{aligned} u_1 &= \frac{c}{E} x_1 x_2, \\ u_2 &= \frac{c}{2E} [\nu(x_3^2 - x_2^2) - x_1^2], \\ u_3 &= -\frac{\nu c}{E} x_2 x_3. \end{aligned} \tag{3.1.164}$$

Let us consider the deformation of a plane, rectangular cross-section of the beam. As in the first example, let us observe the difference between ξ_i and x_i. Thus, we write

$$u_1 = \frac{c}{E} \xi_1 \xi_2, \; u_2 = \frac{c}{2E} [\nu(\xi_3^2 - \xi_2^2) - \xi_1^2], \; u_3 = -\frac{\nu c}{E} \xi_2 \xi_3,$$

and

$$x_1 = \xi_i + u_i,$$

in order to calculate the form which any plane, rectangular cross-section $\xi_1 = K = \text{const}$ will assume after deformation.

For the plane lateral surface $\xi_3 = B$ of the beam we obtain

$$x_3 = B\left(1 - \frac{\nu c}{E}\xi_2\right), \quad x_1 = \xi_1\left(1 + \frac{c}{E}\xi_2\right),$$

$$x_2 = \xi_2 + \frac{c}{2E}[\nu(B^2 - \xi_2^2) - \xi_1^2].$$

Let us assume c/E to be small of higher order, which is the same as having little curvature in the deformed beam's axis.

We then have $x_1 \approx \xi_1$, $x_2 \approx \xi_2$ and

$$x_3 \approx B\left(1 - \frac{\nu c}{E}x_2\right),$$

which shows that the trace of the deformed lateral surface in the x_2, x_3-plane is a tilting straight line. The same is true for the other lateral surface. For the upper and lower lateral surfaces $\xi_2 = \pm A$ of the beam we write

$$x_1 = \xi_1\left(1 \pm \frac{c}{E}A\right), \quad x_2 = \pm A + \frac{c}{2E}[\nu(\xi_3^2 - A^2) - \xi_1^2],$$

$$x_3 = \xi_3\left(1 \pm \frac{c}{E}A\right).$$

Again, assuming the same smallness of c/E, this time $x_1 \approx \xi_1, x_3 \approx \xi_3$ is true, and we obtain

$$x_2 \approx \pm A + \frac{c}{2E}[\nu(x_3^2 - A^2) - x_1^2].$$

This represents the equation for a hyperbolic paraboloid. We have $x_1 \approx \xi_1 = K$ for the assumed cross-section, and hence traces of the deformed upper and lower lateral surfaces in the x_2, x_3 plane, can be represented by the parabolae

$$x_2 = \pm A + \frac{c}{2E}[\nu(x_3^2 - A^2) - K^2].$$

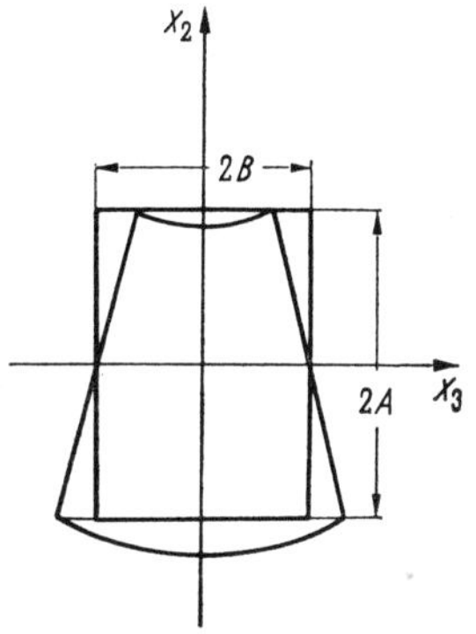

Figure 30

A view of the deformed beam cross-section in the x_2, x_3 plane is shown in Figure 30.

For $\xi_1 = K$, we find

$$x_1 \approx K\left(1 + \frac{c}{E}x_2\right)$$

by way of the previously used condition for c/E. Thus, we determine that plane cross-sections remain approximately plane despite deformations. Therefore, Bernoulli's hypothesis of the technical bending theory of beams has been adequately justified.

3.2 Torsion

3.2.1 *Prandtl's stress function*

We shall discuss the torsion of cylindrical rods having any arbitrarily shaped cross-section (Figure 31). The fundamentals of this theory are traced back to Coulomb and Navier. However, only Saint-Vénant has been able to arrive at the right solution of the torsion problem by assuming that owing to torsion, the rod's cross-section would warp. We shall follow Saint-Vénant's approach in this section, in that we shall use – following his example – the semi-inverse method, and allowing the possibility of cross-sectional warping. We shall, however, deviate by taking up a proposal by Prandtl, who introduced a so-called *stress function* Φ. Thus, we have encountered a fundamentally important concept, which we shall use continually in connection with the mathematical theory of elasticity. The application of stress functions has proved itself very worthwhile in the solutions of many problems. This will be demonstrated by a torsion problem immediately following, because Prandtl's function is nothing more than a special stress function.

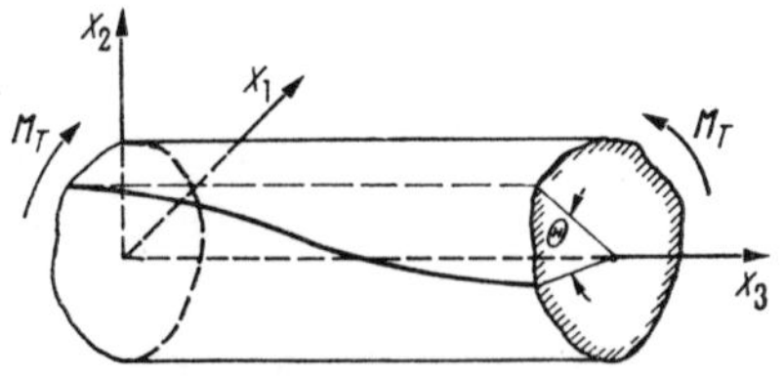

Figure 31

For the stress tensor, let us start with

$$S = \begin{pmatrix} 0 & 0 & \sigma_{13} \\ 0 & 0 & \sigma_{23} \\ \sigma_{13} & \sigma_{23} & 0 \end{pmatrix}. \tag{3.2.1}$$

We ignore all volume forces and write down $\boldsymbol{K} \equiv 0$. Thus, we obtain from the equilibrium conditions (3.1.1)

$$\sigma_{13,3} = 0, \ \sigma_{23,3} = 0 \tag{3.2.2}$$

and

$$\sigma_{13,1} + \sigma_{23,2} = 0. \tag{3.2.3}$$

From (3.2.2) we read off

$$\sigma_{13} = \sigma_{13}(x_1, x_2), \ \sigma_{23} = \sigma_{23}(x_1, x_2). \tag{3.2.4}$$

Using the physical equations (3.1.3b), we find the strain tensor components, and using $E = 2(1+\nu)G$ we find

$$\varepsilon_{11} = \varepsilon_{22} = \varepsilon_{33} = \varepsilon_{12} = 0,\ \varepsilon_{13} = \frac{1}{2G}\,\varepsilon_{13},\, \varepsilon_{23} = \frac{1}{2G}\,\varepsilon_{23}. \tag{3.2.5}$$

The geometric equations yield together with (3.2.5)

$$\begin{aligned} &u_{1,3}+u_{3,1} = \frac{1}{G}\,\sigma_{13},\, u_{2,3}+u_{3,2} = \frac{1}{G}\,\sigma_{23}, \\ &u_{1,1} = u_{2,2} = u_{3,3} = u_{1,2}+u_{2,1} = 0, \end{aligned} \tag{3.2.6}$$

which leads to

$$\begin{aligned} u_1 &= u_1(x_2, x_3), \\ u_2 &= u_2(x_1, x_3), \\ u_3 &= u_3(x_1, x_2). \end{aligned} \tag{3.2.7}$$

Prandtl's function $\Phi = \Phi(x_1, x_2)$ is introduced by

$$\sigma_{13} = \Phi_{,2},\, \sigma_{23} = -\Phi_{,1}. \tag{3.2.8}$$

By applying (3.2.8) in (3.2.3), we notice that the equilibrium conditions have been satisfied. A characteristic of all stress functions is that their selection should be such that they satisfy equilibrium conditions. For the stress tensor (3.2.1) is $\sigma_{kk} \equiv 0$. As no volume forces are present, calculation is carried out by using Beltrami's equations (3.1.26). Because we stated that $\sigma_{kk} = 0$, then only $\nabla^2\sigma_{ij} = 0$ remains. In this way, and by adhering to (3.2.8), as well as by changing the order of the differentiation, we obtain the relation

$$\begin{aligned} &\nabla^2\sigma_{13} = \nabla^2\Phi_{,2} = (\nabla^2\Phi)_{,2} = 0, \\ &\nabla^2\sigma_{23} = -\nabla^2\Phi_{,1} = -(\nabla^2\Phi)_{,1} = 0,\ (\nabla^2\Phi)_{,1} = 0. \end{aligned} \tag{3.2.9}$$

Since, by assumption, Φ is only a function of x_1 and of x_2, then $\nabla^2\Phi$ can at the most be dependent of x_1 and x_2. Thus, from (3.2.9) is

$$\nabla^2\Phi = \text{const.} \tag{3.2.10}$$

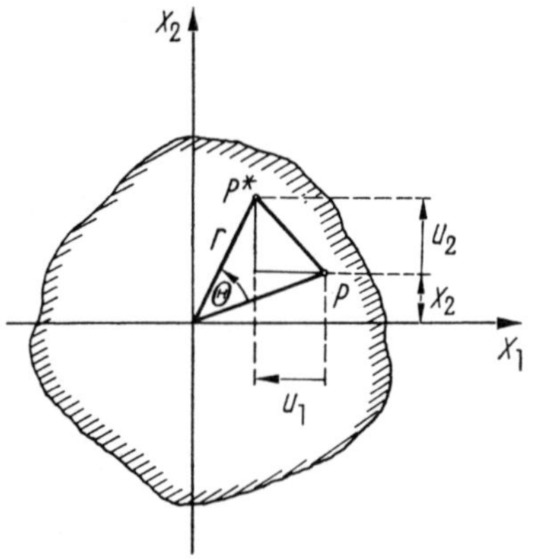

Figure 32

Let us look at the rod's cross-section (Figure 32). At the moment of torsion, point P changes into P^*. Let $\overline{PP*} = r\Theta$. Additionally, we read off the figure

$$\frac{|u_1|}{\overline{PP*}} = \frac{x_2}{r},$$

from which we obtain by noticing the plus or minus signs

$$u_1 = -\frac{\overline{PP*}}{r} x_2 = -\Theta x_2 . \tag{3.2.11}$$

Accordingly, we find

$$u_2 = \Theta x_1 . \tag{3.2.12}$$

Because the deformation of the rod is small, it is assumed that the total angle of twist Θ is proportional to the distance of the particular cross-section of the rod's end. Thus,

$$\Theta = \vartheta x_3 \tag{3.2.13}$$

is valid. The proportional factor ϑ is called twist per unit length. Through (3.2.13) we obtain from (3.2.11) and (3.2.12)

$$u_1 = -\vartheta x_2 x_3,$$
$$u_2 = \vartheta x_1 x_3, \tag{3.2.14}$$

additionally we write down, obeying Saint-Vénant's hypothesis,

$$u_3 = \vartheta\varphi(x_1, x_2), \tag{3.2.15}$$

when $\varphi(x_1, x_2)$ represents the *torsion function.*

While Saint-Vénant deals mainly with evaluating the torsion function φ, we shall proceed differently by calculating the stress function Φ. Using this function, all other quantities of interest are deducible.

From (3.2.6), (3.2.8), and (3.2.14) results

$$\begin{aligned} u_{3,1}+u_{1,3} &= u_{3,1}-\vartheta x_2 = \frac{1}{G}\sigma_{13} = \frac{1}{G}\Phi_{,2}, \\ u_{3,2}+u_{2,3} &= u_{3,2}+\vartheta x_1 = \frac{1}{G}\sigma_{23} = -\frac{1}{G}\Phi_{,1}, \end{aligned} \tag{3.2.16}$$

and thus,

$$u_{3,1} = \frac{1}{G}\Phi_{,2}+\vartheta x_2, \; u_{3,2} = -\left(\frac{1}{G}\Phi_{,1}+\vartheta x_1\right).$$

Through further differentiation we obtain at the same time

$$u_{3,12} = \frac{1}{G}\Phi_{,22}+\vartheta \text{ and } u_{3,21} = -\left(\frac{1}{G}\Phi_{,11}+\vartheta\right),$$

which must agree. This is the case if

$$\frac{1}{G}\Phi_{,22}+\vartheta = -\left(\frac{1}{G}\Phi_{,11}+\vartheta\right)$$

is valid. Thus we arrive at

$$\Phi_{,11}+\Phi_{,22} = -2G\vartheta,$$

i.e., in accordance with (3.2.10) it is

$$\nabla^2\Phi = -2G\vartheta. \tag{3.2.17}$$

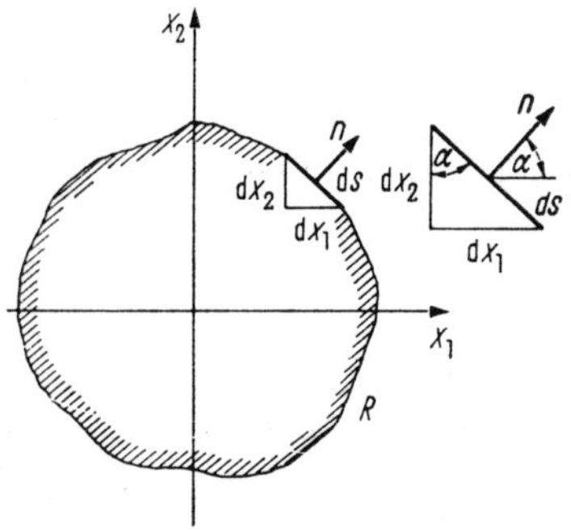

Figure 33

Hence, the stress function Φ represents a solution of a Poisson's differential equation.

Let us find the boundary conditions for Φ, and start by looking at Figure 33. At boundary R of the cross-section, $\boldsymbol{p} = 0$ is valid. According to (3.1.4), at the boundary must therefore be $\sigma_{ij}n_j = 0$. For the normals $\boldsymbol{n}$ of the boundary we read off from Figure 33

$$n_1 = \cos\alpha = \frac{dx_2}{ds}, \; n_2 = \sin\alpha = -\frac{dx_1}{ds}, \; n_3 = 0 \tag{3.2.18}$$

From $\sigma_{ij}n_j = 0$, and because of (3.2.1) and (3.2.18), together with (3.2.8), the one condition

$$\frac{\partial\Phi}{\partial x_2}\frac{dx_2}{ds} + \frac{\partial\Phi}{\partial x_1}\frac{dx_1}{ds} = \frac{d\Phi}{ds} = 0 \tag{3.2.19}$$

is derived. For this reason $\Phi(R) = \text{const}$ is required. Since in the application of Φ only the derivatives of this function are important, we can simply take as the boundary condition for the stress function

$$\Phi(R) = 0. \tag{3.2.20}$$

Evaluation of Φ eventually leads to the solution of the boundary value problem (3.2.17) and (3.2.20). We shall leave aside the various possibilities of a solution at this point, but assume that Φ has been calculated. Assuming that Φ is known, let us calculate the torque M_T. From Figure 34 we read off

$$\overleftarrow{M}_{\mathrm{T}} = \iint_F (\sigma_{32} x_1 - \sigma_{31} x_2)\,\mathrm{d}x_1\,\mathrm{d}x_2 \tag{3.2.21}$$

which, because of (3.2.8), changes to

$$\overleftarrow{M}_{\mathrm{T}} = -\iint_F (x_1 \Phi_{,1} + x_2 \Phi_{,2})\,\mathrm{d}x_1\,\mathrm{d}x_2 .$$

If the surface integral is written out for the integration limits, we obtain

$$\begin{aligned} \overleftarrow{M}_{\mathrm{T}} = & -\int_{x_2^U}^{x_2^O} \left[\int_{x_1^l(x_2)}^{x_1^r(x_2)} x_1 \Phi_{,1}\,\mathrm{d}x_1\right] \mathrm{d}x_2 \\ & -\int_{x_1^L}^{x_1^R} \left[\int_{x_2^U(x_1)}^{x_2^O(x_1)} x_2 \Phi_{,2}\,\mathrm{d}x_2\right] \mathrm{d}x_1 . \end{aligned} \tag{3.2.22}$$

Through integration by parts we find

$$\int_{x_1^l(x_2)}^{x_1^r(x_2)} x_1 \Phi_{,1}\,\mathrm{d}x_1 = [x_1 \Phi]_{x_1^l}^{x_1^r} - \int_{x_1^l(x_2)}^{x_1^r(x_2)} \Phi\,\mathrm{d}x_1 .$$

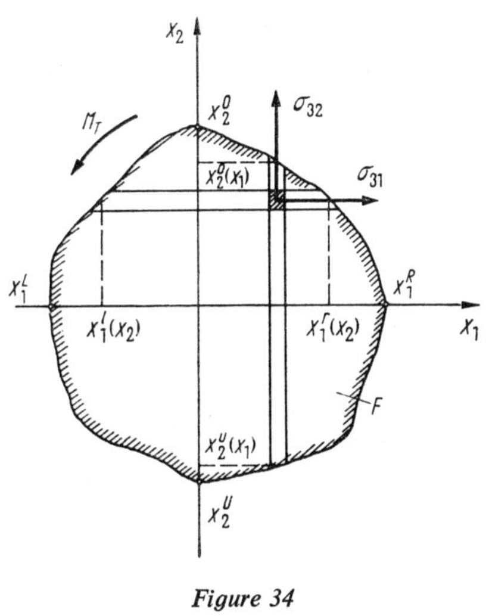

Figure 34

But x_1^r, x_1^l are boundary points. At the boundary, according to (3.2.20), Φ equals nought. Thus, the first term disappears on the right side, and there remains

$$\int_{x_1{}^l(x_2)}^{x_1{}^r(x_2)} x_1 \Phi_{,1} \, dx_1 = -\int_{x_1{}^l(x_2)}^{x_1{}^r(x_2)} \Phi \, dx_1 . \tag{3.2.23}$$

Similarly, we find

$$\int_{x_2{}^U(x_1)}^{x_2{}^O(x_1)} x_2 \Phi_{,2} \, dx_2 = -\int_{x_2{}^U(x_1)}^{x_2{}^O(x_1)} \Phi \, dx_2 . \tag{3.2.24}$$

If (3.2.23) and (3.2.24) are substituted in (3.2.22), then we obtain

$$\overleftarrow{M_T} = \int_{x_2{}^U}^{x_2{}^O} \int_{x_1{}^l}^{x_1{}^r} \Phi \, dx_1 \, dx_2 + \int_{x_1{}^L}^{x_1{}^R} \int_{x_2{}^U}^{x_2{}^O} \Phi \, dx_2 \, dx_1 ,$$

this being nothing more than

$$\overleftarrow{M_T} = 2 \iint_F \Phi \, dx_1 \, dx_2 , \tag{3.2.25}$$

i.e., the torque $\overleftarrow{M_T}$ is equal to twice the volume of the *stress hill*. This is the volume enclosed by surface $\Phi(x_1, x_2)$ and the rod's cross-section.

Another useful expression for M_T is as follows: from (3.2.21) and (3.2.16) is

$$\overleftarrow{M_T} = \iint_F [x_1 G(u_{3,2} + \vartheta x_1) + x_2 G(-u_{3,1} + \vartheta x_2)] \, dx_1 \, dx_2 .$$

This can be changed into

$$\overleftarrow{M_T} = G\vartheta \iint_F (x_1^2 + x_2^2) \, dx_1 \, dx_2 + G \iint_F (x_1 u_{3,2} - x_2 u_{3,1}) \, dx_1 \, dx_2 . \tag{3.2.26}$$

Let

$$J_p = \iint_F (x_1^2 + x_2^2) \, dx_1 \, dx_2$$

be the polar moment of inertia of the rod's cross-section. Then, we can also write

$$\overleftarrow{M}_{\mathrm{T}} = G\vartheta J_p + G\int_F\int (x_1 u_{3,2} - x_2 u_{3,1})\,\mathrm{d}x_1\,\mathrm{d}x_2\,.$$

This formula clearly indicates the difference between the Saint-Vénant theory, and the elementary torsion of a rod with a circular cross-section theory. For the latter, $M_T = G\vartheta J_p$ is usually applied. The above formula contains yet a correction term, which, when applied to any cross-section where warping $u_3 \neq 0$ occurs, will differ from nought. For the twisted circular cylindrical rod, $u_3 \equiv 0$ is true instead, since no warping occurs. Then the correction term vanishes, and the above formula will result in the correct calculation of the rod with circular cross-section. Summarizing the stages of the calculation, there is as follows – when the twist is given, then from (3.2.17) and (3.2.20), the stress function Φ is calculated. Warping u_3 results from (3.2.16) or from

$$u_{3,1} = \frac{1}{G}\Phi_{,2} + x_2\vartheta,\ u_{3,2} = -\left(\frac{1}{G}\Phi_{,1} + \vartheta x_1\right)$$

through integration. The moment M_T results from (3.2.25), and the shearing stress σ_{13}, σ_{23} can be found according to (3.2.8). We shall exemplify this process at a latei stage.

3.2.2 *Saint-Vénant's torsion function*

In the preceding chapter we have only discussed the calculation of the stress function. It is, however, possible to work equally well by using the torsion function $\varphi(x_1 x_2)$, which was introduced by (3.2.15). In this way we can acquaint ourselves with an important method in the mathematical theory of elasticity, i.e., in dealing with the calculation by means of complex variables.

If (3.2.14) and (3.2.15) are put in (3.2.6), and we solve with respect to the stresses, then

$$\sigma_{13} = G\vartheta(\varphi_{,1} - x_2),\ \sigma_{23} = G\vartheta(\varphi_{,2} + x_1) \tag{3.2.27}$$

is obtained. When using this for the equilibrium condition (3.2.3), we obtain

$$\varphi_{,11} + \varphi_{,22} \equiv \nabla^2\varphi = 0. \tag{3.2.28}$$

This represents Laplace's differential equation for φ.

As for the boundary conditions, they follow on from $\sigma_{ij}n_j = 0$, which, because of (3.2.27) and (3.2.18), leads to

$$(\varphi_{,1}-x_2)\frac{dx_2}{ds} - (\varphi_{,2}+x_1)\frac{dx_1}{ds} = 0, \text{ on } R.$$

This relation is transformed into

$$\frac{\partial\varphi}{\partial x_1}\frac{dx_2}{ds} - \frac{\partial\varphi}{\partial x_2}\frac{dx_1}{ds} = \frac{d\varphi}{dn} = x_2\frac{dx_2}{ds} + x_1\frac{dx_1}{ds} \text{ on } R. \tag{3.2.29}$$

Through (3.2.28) and (3.2.29), it becomes evident that for the torsion function φ we are dealing with Neumann's second boundary value problem of the potential theory. If this is solved, and φ is subsequently known, then for the given twist per unit length ϑ all other quantities of interest can be deduced from φ, as it has correspondingly been done in 3.2.1 using Φ:

The equation (3.2.15) yields the warping component u_3, and hence

$$u_{3,1} = \vartheta\varphi_{,1}, u_{3,2} = \vartheta\varphi_{,2}. \tag{3.2.30}$$

Inserting this in (3.2.26), we obtain

$$M_T = G\vartheta\int_F\int(x_1^2+x_2^2+x_1\varphi_{,2}-x_2\varphi_{,1})\,dx_1\,dx_2, \tag{3.2.31}$$

which enables us to find the torque M_T. It is quite usual instead of (3.2.31), to write

$$\overleftarrow{M_T} = GJ_t\vartheta$$

using

$$J_t = \int_F\int(x_1^2+x_2^2+x_1\varphi_{,2}-x_2\varphi_{,1})\,dx_1\,dx_2, \tag{3.2.32}$$

and to define the quantity GJ_t by *torsional rigidity*. J_t itself is defined as the *torsional constant* which is identical with the polar moment of inertia for a rod having a circular cross section. Finally, there are the stresses from (3.2.27). Thus, all wanted quantities have been found by means of φ.

Let us deal with the actual calculation of φ, showing the use of the theory of complex functions. Let us introduce the analytical, complex function

$$F(z) = \varphi + i\psi. \tag{3.2.33}$$

Its real part is the torsion function φ. The imaginary part ψ of F is linked with φ by Cauchy-Riemann's differential equations:

$$\varphi_{,1} = \psi_{,2} \text{ and } \varphi_{,2} = -\psi_{,1}. \tag{3.2.34}$$

Therefore, as the boundary condition for ψ from (3.2.29), we have

$$\frac{\partial\psi}{\partial x_2}\frac{dx_2}{ds} + \frac{\partial\psi}{\partial x_1}\frac{dx_1}{ds} = \frac{d\psi}{ds} = x_2\frac{dx_2}{ds} + x_1\frac{dx_1}{ds} \text{ on } R.$$

This yields through integration

$$\psi = \tfrac{1}{2}(x_1^2 + x_2^2) + c \text{ on } R, \tag{3.2.35}$$

containing the constant of integration c.

These relations can be used for solving problems by reverting them, and by using the following method of inversion of Saint-Vénant. For this purpose we shall prescribe $F(z)$ so that φ and ψ are known. If the result of (3.2.35) using this ψ is a *closed curve*, then the boundary of the cross-section follows from (3.2.35). For the cross-section determined in this way, the following quantities can be found; the warping component from (3.2.15) stresses from (3.2.27) and the torque from (3.2.31). This will yet be shown by an example.

Let us proceed with the theory. Because of $x_1^2 + x_2^2 = z\bar{z}$ [1] we have

$$\psi = \tfrac{1}{2}z\bar{z} + c \text{ on } R. \tag{3.2.36}$$

Let the cross-section of the rod, given with respect to the z plane, be conformally mapped on the unit circle in a ζ-plane by

[1] Barred quantities are conjugate complex.

$$z = f(\zeta). \tag{3.2.37}$$

Through (3.2.37), $F(z)$ is changed into $F_1(\zeta)$, and the boundary condition (3.2.36) changes into

$$\psi = \tfrac{1}{2}f(\zeta)\bar{f}(\bar{\zeta})+c \text{ on } |\zeta| = 1. \tag{3.2.38}$$

Let

$$\begin{aligned} \varphi &= \tfrac{1}{2}[F_1(\zeta)+\bar{F}_1(\bar{\zeta})], \\ \mathrm{i}\psi &= \tfrac{1}{2}[F_1(\zeta)-\bar{F}_1(\bar{\zeta})]. \end{aligned} \tag{3.2.39}$$

Thus, (3.2.38) changes to

$$F_1(\zeta)-\bar{F}_1(\bar{\zeta}) = \mathrm{i}f(\zeta)\bar{f}(\bar{\zeta})+c \text{ on } |\zeta| = 1. \tag{3.2.40}$$

Let us assume the mapping function being given as $f(\zeta)$. The points on the unit circle are $\zeta = e^{\mathrm{i}\alpha}$, hence, for the mapping function on the unit circle, using a Fourier expansion, the following is true,

$$f(e^{\mathrm{i}\alpha}) = \sum_{-\infty}^{\infty} A_n e^{\mathrm{i}n\alpha}, \quad A_n = \frac{1}{2\pi}\int_0^{2\pi} f(e^{\mathrm{i}\alpha})e^{-\mathrm{i}n\alpha}\mathrm{d}\alpha. \tag{3.2.41}$$

Let us assume further that $F_1(\zeta)$ has the series expansion

$$F_1(\zeta) = \sum_0^{\infty} B_n\zeta^n, \tag{3.2.42}$$

then the following is true for the circle of unit radius,

$$F_1(e^{\mathrm{i}\alpha}) = \sum_0^{\infty} B_n e^{\mathrm{i}n\alpha}, \ \bar{F}_1\overline{(e^{\mathrm{i}\alpha})} = \sum_0^{\infty} \bar{B}_n e^{-\mathrm{i}n\alpha}, \tag{3.2.43}$$

so that (3.2.40) changes into

$$\sum_0^{\infty} B_n e^{\mathrm{i}n\alpha} - \sum_0^{\infty} \bar{B}_n e^{-\mathrm{i}n\alpha} = \mathrm{i}(\sum_0^{\infty} A_n e^{\mathrm{i}n\alpha})(\sum_0^{\infty} \bar{A}_n e^{-\mathrm{i}n\alpha})+c. \tag{3.2.44}$$

Because the A_n, as well as the $\bar{A}_n$, are known, then the coefficients B_n are found through (3.2.44), and the wanted $F_1(\zeta)$ according to (3.2.42) is determined. Thus, an important part of the calculation has been found. When comparing the coefficients of $e^{in\alpha}$ for all n, with the exception of $n = 0$, we find

$$B_n = iC_n,\ \bar{B}_n = i\bar{C}_n,\ n \neq 0,$$
$$C_n = \sum_{m=-\infty}^{\infty} A_{n+m}\bar{A}_m,\ \bar{C}_n = C_{-n},\ n = 0, 1, 2, 3, \ldots. \tag{3.2.45}$$

Hence, we finally obtain

$$F_1(\zeta) = i\sum_{n=1}^{\infty} C_n\zeta^n + \text{const.} \tag{3.2.46}$$

Using

$$x_1 = \tfrac{1}{2}[f(\zeta)+\bar{f}(\bar{\zeta})],\ x_2 = \frac{1}{2i}[f(\zeta)-\bar{f}(\bar{\zeta})],$$

and the first line from (3.2.39), we are able to calculate from (3.2.31)

$$\overset{\leftarrow}{M_T} = G\vartheta\left\{\frac{1}{4i}\oint [\bar{f}(\bar{\zeta})]^2 f(\zeta)\frac{df(\zeta)}{d\zeta}\,d\zeta\right.$$
$$\left.-\frac{1}{4}\oint [F_1(\zeta)+\bar{F}_1(\bar{\zeta})][f(\zeta)d\bar{f}(\bar{\zeta})+\bar{f}(\bar{\zeta})df(\zeta)]\right\}. \tag{3.2.47}$$

The integrals in (3.2.47) are to be calculated along the boundary of the unit circle of the ζ-plane. Since the functions $f(\zeta)$ and $F_1(\zeta)$ are assumed as given, i.e., as calculated, then (3.2.47) yields the torque M_T.

The warping component is calculated according to (3.2.15) and (3.2.39), and reads

$$u_3 = \tfrac{1}{2}\vartheta[F_1(\zeta)+\bar{F}_1(\bar{\zeta})]. \tag{3.2.48}$$

Let us finally find the relations by means of which the stresses can be determined. According to (3.2.27),

$$\sigma_{13}-i\sigma_{23} = G\vartheta(\varphi_{,1}-i\varphi_{,2}-x_2-ix_1).$$

Through (3.2.34) it changes to

$$\sigma_{13}-i\sigma_{23} = G\vartheta[\varphi_{,1}+i\psi_{,1}-i(x_1-ix_2)].$$

But

$$\varphi_{,1}+i\psi_{,1} = F'(z) = F_1'(\zeta)\cdot[f'(\zeta)]^{-1} \text{ and } x_1-ix_2 = \bar{z} = \bar{f}(\bar{\zeta}).$$

Therefore,

$$\sigma_{13}-i\sigma_{23} = G\vartheta\left[\frac{F_1'(\zeta)}{f'(\zeta)}-i\bar{f}(\bar{\zeta})\right] \tag{3.2.49}$$

is also true.

(3.2.49) yields the wanted stresses σ_{13} and σ_{23} by comparison of the real and imaginary parts on both sides of the equation.

Thus, the torsion problem for prismatic rods having a simply connected cross-section has been completely solved, by using complex variables.

3.2.3 *Further prospects and examples*

Certain analogies are usable for the practical solution of the torsion problem. Let us start by examining *Prandtl's membrane analogy*.

As has been shown in 3.2.1, the stress function Φ is a key function for finding all required quantities. It is derived from the boundary problem

$$\nabla^2\Phi = -2G\vartheta, \quad \Phi(R) = 0. \tag{3.2.50}$$

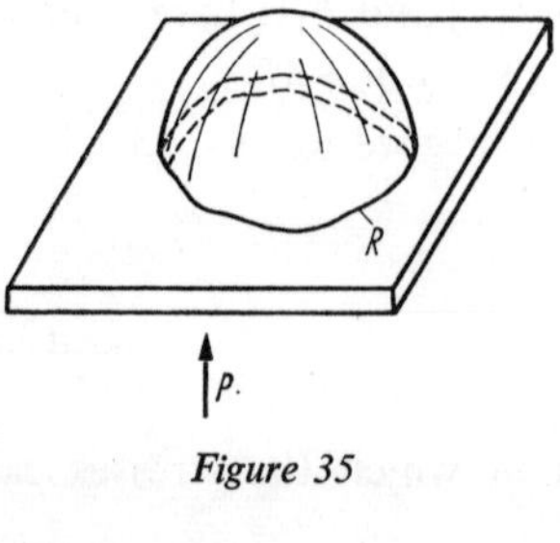

Figure 35

Let us consider a soap film, stretched as a membrane over a hole shaped like the cross-section with contour R. Let it bulge as a result of a normal pressure p, showing the deflection w (Figure 35). For w there is the boundary value problem

$$\nabla^2 w = -\frac{p}{S}, \quad w(R) = 0.$$

S is the tension in the membrane per unit length. Evidently, the problems for Φ and for w are mathematically identical, if the contour R is the same in both cases. It can be assumed that the solution for Φ may be derived from a solution of w when appropriate conversion factors have been considered. The solution for w is found experimentally, i.e., the deflection of the soap film is measured [19]. By using the results of these measurements, the desired data for Φ are then calculated. *Hydrodynamic analogies* are either used for solving torsion problems experimentally or for explaining them qualitatively. They stem, for example, from Lord Kelvin, Boussinesq and Greenhill. Some details are given by Timoshenko-Goodier [20] and by Föppl [21]. It might also be of value to mention *electrical analogies* which are useful for finding solutions by experiments (these are discussed in [19]). The torsion problem of rods can be extended in various ways. For example, it is possible to explore the influences of *notches* and *grooves* or of *restrained warping*. Furthermore, torsion of rods having *varying* or *thin-walled, open cross-sections*, can also be investigated. Finally one may be dealing with the torsion of *hollow rods* having *multiple connected cross-section*, as well as the related special case of torsion of *thin-walled tubes*. Solutions of special interest to the engineer are discussed in [20].

To illustrate 3.2.1 and 3.2.2, the following examples are given:

1. *Example.* Let the rod under torsion be of an elliptical cross-section with semi-axes a and b, and let the coordinates x_{iR} of its contour points satisfy the equation

$$\frac{x_{1R}^2}{a^2} + \frac{x_{2R}^2}{b^2} = 1. \tag{3.2.51}$$

Let us solve the problem by using the stress function Φ which has to satisfy the boundary value problem (3.2.50). Let us use the semi-inverse method, starting with

$$\Phi = c\left(\frac{x_1^2}{a^2} + \frac{x_2^2}{b^2} - 1\right). \tag{3.2.52}$$

For adaption purposes to (3.2.50) this expression contains value c.

We obtain $\Phi(R)$ by substitution of x_{iR} for x_i in (3.2.52), and we notice that, because of (3.2.51), $\Phi(R) = 0$ has already been satisfied. Let us consider the differential equation $\nabla^2\Phi = -2G\vartheta$. In carrying out all differentiations in (3.2.52),

$$\nabla^2\Phi = 2c\left(\frac{1}{a^2} + \frac{1}{b^2}\right) \tag{3.2.53}$$

is obtained.

In order to fulfill the differential equation by (3.2.52), we write

$$2c\left(\frac{1}{a^2} + \frac{1}{b^2}\right) = -2G\vartheta \text{ i.e., } c = -\frac{a^2b^2G\vartheta}{a^2+b^2}. \tag{3.2.54}$$

Thus, the constant c has been determined, and the remaining part of the problem is calculated by using

$$\Phi = -G\vartheta\frac{a^2b^2}{a^2+b^2}\left(\frac{x_1^2}{a^2} + \frac{x_2^2}{b^2} - 1\right). \tag{3.2.55}$$

For u_3, using (3.2.55), we find

$$u_{3,1} = \frac{1}{G}\Phi_{,2} + \vartheta x_2 = -\frac{a^2-b^2}{a^2+b^2}\vartheta x_2,$$

$$u_{3,2} = -\left(\frac{1}{G}\Phi_{,1} + \vartheta x_1\right) = -\frac{a^2-b^2}{a^2+b^2}\vartheta x_1.$$

Through integration we get

$$u_3 = -\frac{a^2-b^2}{a^2+b^2}\vartheta x_2 x_1 + f(x_2)$$

as well as

$$u_3 = -\frac{a^2-b^2}{a^2+b^2}\vartheta x_1 x_2 + g(x_1).$$

Since the solutions of the displacements are unique, then both the solutions for u_3 must be equal. This can only be the case when $f(x_2) \equiv 0$, $g(x_1) \equiv 0$ is assumed for the arbitrary function of integration. Thus,

$$u_3 = -\frac{a^2-b^2}{a^2+b^2}\vartheta x_1 x_2,$$

i.e., the originally plane cross-section of the rod changes through torsion into a hyperbolic paraboloid.

According to (3.2.25), the calculation of M_T is carried out using (3.2.55) and yields

$$M_\mathrm{T} = -2G\vartheta\frac{a^2b^2}{a^2+b^2}\int_F\int\left(\frac{x_1^2}{a^2}+\frac{x_2^2}{b^2}-1\right)\mathrm{d}x_1\,\mathrm{d}x_2 = G\vartheta\pi\frac{a^3b^3}{a^2+b^2}.$$

Distribution of stresses results in (3.2.55) from (3.2.8). It is

$$\sigma_{13} = \Phi_{,2} = -2G\vartheta\frac{a^2}{a^2+b^2}x_2 = -\frac{2M_\mathrm{T}}{\pi ab^3}x_2,$$

$$\sigma_{23} = -\Phi_{,1} = 2G\vartheta\frac{b^2}{a^2+b^2}x_1 = \frac{2M_\mathrm{T}}{\pi a^3 b}x_1.$$

Maximal values of torsional shearing stresses occur at the extremities of the small elliptical axis, i.e., at the boundary points having coordinates $x_1 = 0$, $x_2 = \pm b$. When these coordinates are substituted for the relations dealing with stresses, then $2M/\pi ab^2$ is found to be the value of the largest shearing stress.

2. *Example.* The following deals with the torsion of a rod with rectangular cross-section (Figure 36), and using torsion function φ. Then the boundary value problem

$$\nabla^2\varphi = 0,\ (\varphi_{,1}-x_2)\frac{\mathrm{d}x_2}{\mathrm{d}s}-(\varphi_{,2}+x_1)\frac{\mathrm{d}x_1}{\mathrm{d}s} = 0 \text{ on } R \tag{3.2.56}$$

applies.

But because of the rectangular cross-section, $\mathrm{d}x_2/\mathrm{d}s = \pm 1$, $\mathrm{d}x_1/\mathrm{d}s = 0$ is for $x_1 = \pm A$, and similarly $\mathrm{d}x_2/\mathrm{d}s = 0$, $\mathrm{d}x_1/\mathrm{d}s = \pm 1$ is for $x_2 = \pm B$, so that the boundary conditions of (3.2.56) change into

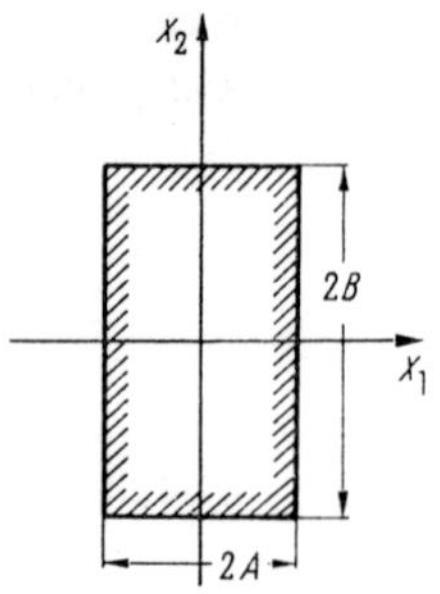

Figure 36

$$\varphi_{,1} = x_2 \text{ for } x_1 = \pm A, \qquad \varphi_{,2} = -x_1 \text{ for } x_2 = \pm B. \tag{3.2.57}$$

To simplify further, the auxiliary function

$$\varphi^* = x_1 x_2 - \varphi \tag{3.2.58}$$

is introduced.

Because of the characteristic of φ, this function is also

$$\nabla^2 \varphi^* = 0, \tag{3.2.59}$$

and because of (3.2.57) and (3.2.50) its boundary conditions are

$$\varphi^*_{,1} = 0 \text{ for } x_1 = \pm A, \qquad \varphi^*_{,2} = 2x_1 \text{ for } x_2 = \pm B. \tag{3.2.60}$$

To obtain the solution of φ from (3.2.59), (3.2.60), the series expansion

$$\varphi^* = \sum_{n=0}^{\infty} \rho_n f_n(x_1) g_n(x_2) \tag{3.2.61}$$

is set up.

Substituting in (3.2.59), we obtain for the coordinate functions f_i, g_i the two ordinary differential equations

$$f_n'' + \kappa_n^2 f_n = 0, \; g_n'' - \kappa_n^2 g_n = 0$$

with the general solutions

$$f_n(x_1) = c_1 \sin \kappa_n x_1 + c_2 \cos \kappa_n x_1 ,$$
$$g_n(x_2) = c_3 \sinh \kappa_n x_2 + c_4 \cosh \kappa_n x_2 . \tag{3.2.62}$$

To determine the constants of integration c_1 to c_4, we use the boundary conditions. On one hand, because of (3.2.61), there is

$$\varphi^*_{,2} = \sum_{n=0}^{\infty} \rho_n f_n(x_1) g'_n(x_2) = 2x_1 \text{ for } x_2 = \pm B.$$

This is only possible if $g'_n(x_2)$ is an even function, and if $f_n(x_1)$ is an uneven function, which means that $c_2 = c_4 = 0$. But on the other hand,

$$\sum_{n=0}^{\infty} \rho_b f'_n(x_1) g_n(x_2) = 0 \text{ for } x_1 = \pm A.$$

This is satisfied if $f'_n(\pm A) = 0$. Thus, from (3.2.62), using $c_2 = 0$, follows the condition

$$f'_n(\pm A) = c_1 \kappa_n \cos \kappa_n(\pm A) = 0.$$

It yields the characteristic values κ_n,

$$\kappa_n = \frac{(2n+1)\pi}{2A}. \tag{3.2.63}$$

Since c_1, c_2 and κ_n are known, we write for (3.2.61), by putting down $r_n = c_1 c_3 \rho_n$,

$$\varphi^* = \sum_{n=0}^{\infty} r_n \sin \frac{(2n+1)}{2A} \pi x_1 \sinh \frac{(2n+1)}{2A} \pi x_2 , \tag{3.2.64}$$

and using the second condition of (3.2.60), we try calculating the new coefficients r_n. Thus as a condition equation for r_n we arrive at

$$\sum_{n=0}^{\infty} r_n \frac{2n+1}{2A} \pi \sin \frac{(2n+1)}{2A} \pi x_1 \cosh \frac{(2n+1)}{2A} \pi B = 2x_1. \tag{3.2.65}$$

Using the Fourier's expansion of $2x_1$ and by comparison of coefficients,

$$r_n = (-1)^n \frac{32A^2}{\pi^3(2n+1)^3} \frac{1}{\cosh \dfrac{2n+1}{2A} \pi B} \tag{3.2.66}$$

is obtained from (3.2.65).

It is now possible to outline torsion-function φ, since from (3.2.58), (3.2.64) and (3.2.66) follows

$$\varphi = x_1 x_2 - \frac{32A^2}{\pi^3} \sum_{n=0}^{\infty} \frac{(-1)^n}{(2n+1)^3} \frac{1}{\cosh \kappa_n B} \sin \kappa_n x_1 \sinh \kappa_n x_2 \tag{3.2.67}$$

when (3.2.63) is substituted for κ_n. Through φ, the remainder is calculated, i.e., the torque M_T from (3.2.31) and (3.2.67), the stresses from (3.2.27) and (3.2.67), and warping from (3.2.15) and (3.2.67). After lengthy calculation we find, for example,

$$M_T = 16G\vartheta A^3 B \left[\frac{1}{3} - \frac{64A}{\pi^5 B} \sum_{n=0}^{\infty} \frac{\tanh \kappa_n B}{(2n+1)^5}\right],$$

$$\sigma_{13} = -16G\vartheta \frac{A}{\pi^2} \sum_{n=0}^{\infty} \frac{(-1)^n}{(2n+1)^2} \frac{\sinh \kappa_n x_2}{\cosh \kappa_n B} \cos \kappa_n x_1,$$

as well as

$$\sigma_{23} = G\vartheta \left[2x_1 - \frac{16A}{\pi^2} \sum_{n=0}^{\infty} \frac{(-1)^n}{(2n+1)^2} \frac{\cosh \kappa_n x_2}{\cosh \kappa_n B} \sin \kappa_n x_1\right].$$

The rectangular cross-section has been chosen in such a way that $B > A$. For this reason, the maximum shearing stresses occur at the boundary points $x_1 = \pm A$, $x_2 = 0$. If, in the above statement, the value of the coordinators $x_1 = A$, $x_2 = 0$ is substituted in the formulae for the stresses, it yields

$$\sigma_{13} = 0,$$

$$\sigma_{23,\max} = 2G\vartheta A \left[1 - \frac{8}{\pi^2} \sum_{n=0}^{\infty} \frac{1}{(2n+1)^2 \cosh \kappa_n B}\right].$$

3. *Example.* In this example we shall use the inverse method, i.e., the method

of reversing the problem. Let us assume the function $F(z)$ having the real and imaginary parts

$$\varphi = \frac{1}{6A}(3x_1^2 x_2 - x_2^3),\ \psi = -\frac{1}{6A}(x_1^3 - 3x_1 x_2^2) + \frac{2A^2}{3} + c. \tag{3.2.68}$$

According to (3.2.35),

$$-\frac{1}{6A}(x_1^3 - 3x_1 x_2^2) + \frac{2A^2}{3} + c = \tfrac{1}{2}(x_1^2 + x_2^2) + c$$

must be true for the boundary of the yet unknown cross-section of the rod. The statement may also be expressed by

$$(x_1 - A)(x_1 - x_2\sqrt{3} + 2A)(x_1 + x_2\sqrt{3} + 2A) = 0.$$

Thus, the cross-sectional boundary is determined by the three lines

$$x_1 = A,\ x_1 = \sqrt{3}x_2 - 2A,\ x_1 = -\sqrt{3}x_2 - 2A$$

yielding as the cross-section of the rod the equilateral triangle in Figure 37.

From 3.2.2 it is known that the function φ given by (3.2.68) represents the torsion function of this particular cross-section. Therefore, through (3.2.31)

$$M_T = \frac{3\sqrt{3}}{5} G\vartheta A^4$$

is found, through (3.2.27)

$$\sigma_{13} = G\vartheta\left(\frac{x_1 x_2}{A} - x_2\right),\ \sigma_{23} = G\vartheta\left(\frac{x_1^2 - x_2^2}{2A} + x_1\right),$$

and according to (3.2.15), the warping of the cross-section is equal to

$$u_3 = \frac{\vartheta}{6A}(3x_1^2 x_2 - x_2^3).$$

The maximum shearing stresses occur in the centre of the sides of the triangle. Therefore, if for example $x_1 = A, x_2 = 0$ is substituted in the equations for stresses, it yields

$$\sigma_{13} = 0, \sigma_{23,\max} = \tfrac{3}{2}G\vartheta A.$$

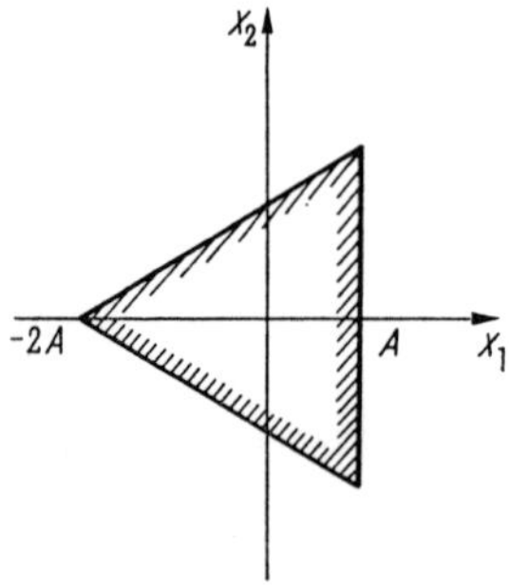

Figure 37

3.3 Planar problems

3.3.1 *The classic case of plane strain and stress*

Plane strain, when related to a particular body, is characterized by the fundamental assumption that displacement u_3 disappears completely, and that the other displacements are functions of x_1, x_2 only, thus being independent from x_3. Hence,

$$u_3 = 0, \; u_1 = u_1(x_1, x_2), \; u_2 = u_2(x_1, x_2). \tag{3.3.1}$$

Based on (3.1.2),

$$\varepsilon_{33} = \varepsilon_{13} = \varepsilon_{23} = 0 \tag{3.3.2}$$

holds, and the remaining distortions are similar to u_1 and u_2, from which they have been derived, functions from x_1, x_2 only. Because of

$$\varepsilon_{11} = \varepsilon_{11}(x_1, x_2), \; \varepsilon_{22} = \varepsilon_{22}(x_1, x_2), \; \varepsilon_{12} = \varepsilon_{12}(x_1, x_2) \tag{3.3.3}$$

and (3.3.2), the equations (3.1.3a) yield for the stresses

$$\begin{aligned}&\sigma_{13} = \sigma_{23} = 0,\\ &\sigma_{11} = \sigma_{11}(x_1, x_2),\ \sigma_{22} = \sigma_{22}(x_1, x_2),\ \sigma_{33} = \sigma_{33}(x_1, x_2),\\ &\sigma_{12} = \sigma_{12}(x_1, x_2),\end{aligned} \tag{3.3.4}$$

i.e., the non-vanishing components of the stress tensor are again functions from x_1, x_2 only. Also let us assume that the volume force $\boldsymbol{K}$ has no component in the x_3 direction, and that the remaining components, like everything preceding it, are independent from x_3:

$$K_3 = 0,\ K_1 = K_1(x_1, x_2),\ K_2 = K_2(x_1, x_2). \tag{3.3.5}$$

Similarly, for the surface forces p:

$$p_3 = 0,\ p_1 = p_1(x_1, x_2),\ p_2 = p_2(x_1, x_2), \tag{3.3.6}$$

so that the general boundary conditions (3.1.4) change to

$$\begin{aligned}&\sigma_{11}n_1 + \sigma_{12}n_2 = p_1,\\ &\sigma_{21}n_1 + \sigma_{22}n_2 = p_2.\end{aligned} \tag{3.3.7}$$

The third condition $\sigma_{33}n_3 = 0$ derived from (3.1.4) is identically satisfied, since it is assumed that all considerations concerning the boundary conditions are solely referred to those parts of the body's surface, whose surface normals will always be vertical to the x_3-direction. But for these it is $n_3 \equiv 0$. The practical significance, and hence the justification of this assumption, will be shown in the next paragraph which will deal with the implementation of planar distortion.

(3.3.1) to (3.3.7) have established that on the whole we are dealing with a problem solely dependent on x_1, x_2, i.e., a two-dimensional or planar problem. To solve it, we shall use the general equations of the theory of elasticity considering the restricting conditions, as given by (3.3.1) to (3.3.7). Since, for reasons of (3.3.7), we are dealing with a problem having specified surface forces, we shall naturally prefer working with the equations of the second boundary value problem as indicated in 3.1.1.

Let us look therefore at (3.1.23) which reads as

$$\left(\frac{\partial^2}{\partial x_1^2} + \frac{\partial^2}{\partial x_2^2} + \frac{\partial^2}{\partial x_3^2}\right)(\sigma_{11} + \sigma_{22} + \sigma_{23}) = -\frac{1+\nu}{1-\nu}\left(\frac{\partial K_1}{\partial x_1} + \frac{\partial K_2}{\partial x_2} + \frac{\partial K_3}{\partial x_3}\right).$$

Because of (3.3.4) and (3.3.5) it changes to

$$\left(\frac{\partial^2}{\partial x_1^2}+\frac{\partial^2}{\partial x_2^2}\right)(\sigma_{11}+\sigma_{22}+\sigma_{33}) = -\frac{1+\nu}{1-\nu}\left(\frac{\partial K_1}{\partial x_1}+\frac{\partial K_2}{\partial x_2}\right). \tag{3.3.8}$$

But from (3.1.3b) and (3.3.2)

$$\varepsilon_{33} = \frac{1+\nu}{E}\sigma_{33} - \frac{\nu}{E}(\sigma_{11}+\sigma_{22}+\sigma_{33}) = 0,$$

so that specially

$$\sigma_{33} = \nu(\sigma_{11}+\sigma_{22}) \tag{3.3.9}$$

is derived. Substituting in (3.3.8) yields

$$\left(\frac{\partial^2}{\partial x_1^2}+\frac{\partial^2}{\partial x_2^2}\right)(\sigma_{11}+\sigma_{22}) = -\frac{1}{1-\nu}\left(\frac{\partial K_1}{\partial x_1}+\frac{\partial K_2}{\partial x_2}\right), \tag{3.3.10}$$

which represents the fundamental equation of plane strain. Let us assume further that for the volume forces the following is true

$$\begin{aligned} &\mathbf{K} = -\operatorname{grad} U, \\ &\text{i.e., } K_1 = -U_{,1},\ K_2 = -U_{,2}. \end{aligned} \tag{3.3.11}$$

This assumption is of great practical significance since it holds for centrifugal forces or the weight. In addition, let us introduce the *stress function* $\Phi = \Phi(x_1, x_2)$, related to the stresses as follows:

$$\sigma_{11} = \Phi_{,22}+U,\ \sigma_{22} = \Phi_{,11}+U,\ \sigma_{12} = -\Phi_{,12}. \tag{3.3.12}$$

These equations satisfy identically the equilibrium conditions (3.1.1) when (3.3.4), (3.3.5) and (3.3.11) are considered. Through (3.3.11) and (3.3.12), (3.3.10) is changed into

$$\left(\frac{\partial^2}{\partial x_1^2}+\frac{\partial^2}{\partial x_2^2}\right)\left(\frac{\partial^2}{\partial x_1^2}+\frac{\partial^2}{\partial x_2^2}\right)\Phi = \frac{2\nu-1}{1-\nu}\left(\frac{\partial^2}{\partial x_1^2}+\frac{\partial^2}{\partial x_2^2}\right)U. \tag{3.3.13}$$

Since the calculation does not go beyond x_1, x_2, thus not risking a mistake, we simply write for

$$\frac{\partial^2}{\partial x_1^2} + \frac{\partial^2}{\partial x_2^2}$$

symbolically ∇^2, then (3.3.13) changes to

$$\nabla^2 \nabla^2 \Phi = \frac{2\nu - 1}{1-\nu} \nabla^2 U. \tag{2.3.14}$$

This being the special form assumed by the basic equation (3.3.10) under the specific assumption (3.3.11).

Another important case leading to further simplification is that of $\boldsymbol{K} \equiv 0$, i.e., when either no volume forces are present or when they are disregarded. Hence $U \equiv 0$. Thus, (3.3.12) changes to

$$\sigma_{11} = \Phi_{,22}, \sigma_{22} = \Phi_{,11}, \sigma_{12} = -\Phi_{,12}, \tag{3.3.15}$$

and (3.3.14) changes into

$$\nabla^2 \nabla^2 \Phi = 0. \tag{3.3.16}$$

Thus, we have arrived back at the biharmonic equation, whose significance has already been pointed out.

Working with problems that deal with plane strain is such that solutions to (3.3.14) and (3.3.16) have to be found, which through (3.3.12) and (3.3.15) lead to stresses satisfying the boundary conditions (3.3.7). In many instances, the state of stress may be of interest only, in which case knowledge of the above solutions also represents the solution to the problem. However, in the event of finding, in addition, either strains or displacements, further calculation becomes necessary, which will be discussed later.

At this point let us discuss the *state of plane stress* which is characterized by the following assumption

$$\begin{aligned} &\sigma_{33} = \sigma_{13} = \sigma_{23} = 0, \\ &\sigma_{11} = \sigma_{11}(x_1, x_2), \ \sigma_{22} = \sigma_{22}(x_1, x_2), \ \sigma_{12} = \sigma_{12}(x_1, x_2). \end{aligned} \tag{3.3.17}$$

Let us leave it open whether there is a possibility of realizing this assumption or not. Additionally, let the assumption which had been made for plane strain concerning volume and surface forces be valid also in this case. Thus (3.3.5) and (3.3.7) have to be used for the following calculations.

Since we are dealing with a boundary value problem of the second kind, let us go back to (3.1.19) expanding that equation for $i = 1, j = 2, k = 1, l = 2$. Let us consider that following the assumptions, all values bearing the Index 3 will vanish. This yields

$$\sigma_{12,12}+\sigma_{12,12}-\sigma_{11,22}-\sigma_{22,11} = \frac{\nu}{1+\nu}(-\sigma_{mm,22}-\sigma_{mm,11}),$$

$$\sigma_{mm} = \sigma_{11}+\sigma_{22}.$$

Rearranging this relation, it becomes

$$2(1+\nu)\sigma_{12,12} = \sigma_{11,22}+\sigma_{22,11}-\nu(\sigma_{22,22}+\sigma_{11,11}). \tag{3.3.18}$$

Differentiating the equilibrium conditions (3.1.1), it yields

$$\sigma_{ik,ki} = -K_{i,i}.$$

Expanding this in full with all assumptions in mind, we obtain by rearranging

$$2\sigma_{12,12} = -(\sigma_{11,11}+\sigma_{22,22}+K_{1,1}+K_{2,2}). \tag{3.3.19}$$

Substituting (3.3.19) in (3.3.18), we obtain after simple calculation

$$\sigma_{11,11}+\sigma_{11,12}+\sigma_{22,11}+\sigma_{22,22} = -(1+\nu)(K_{1,1}+K_{2,2}),$$

which is also expressed as

$$\left(\frac{\partial^2}{\partial x_1^2}+\frac{\partial^2}{\partial x_2^2}\right)(\sigma_{11}+\sigma_{22}) = -(1+\nu)\left(\frac{\partial K_1}{\partial x_1}+\frac{\partial K_2}{\partial x_2}\right). \tag{3.3.20}$$

Thus, the fundamental equations of a plane stress is obtained, which represents the counterpart of (3.3.10), the fundamental equation of plane strain.

Following assumption (3.3.11) and equations (3.3.12), we obtain from (3.3.20)

$$\nabla^2\nabla^2\Phi = -(1-\nu)\nabla^2 U, \tag{3.3.21}$$

which corresponds to equation (3.3.14) of plane strain. Disregarding the different values of the constants on the right side of (3.3.14) and (3.3.21), the problems of plane strain and plane stress are mathematically equivalent, since both have the differential equation $\nabla^2\nabla^2\Phi = \text{const}\cdot\nabla^2 U$, and boundary conditions (3.3.7) in common.

Congruency increases when $\boldsymbol{K} \equiv 0$, $U \equiv 0$ are assumed to hold, i.e., when all volume forces in the calculation are omitted. Then (3.3.21) changes into the biharmonic equation (3.3.16), and (3.3.15) becomes valid. At this point, the problems of plane strain and plane stress have become identical. All solutions of the boundary value problem (3.3.7), (3.3.15) and (3.3.16) yield a state of stress which applies to one as well as to the other kind of planar problem. The difference between the problems of plane strain and plane stress becomes only apparently when strains and displacements are actually calculated.

This will be shown in the following. To find for example the components of plane strain, we shall start by using the already calculated stresses $\sigma_{11}, \sigma_{22}, \sigma_{12}$ and find, based on (3.3.9) and (3.1.3b), the relations

$$\varepsilon_{11} = \frac{1-\nu^2}{E}\left(\sigma_{11}-\frac{\nu}{1-\nu}\sigma_{22}\right), \quad \varepsilon_{22} = \frac{1-\nu^2}{E}\left(\sigma_{22}-\frac{\nu}{1-\nu}\sigma_{11}\right),$$
$$\varepsilon_{12} = \frac{\sigma_{12}}{2G}, \quad \varepsilon_{33} = \varepsilon_{13} = \varepsilon_{23} = 0 \tag{3.3.22}$$

which yield the wanted strains.

In the case of plane stress, the components of the strain tensor are obtained from the known stresses $\sigma_{11}, \sigma_{22}, \sigma_{12}$ through (3.1.3b), and because of $\sigma_{33} = 0$, they read

$$\begin{aligned} \varepsilon_{11} &= \frac{1}{E}(\sigma_{11}-\nu\sigma_{22}), \quad \varepsilon_{22} = \frac{1}{E}(\sigma_{22}-\nu\sigma_{11}), \\ \varepsilon_{12} &= \frac{\sigma_{12}}{2G}, \quad \varepsilon_{33} = -\frac{\nu}{E}(\sigma_{11}+\sigma_{22}), \quad \varepsilon_{13} = \varepsilon_{23} = 0. \end{aligned} \tag{3.3.23}$$

Comparison of (3.3.22) with (3.3.23) clearly indicates the difference between the two problems. But even then we can still sustain the congruency in the formal mathematical sense. All that must be done is to put into

(3.3.22) the ideal elasticity modulus E^* for $E/(1-\nu^2)$, the ideal Poisson's ratio ν^* for $\nu/(1-\nu)$, and put into (3.3.23) approximately $\varepsilon_{33} \approx 0$. Then (3.3.22) and (3.3.23) are equal for formal calculation purposes. Putting the approximation $\varepsilon_{33} \approx 0$ into (3.3.23) will be justified later.

Let us assume that the displacements must be found. This can be done in two ways, either by integrating the known strains from (3.3.22) and (3.3.23), or by derivation from the known stress function. Let us discuss the latter way, assuming that no mass forces are likely to occur.

Substituting (3.3.15) in (3.1.3a), we obtain

$$\Phi_{,22} = \lambda\varepsilon_{kk}+2\mu\varepsilon_{11}, \; \Phi_{,11} = \lambda\varepsilon_{kk}+2\mu\varepsilon_{22}. \tag{3.3.24}$$

Let us assume that $\varepsilon_{33} = 0$ is true for both planar problems. While in one case this is precisely so, it is only approximately true in the other case. Hence, following our assumption, $\varepsilon_{kk} = \varepsilon_{11}+\varepsilon_{22}$ is valid, and from (3.3.24), by adding both equations and by notating $\Phi_{,11}+\Phi_{,22} = \nabla^2\Phi$, we obtain

$$\varepsilon_{kk} = \frac{1}{2(\mu+\lambda)}\nabla^2\Phi. \tag{3.3.25}$$

Substituting (3.3.25) in (3.3.24), let us solve for ε_{11}, ε_{22} respectively, thus obtaining

$$2\mu\varepsilon_{11} = \Phi_{,22}-\frac{\lambda}{2(\mu+\lambda)}\nabla^2\Phi, \; 2\mu\varepsilon_{22} = \Phi_{,11}-\frac{\lambda}{2(\mu+\lambda)}\nabla^2\Phi. \tag{3.3.26}$$

But $\Phi_{,22} = \nabla^2\Phi-\Phi_{,11}$ and $\Phi_{,11} = \nabla^2\Phi-\Phi_{,22}$. Using this in (3.3.26), inserting $\varepsilon_{11} = u_{1,1}$, $\varepsilon_{22} = u_{2,2}$, and rearranging the whole slightly, we arrive at

$$\begin{aligned} 2\mu u_{1,1} &= -\Phi_{,11}+\frac{2\mu+\lambda}{2(\mu+\lambda)}\nabla^2\Phi, \\ 2\mu u_{2,2} &= -\Phi_{,22}+\frac{2\mu+\lambda}{2(\mu+\lambda)}\nabla^2\Phi. \end{aligned} \tag{3.3.27}$$

Since Φ represents a solution of $\nabla^2\nabla^2\Phi = 0$, then $\nabla^2\Phi$ is itself a solution of Laplace's differential equation, and hence a harmonic function p. Let q

be the conjugate harmonic function of $\nabla^2\Phi \equiv p$ which, except for one arbitrary constant, can be set up if $\nabla^2\Phi$ is known. For the known Φ (and hence $\nabla^2\Phi$), the complex function $f(z) = p+iq$ is constructed which leads to $F(z) = P+iQ = \frac{1}{4}\int f(z)dz$. Differentiating $F(z)$ and applying the Cauchy-Riemann differential equations on P and Q, we find that

$$\nabla^2\Phi = 4P_{,1} = 4Q_{,2} = p. \tag{3.3.28}$$

Using (3.3.28), (3.3.27) changes into

$$2\mu u_{1,1} = -\Phi_{,11} + \frac{2(2\mu+\lambda)}{\mu+\lambda} P_{,1}, \; 2\mu u_{2,2} = -\Phi_{,22} + \frac{2(2\mu+\lambda)}{\mu+\lambda} Q_{,2},$$

offering the possibility to integrate. We ignore all functions of integration pertaining to the rigid body motion, and get by integration

$$2\mu u_1 = -\Phi_{,1} + \frac{2(2\mu+\lambda)}{\mu+\lambda} P, \; 2\mu u_2 = -\Phi_{,2} + \frac{2(2\mu+\lambda)}{\mu+\lambda} Q. \tag{3.3.29}$$

The formulae (3.3.29) enable us to find both the displacements u_1 and u_2, which are the essential ones in a planar case, from the already known stress function Φ, if the particularly body is simply connected.

If, on the other hand, we are only interested in the displacements, and not in the stresses, then from the start, both planar problems can be set up in the displacements. Let us follow this in the case of plane strain. Let (3.3.1), (3.3.5) and (3.3.7) be true for this case. Therefore, observing the above conditions, special equation

$$\begin{gathered}(\lambda+\mu)\Theta_{,\alpha} + \mu\nabla^2 u_\alpha + K_\alpha = 0, \; \alpha = 1, 2, \\ \Theta = u_{1,1} + u_{2,2}, \; \nabla^2 = \frac{\partial^2}{\partial x_1^2} + \frac{\partial^2}{\partial x_2^2}, \; K_\alpha = K_\alpha(x_1, x_2)\end{gathered} \tag{3.3.30}$$

is obtained from the general equation (3.1.9). If specification of conditions for the displacements occurring at the contour of the cross-section in the x_1x_2-plane, instead of boundary conditions (3.3.7) notated in the stresses, can be given, then the problem of plane strain has been changed into a boundary value problem of the first kind. Its solution yields the wanted displacements.

Let us discuss the plane state of stress. Then $\sigma_{33} = 0$, and hence, because of (3.1.3a) and $\varepsilon_{i(i)} = u_{i(i)}$,

$$\lambda(u_{1,1}+u_{2,2}+u_{3,3})+2\mu u_{3,3} = 0,$$

from which

$$u_{3,3} = -\frac{\lambda}{\lambda+2\mu}(u_{1,1}+u_{2,2}) \tag{3.3.31}$$

is obtained. But also $\sigma_{13} = 0$, and according to (3.1.3a) also $\varepsilon_{13} = 0$, which according to (3.1.2) is followed by $u_{1,3} = -u_{3,1}$. Differentiating further with respect to x_3, and rearranging the order of the differentiation as well as using (3.3.31), yields

$$u_{1,33} = -u_{3,13} = -u_{3,31} = \frac{\lambda}{\lambda+2\mu}\frac{\partial}{\partial x_1}(u_{1,1}+u_{2,2}). \tag{3.3.32}$$

When relationship $\Theta = u_{1,1}+u_{2,2}$ is used, then the shortened version is

$$u_{3,3} = -\frac{\lambda\Theta}{\lambda+2\mu}, \quad u_{1,33} = \frac{\lambda\Theta_{,1}}{\lambda+2\mu}. \tag{3.3.33}$$

Expanding (3.1.9) with respect to $j = 1$, we obtain

$$(\lambda+\mu)u_{1,11}+u_{2,21}+u_{3,31})+\mu(u_{1,11}+u_{1,22}+u_{1,33})+K_1 = 0.$$

Applying Θ, ∇^2 according to (3.3.30) and (3.3.33), this changes to

$$(\lambda+\mu)\left(\Theta_{,1}-\frac{\lambda\Theta_{,1}}{\lambda+2\mu}\right)+\mu\left(\nabla^2 u_1+\frac{\lambda}{\lambda+2\mu}\Theta_{,1}\right)+K_1 = 0,$$

and

$$\left(\frac{2\lambda\mu}{\lambda+2\mu}+\mu\right)\Theta_{,1}+\mu\nabla^2 u_1+K_1 = 0.$$

A corresponding equation for the index $j = 2$ is easily set up, and we write in general

$$\left(\frac{2\lambda\mu}{\lambda+2\mu}+\mu\right)\Theta_{,\alpha}+\mu\nabla^2 u_\alpha+K_\alpha = 0,\ \alpha = 1,2,$$
$$\Theta = u_{1,1}+u_{2,2},\ \nabla^2 = \frac{\partial^2}{\partial x_1^2}+\frac{\partial_2^2}{\partial x_2},\quad K_\alpha = K_\alpha(x_1, x_2). \tag{3.3.34}$$

Thus the fundamental plane stress equation in the displacements, analogous to (3.3.30), has been determined. In the case of being able to express the boundary conditions (3.3.7) in terms of the displacements, we have again obtained a boundary value problem of the first kind. The comparison of (3.3.30) with (3.3.34) shows that also in this formulation both problems are fundamentally different because of the different coefficients in the differential equation. It is possible, however, as far as formal mathematics are concerned, to make them equal by inserting $\lambda^* = 2\lambda\mu/(\lambda+2\mu)$ in (3.3.34). The equations (3.3.30) and (3.3.34) can then be treated uniformly when a solution is wanted. We must remember, though, when applying the solution, that in one case, instead of λ, λ^* with its special meaning must be inserted.

At this point, let us discuss that $\boldsymbol{K} \equiv 0$. Then the differential equations of both planar problems can be summed up by

$$\begin{aligned}&(\lambda^*+\mu)\Theta_{,\alpha}+\mu\nabla^2 u_\alpha = 0,\ \alpha = 1, 2,\\ &\lambda^* = \lambda \text{ for plane strain,}\\ &\lambda^* = \frac{2\lambda\mu}{\lambda+2\mu} \text{ for plane stress.}\end{aligned} \tag{3.3.35}$$

Let us change to symbolic notation of (3.3.35) by temporarily assuming all to be three-dimensional and work with vectors

$$\nabla = (\partial/\partial x_1, \partial/\partial x_2, \partial/\partial x_3) \text{ and } \boldsymbol{u} = (u_1, u_2, 0).$$

Thus, (3.3.35) changes into

$$(\lambda^*+\mu) \operatorname{grad} \operatorname{div} \boldsymbol{u}+\mu\nabla^2\boldsymbol{u} = 0. \tag{3.3.36}$$

Let us investigate now whether it is possible to set up a solution for (3.3.36) by using a *displacement function* Ψ. This displacement function would then be analogous to the known stress function, and has actually been introduced by Love. However, instead, let us follow Marguerre's method and start with

$$\boldsymbol{u} = \operatorname{grad}\varphi+\operatorname{curl}\boldsymbol{a}, \tag{3.3.37}$$

when φ is a scalar quantity and $\boldsymbol{a}$ is vector

$$\boldsymbol{a} = (0, 0, \psi).$$

Then,

$$\operatorname{curl} \boldsymbol{a} = (\psi_{,2}, -\psi_{,1}, 0). \tag{3.3.38}$$

Introducing (3.3.37) in (3.3.36), and because of $\operatorname{div} \boldsymbol{u} = \nabla^2\varphi$, we obtain from (3.3.36) after simple rearrangement

$$\nabla^2[(\lambda^* + 2\mu) \operatorname{grad} \varphi + \mu \operatorname{curl} \boldsymbol{a}] = 0. \tag{3.3.39}$$

Let us assume at this point that (3.3.39) is once more two-dimensional, which, according to (3.3.38) curl $\boldsymbol{a}$ is anyway. For grad φ, i.e., ∇^2, let us write

$$\operatorname{grad} \varphi = (\varphi_{,1}, \varphi_{,2}, 0), \text{ i.e., } \nabla^2 = \frac{\partial^2}{\partial x_1^2} + \frac{\partial^2}{\partial x_2^2}. \tag{3.3.40}$$

We shall then obtain the two scalar equations

$$\nabla^2[(\lambda^* + 2\mu)\varphi_{,1} + \mu\psi_{,2}] = 0, \; \nabla^2[(\lambda^* + 2\mu)\varphi_{,2} - \mu\psi_{,1}] = 0 \tag{3.3.41}$$

from (3.3.39).

As for the displacement function Ψ we start assuming

$$\varphi = \frac{1}{\lambda^* + 2\mu} \Psi_{,2}, \psi = -\frac{1}{\mu} \Psi_{,1}. \tag{3.3.42}$$

When using it for (3.3.41), the first equation (3.3.41) is identically satisfied, and from the second we obtain

$$\nabla^2\nabla^2\Psi = 0, \tag{3.3.43}$$

i.e., the displacement function must also satisfy the biharmonic equation as did the stress function for vanishing volume forces.

Let us calculate from (3.3.37), (3.3.38), (3.3.40) and (3.3.42) that

$$u_1 = \varphi_{,1}+\psi_{,2} = -\frac{\lambda^*+\mu}{\mu(\lambda^*+2\mu)}\Psi_{,12},$$
$$u_2 = \varphi_{,2}-\psi_{,1} = -\frac{\lambda^*+\mu}{\mu(\lambda^*+2\mu)}\Psi_{,22}+\frac{1}{\mu}\nabla^2\Psi. \qquad (3.3.44)$$

Hence, we can derivate the displacements from the previously found displacement function Ψ.

Let us finally consider the *generalized state of plane strain* which occurs when displacement u_3 does not equal nought but is a linear function of x_3. The ε_{33} does not equal nought as in a classic case, but is constant. For example, let $\varepsilon_{33} = c = \text{const}$. Hence, instead of (3.3.9), we obtain

$$\sigma_{33} = \nu(\sigma_{11}+\sigma_{22})+E\cdot\varepsilon_{33} = \nu(\sigma_{11}+\sigma_{22})+cE. \qquad (3.3.45)$$

By putting (3.3.45) in (3.3.8), we arrive back at (3.3.10) and (3.3.14), respectively. Therefore, the general case does not differ from the classic case in the course of the further calculation. Except when calculating the strains does a difference occur, because instead of (3.3.22)

$$\varepsilon_{11} = \frac{1-\nu^2}{E}\left(\sigma_{11}-\frac{\nu}{1-\nu}\sigma_{22}\right)-\nu c,$$
$$\varepsilon_{22} = \frac{1-\nu^2}{E}\left(\sigma_{22}-\frac{\nu}{1-\nu}\sigma_{11}\right)-\nu c, \qquad (3.3.46)$$
$$\varepsilon_{12} = \frac{\sigma_{12}}{2G}, \quad \varepsilon_{13} = \varepsilon_{23} = 0, \quad \varepsilon_{33} = c$$

is to be taken.

3.3.2 *On realizing planar conditions*

A characteristic example of plan strain occurs with the long cylinder (Figure 38), whose lateral surface along its generating line is subjected to a uniformly distributed load which does not have any components in the direction of the generating line. By suitably supporting the ends of the cylinder it is possible to satisfy conditions (3.3.1) for the displacements. The boundary conditions (3.3.7) are valid for the cylinder's lateral surface, on which, in actual fact, $n_3 = 0$, is true. We must, however, realize that, according to the calculation on both bases of the cylinder, stress $\sigma_{33} = \nu(\sigma_{11}+\sigma_{12})$ does occur. In order to realize the classic case of plain strain, we should see to it that in supporting the cylinder's ends, such forces are put into operation which enable the stresses σ_{33} to be present. In this way, boundary conditions will be obtained which then are added to those already valid for the lateral surface of the cylinder, and which either cannot be, or should not be, realized. Hence, the question arises whether the discussion of a classic state of plane strain is of any significance.

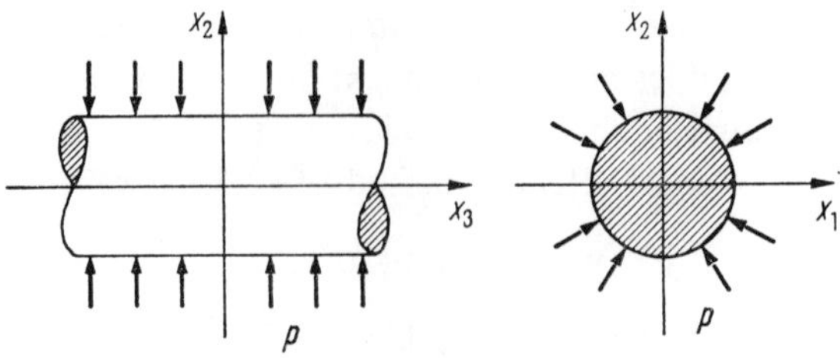

Figure 38

We can safely say that it is so, as the solution of the classic case does, at least, represent a first approximation which applies quite well to the middle part of the cylinder – which is at a good distance from both ends of the cylinder. Let us suppose further that we wanted to discuss a cylinder which is loaded along its lateral surface as in the classic case, yet whose bases are free of any forces. In this case, we set up the classic solution for the cylinder which yields a certain stress distribution σ_{33} for the bases, but offends the boundary condition there. Following this, we take the same cylinder, but this time with no stress on its lateral surface, and the bases stressed with $-\sigma_{33}$. A solution to this new case of stress is easily found by using Saint-Vénant's principle, i.e., by substituting the actual distribution $-\sigma_{33}$ by a more simple one, yet statistically equivalent. By superimposing both solutions, we obtain, according to the superposition principle, the true solution for the cylinder when it is loaded on the lateral surface only, and not on the bases, as had been specified originally.

The state of plane stress is approximately encountered in a plan slab of constant thickness (Figure 39), which is affected by loads applied to its contour which are uniformly distributed over the slab's thickness and are parallel to the slab's middle surface. As already mentioned, on a slab of this kind, plane stress cannot be exact but only approximate. The condition (3.3.17), in particular, does not apply, since the stresses σ_{i3}, $i = 1, 2, 3$, do not equal nought, and since the remaining stresses are actually dependent on x_3. Thus, as far as the slab is concerned, we are faced with a three-dimensional problem.

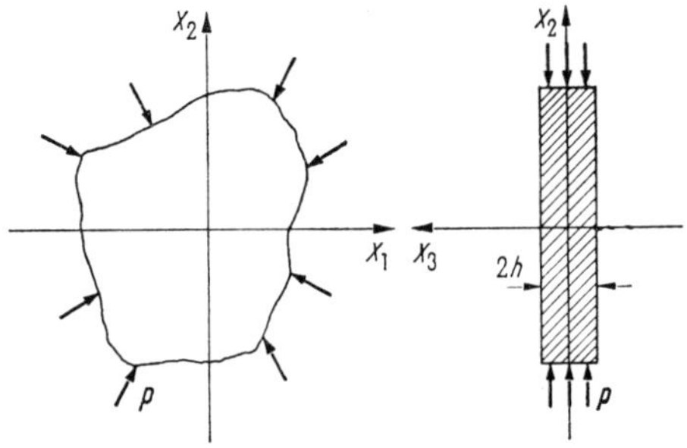

Figure 39

There are two ways of setting our assumptions on plane stress right. One goes back to Filon, and tells us how to set up the slab problem making it artificially two-dimensional. This is achieved by averaging all quantities depending on x_3, i.e., using quantities from which the dependence of x_3 has been removed by integration over x_3. It is said that for the problem we have now obtained, the *generalized state of plane stress* is given. The other way extends the calculation of the classic state of plane stress by considering the dependence of all possible quantities of x_3. Let us start by discussing the generalized state of plane stress, and assume that because of the minimal thickness of the slab, and because of the kind of load acting on the slab, the displacement u_3 is very small, so that for its average we can write

$$\bar{u}_3(x_1, x_2) = \frac{1}{2h}\int_{-h}^{+h} u_3(x_1, x_2, x_3)\mathrm{d}x_3 = 0.$$

This fact proves the justification of approximation in (3.3.2), that $\varepsilon_{33} \approx 0$ is true. Since the slab is only stressed along its contour, then

$$\sigma_{i3}(x_1, x_2, \pm h) = 0, \; i = 1, 2, 3, \tag{3.3.47}$$

must be valid. But it is $\sigma_{13,1}(x_1, x_2, \pm h) = 0$, $\sigma_{23,2}(x_1, x_2, \pm h) = 0$, and since by definition, $K_3 = 0$, then, from the third equilibrium condition

$$\sigma_{13,1}+\sigma_{23,2}+\sigma_{33,3} = 0,$$

which is also valid for $x_3 = \pm h$, $\sigma_{33,3}(x_1, x_2, \pm h) = 0$ is obtained. Thus, we have that not only σ_{33} but also $\sigma_{33,3}$ for $x_3 = \pm h$ is equal to nought. We can deduce that σ_{33} differs only slightly from nought, and we are justified in writing that $\sigma_{33} \equiv 0$.

Let us expend (3.1.19) for $i = 1, j = 2, k = 2, l = 2$, taking into account that $\sigma_{33} = 0$, and using the average of all remaining quantities. Thus we are formally back to equation (3.3.18), the only difference being that throughout we find averages of all quantities actually dependent on x_1 and x_2 only:

$$2(1+\nu)\bar{\sigma}_{12,12} = \bar{\sigma}_{11,22}+\bar{\sigma}_{22,11}-\nu(\bar{\sigma}_{22,22}+\bar{\sigma}_{11,11}),$$
$$\bar{\sigma}_{ij}(x_1, x_2) = \frac{1}{2h}\int_{-h}^{h} \sigma_{ij}(x_1, x_2, x_3)\,dx_3. \tag{3.3.48}$$

In addition

$$\overline{\sigma_{i3,3}} = \frac{1}{2h}\int_{-h}^{h} \sigma_{i3,3}\,dx_3 = \frac{1}{2h}[\sigma_{i3}(h)-\sigma_{i3}(-h)] = 0, \; i = 1, 2, 3, \tag{3.3.49}$$

is also obtained from (3.3.47). Considering this, and $\sigma_{33} \equiv 0$, $K_3 \equiv 0$, we obtain, after taking the mean value of equation $\sigma_{ik,ki} = -K_{i,i}$, formally (3.3.19), again with the difference that averages are found throughout:

$$2\bar{\sigma}_{12,12} = -(\bar{\sigma}_{11,11}+\bar{\sigma}_{22,22}+\bar{K}_{1,1}+\bar{K}_{2,2}). \tag{3.3.50}$$

(3.3.48) and (3.3.50) correspond to (3.3.18) and (3.3.19) of the classic case, and are two-dimensional like those. Hence, completely analogous to them we obtain:

$$\left(\frac{\partial^2}{\partial x_1^2}+\frac{\partial^2}{\partial x_2^2}\right)(\bar{\sigma}_{11}+\bar{\sigma}_{22}) = -(1+\nu)\left(\frac{\partial \bar{K}_1}{\partial x_1}+\frac{\partial \bar{K}_2}{\partial x_2}\right), \tag{3.3.51}$$

representing the counterpart to differential equation (3.3.20) of the classic case.

Let us take the averages for the boundary condition (3.3.7), thus yielding

$$\begin{aligned}\bar{\sigma}_{11} n_1+\bar{\sigma}_{12} n_2 &= p_1,\\ \bar{\sigma}_{21} n_1+\bar{\sigma}_{22} n_2 &= p_2.\end{aligned} \tag{3.3.52}$$

Comparison of (3.3.51) and (3.3.52) with (3.3.20) and (3.3.7) indicates that each solution of the classic problem represents formally a solution of the generalized problem, i.e., the classic problem of plane stress correctly gives those *averages* of the stresses which occur in a plane slab loaded along its contour. In this way, the practical significance of plane stress has been proven.

For the other way, we notate, as in (3.3.17),

$$\sigma_{33} = \sigma_{13} = \sigma_{23} = 0,$$

which for σ_{33} is justified anyway as well as for σ_{13}, σ_{23} because of (3.3.47) and (3.3.49); but contrary to (3.3.17) assume a certain dependence for σ_{11}, σ_{22} and σ_{12} from x_3. Thus, we are faced by a three-dimensional problem, which yields the solutions for the wanted stresses $\sigma_{11}, \sigma_{22}, \sigma_{12}$ from

$$\begin{aligned} &\sigma_{11} = \Phi_{,22}+U,\ \sigma_{22} = \Phi_{,11}+U,\ \sigma_{12} = -\Phi_{,12}, \\ &\Phi = \Phi^* + \frac{\nu(4h^2-12x_3^2)}{24(1+\nu)}\left[\nabla^2\Phi^*+2U-\frac{1+\nu}{2(1-\nu)}(x_1^2+x_2^2)\nabla^2 U\right], \\ &\Phi^* = \Phi^*(x_1, x_2),\ \nabla^2\nabla^2\Phi^* = -(1-\nu)\nabla^2 U,\ \nabla^2 U = \text{const}, \end{aligned} \tag{3.3.53}$$

when $K_{1,1}+K_{2,2} = -\nabla^2 U$ is valid for the volume forces. We can omit the derivation of (3.3.53).

Let $4h^2-12x_3^2 \approx 0$ be valid for thin slabs. Then (3.3.53) refers back to the known relations of the classic plane stress. Thus, it has again been proven that solutions of the classic problem represent adequate approximations for thin slabs.

As a final remark, let us point out that solutions of (3.3.53) can only satisfy boundary conditions (3.3.52), and not the exact boundary conditions (3.3.7). This is quite evident since (3.3.53) yields $\sigma_{11}, \sigma_{22}, \sigma_{12}$ as functions also of x_3, while p_1 and p_2 by assumption are functions only of x_1, x_2. Following this violation of the boundary conditions, the very solutions of (3.3.53) are not fully satisfying. We must, therefore, as we have already pointed out in the cylinder problem, superpose the solution of a second slab problem as to compensate for the violation of boundary condition (3.3.7). A detailed discussion of this fact has been made by Durelli-Phillips-Tsao [22], from which (3.3.53) is derived.

3.3.3 *The biharmonic equation*

On permitting omission of the volume forces, we are led to the biharmonic equation (3.3.16) and to the boundary conditions (3.3.7) via the two classic planar problems. The boundary conditions may be recalculated as follows: Let R be the contour of the body's cross-section in the x_1, x_2-plane, let s be the arc length of this contour, and let the possibility exist to represent the contour forces p_1, p_2 as a function of the arc length. If, additionally, we use (3.3.15), then (3.3.7) changes to

$$\left.\begin{aligned} \Phi_{,22} n_1 - \Phi_{,12} n_2 &= p_1(s), \\ -\Phi_{,12} n_1 + \Phi_{,11} n_2 &= p_2(s) \end{aligned}\right\} \text{ on } R. \tag{3.3.54}$$

But according to Figure 33, $n_1 = \mathrm{d}x_2/\mathrm{d}s$, $n_2 = -\mathrm{d}x_1/\mathrm{d}s$, so that (3.3.54) changes to

$$\frac{\mathrm{d}}{\mathrm{d}s}(\Phi_{,2}) = p_1(s), \quad -\frac{\mathrm{d}}{\mathrm{d}s}(\Phi_{,1}) = p_2(s), \text{ on } R.$$

This yields by integration,

$$\Phi_{,1} = -\int p_2(s)\,\mathrm{d}s = f_1(s) + \text{const}, \quad \Phi_{,2} = \int p_1(s)\,\mathrm{d}s = f_2(s) + \text{const}.$$

Hence, the boundary value problem

$$\begin{aligned} &\nabla^2\nabla^2\Phi = 0, \\ &\Phi_{,\alpha} = f_\alpha(s) + \text{const, on } R, \ \alpha = 1, 2, \end{aligned} \tag{3.3.55}$$

is fundamental to the classic planar problems, when the volume forces are neglected.

The biharmonic equation does not only occur in connection with the stress function Φ but also with the displacement function Ψ [see (3.3.43)]. Furthermore, when certain conditions are satisfied, the equations of Lamé-Navier and Beltrami may be reduced to the biharmonic equation [see (3.1.18) and (3.1.27)]. All these facts indicate the great practical significance of this differential equation.

Only in the rarest of cases will it be possible to solve directly the boundary value problems connected with it. In most cases it will become necessary to use the inverse or semi-inverse method for a solution of the problems. It is of value, therefore, to have a certain general knowledge of the manifold of its solutions.

First of all it is realized that every *harmonic function* is a solution of the biharmonic equation. Since, for example, if φ is harmonic, then it satisfies Laplace's differential equation $\nabla^2\varphi = 0$. But then, at the same time $\nabla^2(\nabla^2\varphi) = \nabla^2 0 = 0$, i.e., the biharmonic equation is satisfied. The *first type of solution* is thus represented by the harmonic functions φ.

Secondly, let us consider functions $x_i\psi$, $i = 1, 2, 3$, containing ψ as a harmonic function. Let us set up $\nabla^2(x_i\psi) = (x_1\psi)_{,kk}$. It is $(x_i\psi)_{,k} = \delta_{ik}\psi + x_i\psi_{,k}$, and hence $(x_i\psi)_{,kk} = (\delta_{ik}\psi + x_i\psi_{,k})_{,k}$. Calculated in full, it yields $(x_i\psi)_{,kk} = 2\delta_{ik}\psi_{,k} + x_i\psi_{,kk}$. Since ψ is harmonic, then following this condition $\psi_{,kk} = 0$, so that the important relation

$$(x_i\psi)_{,kk} \equiv \nabla^2(x_i\psi) = 2\psi_{,i} \tag{3.3.56}$$

remains. Let us set up $\nabla^2(\nabla^2 x_i\psi)$. According to (3.3.56) it yields $\nabla^2(\nabla^2 x_i\psi) = 2\nabla^2\psi_{,i} = 2(\nabla^2\psi)_{,i} = 0$ since, by definition, $\nabla^2\psi = 0$ is true. But it follows from $\nabla^2\nabla^2 x_i\psi = 0$ that functions of the kind $x_i\psi$, whose ψ is harmonic, satisfy the biharmonic equation, thus forming the *second type of solution*.

Since the sum of solutions is itself a solution because of the linearity of the biharmonic equation, then we obtain the *third type of solution* if the solutions of the first and second type are added:

$$f = \varphi + x_i\psi, \quad i = 1, 2, 3.$$

Let us now consider functions $r^2\psi$ containing $r^2 = x_i x_i$ and the harmonic ψ. Let $(x_i x_i)_k = \delta_{ik}x_i + \delta_{ik}x_i = 2x_k$, i.e., $r^2_{,k} = 2x_k$. Hence, we find $(r^2\psi)_{,k} = 2x_k\psi + r^2\psi_{,k}$ and because of $\psi_{,kk} = 0$ (by assumption),

$$(r^2\psi)_{,kk} = (2x_k\psi + r^2\psi_{,k})_{,k} = 2\delta_{kk}\psi + 4x_k\psi_{,k} + r^2\psi_{,kk} = 6\psi + 4x_k\psi_{,k}. \tag{3.3.57}$$

Applying (3.3.56), and using the property of ψ,

$$\nabla^2(x_k\psi_{,k}) = 2\psi_{,kk} = 0 \tag{3.3.58}$$

is obtained, from which we deduce that $x_k\psi_{,k}$ is harmonic. Following this, however, on the right side of (3.3.57) is the sum of two harmonic functions. Thus, $(r^2\psi)_{,kk}$ is itself harmonic so that

$$[(r^2\psi)_{,kk}]_{,ii} = \nabla^2(\nabla^2 r^2\psi) = 0$$

must be harmonic. Hence, $r^2\psi$ must be a solution of the biharmonic equation. Solutions of this kind yield the *fourth type.*

When ψ is harmonic, then it also applies to $\psi_{,k}$. Thus, $r^2\psi_{,k}$ belongs to the fourth type, and is biharmonic. Therefore it is possible to set up the solutions $\varphi + r^2\psi_{,k}$ of the *fifth type* by summation of solutions of the first type with solutions of type $r^2\psi_{,k}$.

If a is any constant and ψ a harmonic function, then $-a^2\psi_{,k}$ represents a solution of the first type. Adding it to the solutions of the fifth type it yields $\varphi + (r^2 - a^2)\psi_{,k}$, which are *solutions of the sixth type.*

In this way it is possible to set up a whole catalogue of solutions for the biharmonic differential equation, which can be used for practical calculation following the inverse and semi-inverse method. A catalogue of this kind is given for example in [19], p. 131.

A special possibility of dealing with the biharmonic equation and the connected boundary value problems is to carry out the calculation using complex functions. The origins of this method are tracked back to Love and Filon. The procedure was completed by Kolosoff and represented in detail by Muskhelishvili [17]. Despite its great importance, let us discuss it only briefly because a detailed discussion of the method would easily fill a book by itself. For a detailed discussion see [17].

As $\nabla^2\nabla^2\Phi = 0$, then $\nabla^2\Phi$ is a harmonic function p. This fact has already been pointed out in 3.3.1. Using function p and its conjugate q, functions P and Q are obtained by integration, as has been shown in 3.3.1, and for which (3.3.28) is valid. P and Q are conjugate harmonic. Thus, they satisfy the Cauchy-Riemann differential equations, and $P_{,1} = Q_{,2}$ is valid. It follows from (3.3.28)

$$P_{,1} = Q_{,2} = p/4. \tag{3.3.59}$$

Let us set up $\nabla^2(\Phi - Px_1 - Qx_2)$. According to (3.3.56), $\nabla^2 Px_1 = 2P_{,1}$, and $\nabla^2 Qx_2 = 2Q_{,2}$. Thus, we obtain

$$\nabla^2(\Phi - Px_1 - Qx_2) = \nabla^2\Phi - 2P_{,1} - 2Q_{,2}.$$

Because of $\nabla^2\Phi \equiv p$ (by definition) and (3.3.59) we find

$$\nabla^2(\Phi - Px_1 - Qx_2) = 0.$$

This means that $\Phi - Px_1 - Qx_2$ is harmonic, e.g., it is equal to $H(x_1, x_2)$. We deduce

$$\Phi = Px_1 + Qx_2 + H(x_1, x_2). \tag{3.3.60}$$

Using $F(z) = P + \mathrm{i}Q$ and $G(z) = H + \mathrm{i}H^*$ (when H^* is the conjugate of H) we finally represent (3.3.60) as

$$\Phi = \tfrac{1}{2}[\bar{z}F(z) + z\overline{F(z)} + G(z) + \overline{G(z)}]. \tag{3.3.60a}$$

Barred quantities are to be understood as being conjugated. From (3.3.15)

$$\begin{aligned} \sigma_{11} + \mathrm{i}\sigma_{12} &= \Phi_{,22} - \mathrm{i}\Phi_{,12} = -\mathrm{i}(\Phi_{,1} + \mathrm{i}\Phi_{,2})_{,2}, \\ \sigma_{22} - \mathrm{i}\sigma_{12} &= \Phi_{,11} + \mathrm{i}\Phi_{,12} = (\Phi_{,1} + \mathrm{i}\Phi_{,2})_{,1} \end{aligned} \tag{3.3.61}$$

is obtained, and from (3.3.60a)

$$\Phi_{,1} + \mathrm{i}\Phi_{,2} = F(z) + z\overline{F'(z)} + \overline{G'(z)}. \tag{3.3.62}$$

Substituting in (3.3.61), we obtain through further differentiation

$$\begin{aligned} \sigma_{11} + \mathrm{i}\sigma_{12} &= F'(z) + \overline{F'(z)} - z\overline{F''(z)} - \overline{G''(z)}, \\ \sigma_{22} - \mathrm{i}\sigma_{12} &= F'(z) + \overline{F'(z)} + z\overline{F''(z)} + \overline{G''(z)}. \end{aligned} \tag{3.3.63}$$

This formula makes it possible to find the stresses by complex calculation.

Corresponding calculations for the displacements have already been done in 3.3.1, leading to formulae (3.3.29). It is, therefore, not necessary to repeat them at this point. Summing up the various equations, we obtain from (3.3.29)

$$2\mu(u_1 + \mathrm{i}u_2) = \left[-(\Phi_{,1} + \mathrm{i}\Phi_{,2}) + \frac{2(2\mu + \lambda)}{\mu + \lambda}(P + \mathrm{i}Q) \right].$$

Let us use (3.3.62), and considering that $P + \mathrm{i}Q = F(z)$, we arrive at

$$2\mu(u_1+\mathrm{i}u_2) = \frac{3\mu+\lambda}{\mu+\lambda}F(z)-z\overline{(F'(z)}-\overline{G'(z)}. \tag{3.3.64}$$

Following simple calculation, in this way we are able to find the displacements using complex calculation. It is worth mentioning that (3.3.64) is exactly valid for plane strain, because in determining (3.3.29), assumption $\varepsilon_{33} = 0$ was made, which applies directly to plane strain. To obtain the corresponding formula for plane stress, let us insert in (3.3.64), as has been done elsewhere, quantity $\lambda^* = 2\lambda\mu/(\lambda+2\mu)$ instead of λ. This leads to

$$2\mu(u_1+\mathrm{i}u_2) = \frac{5\lambda+6\mu}{3\lambda+2\mu}F(z)-z\overline{F'(z)}-\overline{G'(z)}. \tag{3.3.65}$$

Neglecting the volume forces, the classic planar problems are thus reduced to determining the complex functions $F(z)$ and $G'(z)$ from the boundary conditions. Limiting ourselves to simple connected cross-sections, which for the sake of simplicity we have done in this chapter and elsewhere, we must find functions, for which according to (3.3.55)

$$\Phi_{,1}+\mathrm{i}\Phi_{,2} = f_1(s)+\mathrm{i}f_2(s)+\text{const on } R$$

is valid. Because of (3.3.62), this is equal to condition

$$F(z)+z\overline{F'(x)}+\overline{G'(z)} = f_1(s)+\mathrm{i}f_2(s)+\text{const on } R, \tag{3.3.66}$$

while the arbitrary constant in (3.3.66) is generally complex. In practical calculations, (3.3.66) is used by expanding both sides of the equation into series, and determining the coefficients $F(z)$ and $G'(z)$ by comparing the hitherto undetermined coefficients. References to this method, as well as to other solution methods, are found in [15] and [17]. The constant on the right side of (3.3.66) bears no consequence to the finding of stresses, and only an insignificant part to the finding of the displacements in that these are determined up to a rigid body motion.

In the preceding paragraph, when outlining complex calculation, we have kept to the boundary value problem of the second type. Matching calculations lead to the solution in the case of the boundary value problem of the first kind, when for contour R of the cross-section, displacements u_α, $\alpha = 1, 2$ are specified through

$$u_\alpha = g_\alpha(s), \ \alpha = 1, 2 \ \text{on} \ R. \tag{3.3.67}$$

Instead of (3.3.66) using (3.3.64) and (3.3.67), we arrive at

$$\frac{3\mu+\lambda}{\mu+\lambda} F(z) - z\overline{F'(z)} - \overline{G'(z)} = 2\mu[g_1(s) + ig_2(s)] \ \text{on} \ R, \tag{3.3.68}$$

from which the wanted functions $F(z)$ and $G'(z)$ are calculated.

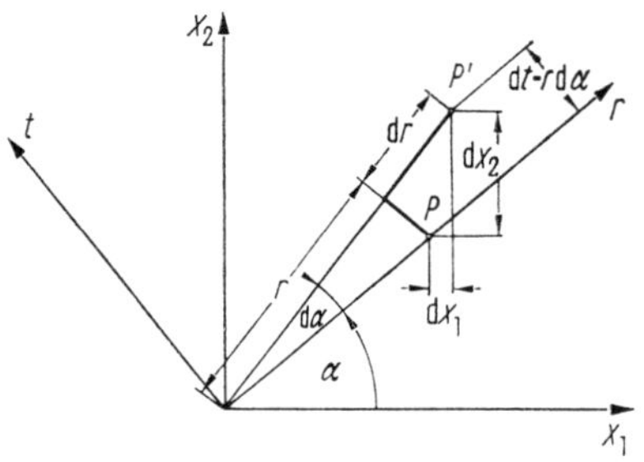

Figure 40

3.3.4 *The planar problem in polar coordinates*

It would prove practical in many cases to go on to curvilinear coordinates as far as the planar problems are concerned, because of the cross-sectional contour in question and its loads, in order to obtain simple boundary conditions. As the polar coordinates play an important part among them, we shall discuss them in detail in this chapter.

In addition to the usual cartesian system x_1, x_2, let us introduce the orthogonal axis system r, t, which in contrast to the original system has been turned about angle α (Figure 40). Let a plane state of stress in relation to the r, t-system be given by tensor

$$\boldsymbol{S} = \begin{pmatrix} \sigma_{rr} & \sigma_{rt} & 0 \\ \sigma_{rt} & \sigma_{tt} & 0 \\ 0 & 0 & 0 \end{pmatrix}.$$

To relate the components of this tensor to those components of the same tensor obtained with regard to the x_1, x_2-system, let us temporarily use the notations

$$\sigma_{rr} = \sigma'_{11}, \sigma_{rt} = \sigma'_{12} = \sigma'_{21}, \sigma_{tt} = \sigma'_{22}. \tag{3.3.69}$$

Let us consider also that the transformation matrix for the change from one coordinate system to the other is

$$\begin{pmatrix} a_{11} = \cos\alpha & a_{12} = \sin\alpha & a_{13} = 0 \\ a_{21} = -\sin\alpha & a_{22} = \cos\alpha & a_{23} = 0 \\ a_{31} = 0 & a_{32} = 0 & a_{33} = 0 \end{pmatrix}. \tag{3.3.70}$$

We are now able to read off the general transformation formula for cartesian tensors, i.e.,

$$\sigma_{ij} = a_{ki}\, a_{lj}\, \sigma'_{kl} \tag{1.1.16}$$

e.g., for $i = j = 1$,

$$\sigma_{11} = a_{11}\, a_{11}\, \sigma'_{11} + a_{11}\, a_{21}\, \sigma'_{12} + a_{21}\, a_{11}\, \sigma'_{21} + a_{21}\, a_{21}\, \sigma'_{22}\,.$$

Adding to this (3.3.69) and (3.3.70), we obtain

$$\sigma_{11} = \cos^2 \alpha\sigma_{rr} - \cos\alpha \sin\alpha\sigma_{rt} - \cos\alpha \sin\alpha\sigma_{rt} + \sin^2 \alpha\sigma_{tt}\,.$$

This expression can be changed to

$$\sigma_{11} = \sigma_{rr} \cos^2 \alpha + \sigma_{tt} \sin^2 \alpha - \sigma_{rt} \sin 2\alpha.$$

Evaluating (1.1.16) for $i = j = 2$ and $i = 1, j = 2$. then the system of equations

$$\begin{aligned} \sigma_{11} &= \sigma_{rr} \cos^2 \alpha + \sigma_{tt} \sin^2 \alpha - \sigma_{rt} \sin 2\alpha, \\ \sigma_{22} &= \sigma_{rr} \sin^2 \alpha + \sigma_{tt} \cos^2 \alpha + \sigma_{rt} \sin 2\alpha, \\ \sigma_{12} &= (\sigma_{rr} - \sigma_{tt}) \sin\alpha \cos\alpha + \sigma_{rt} \cos 2\alpha \end{aligned} \tag{3.3.71}$$

is obtained.

Through reversal of (1.1.16), i.e., from $\sigma'_{kl} = a_{ki} a_{lj} \sigma_{ij}$, we obtain in the same manner

$$\begin{aligned} \sigma_{rr} &= \sigma_{11} \cos^2 \alpha + \sigma_{22} \sin^2 \alpha + \sigma_{12} \sin 2\alpha, \\ \sigma_{tt} &= \sigma_{11} \sin^2 \alpha + \sigma_{22} \cos^2 \alpha - \sigma_{12} \sin 2\alpha, \\ \sigma_{rt} &= (\sigma_{22} - \sigma_{11}) \sin\alpha \cos\alpha + \sigma_{12} \cos 2\alpha. \end{aligned} \tag{3.3.72}$$

The equilibrium conditions are derived from those already mentioned in connection with the cylinder coordinates, if in those all quantities with index z and all derivatives with respect to z are made to equal nought. Thus, we obtain from (3.1.87) for $\boldsymbol{K} \equiv 0$

$$\begin{aligned} &\frac{\partial}{\partial r}(r\sigma_{rr}) - \sigma_{tt} + \frac{\partial \sigma_{rt}}{\partial \alpha} = 0, \\ &\frac{\partial}{\partial r}(r\sigma_{rt}) + \sigma_{rt} + \frac{\partial \sigma_{tt}}{\partial \alpha} = 0, \end{aligned} \tag{3.3.73}$$

if the remaining terms are slightly rearranged and *angle* φ is changed to α and *indices* φ are changed to t.

Similarly[1] from (3.1.82)

$$\begin{aligned} &\varepsilon_{rr} = \frac{\partial u_r}{\partial r}, \quad \varepsilon_{tt} = \frac{1}{r}\frac{\partial u_t}{\partial \alpha} + \frac{u_r}{r}, \quad \varepsilon_{zz} = \frac{\partial u_z}{\partial z}, \\ &\varepsilon_{rt} = \frac{1}{2}\left(\frac{1}{r}\frac{\partial u_r}{\partial \alpha} + \frac{\partial u_t}{\partial r} - \frac{u_t}{r}\right) \end{aligned} \tag{3.3.74}$$

is obtained for the strains, and from (3.1.85) for Hooke's Law

$$\begin{aligned} &\varepsilon_{rr} = \frac{1+\nu}{E}\sigma_{rr} - \frac{\nu}{E}(\sigma_{rr}+\sigma_{tt}+\sigma_{zz}), \\ &\varepsilon_{tt} = \frac{1+\nu}{E}\sigma_{tt} - \frac{\nu}{E}(\sigma_{rr}+\sigma_{tt}+\sigma_{zz}), \\ &\varepsilon_{rt} = \frac{1+\nu}{E}\sigma_{rt}, \quad \varepsilon_{zz} = \frac{1+\nu}{E}\sigma_{zz} - \frac{\nu}{E}(\sigma_{rr}+\sigma_{tt}+\sigma_{zz}). \end{aligned} \tag{3.3.75}$$

Assuming that the volume forces will vanish, then both planar problems are governed by the biharmonic equation $\nabla^2\nabla^2\Phi = 0$. For rewriting this differential equation in terms of polar coordinates, then $\nabla^2\Phi$ from (3.1.107) must be established. For this purpose $h_1 = h_3 = 1$, $h_2 = r$, $q_1 = r$, $q_2 = \alpha$, has to be used as well as all derivatives with respect to q_3 must be equalled to nought. This yields

[1] Once more all quantities having indices z and all derivatives with respect to z are set equal to nought, except for ε_{zz}, σ_{zz}, $\partial u_z/\partial z$.

$$\nabla^2\Phi = \left(\frac{\partial^2}{\partial r^2} + \frac{1}{r}\frac{\partial}{\partial r} + \frac{1}{r^2}\frac{\partial^2}{\partial \alpha^2}\right)\Phi. \tag{3.3.76}$$

Using operator ∇^2 once more, then (3.3.76) leads to

$$\nabla^2\nabla^2\Phi = \left(\frac{\partial^2}{\partial r^2} + \frac{1}{r}\frac{\partial}{\partial r} + \frac{1}{r^2}\frac{\partial^2}{\partial \alpha^2}\right)^2\Phi = 0. \tag{3.3.77}$$

The planar problem is being given in terms of polar coordinates through (3.3.77), and the respective boundary conditions formulated in terms of coordinates r, α. We still need details about stress function Φ. Analogous to (3.3.15), let us assume that

$$\sigma_{rr} = \frac{\partial^2\Phi}{\partial t^2}, \quad \sigma_{tt} = \frac{\partial^2\Phi}{\partial r^2}, \quad \sigma_{rt} = \frac{\partial^2\Phi}{\partial r\,\partial t} \tag{3.3.78}$$

holds, and from Figure 40 we read off relation

$$\partial t = r\,\partial\alpha, \quad \frac{\partial}{\partial t} = \frac{\partial}{r\,\partial\alpha}, \text{ respectively.} \tag{3.3.79}$$

Using it for σ_{rt}, we find first

$$\sigma_{rt} = -\frac{\partial}{\partial r}\left(\frac{1}{r}\frac{\partial\Phi}{\partial\alpha}\right). \tag{3.3.80}$$

Let us take into account the fact that $\sigma_{11}+\sigma_{22} = \sigma_{rr}+\sigma_{tt}$ is valid, since it is an invariant of the stress tensor. But according to (3.3.15), $\sigma_{11}+\sigma_{22} = \nabla^2\Phi$ is true. Therefore, $\sigma_{rr}+\sigma_{tt} = \nabla^2\Phi$ must also be true, which because of (3.3.76) yields

$$\sigma_{rr}+\sigma_{tt} = \frac{\partial^2\Phi}{\partial r^2} + \left(\frac{1}{r}\frac{\partial}{\partial r} + \frac{1}{r^2}\frac{\partial^2}{\partial\alpha^2}\right)\Phi.$$

But because of (3.3.78), σ_{tt} has already been determined, and therefore

$$\sigma_{rr} = \left(\frac{1}{r}\frac{\partial}{\partial r} + \frac{1}{r^2}\frac{\partial^2}{\partial\alpha^2}\right)\Phi \tag{3.3.81}$$

must then be true. Summarizing (3.3.78), (3.3.80) and (3.3.81), then as a relation between stresses and stress function in polar coordinates we have the relations

$$\sigma_{rr} = \left(\frac{1}{r}\frac{\partial}{\partial r} + \frac{1}{r^2}\frac{\partial^2}{\partial \alpha^2}\right)\Phi, \quad \sigma_{tt} = \frac{\partial^2 \Phi}{\partial r^2}, \quad \sigma_{rt} = -\frac{\partial}{\partial r}\left(\frac{1}{r}\frac{\partial \Phi}{\partial \alpha}\right). \tag{3.3.82}$$

Let us check whether stress function Φ satisfies the basic requirement: Let us insert (3.3.82) into equilibrium conditions (3.3.73). These are being identically satisfied, and the relations (3.3.82) are therefore correct.

The calculation process is now as follows: Let us, by considering the boundary conditions, set up a solution Φ for (3.3.77), obtaining from it stresses through (3.3.82). If we are dealing with a plane state of stress, then $\sigma_{zz} \equiv 0$, and according to (3.3.75), the strains are calculated from

$$\begin{aligned} \varepsilon_{rr} &= \frac{1}{E}(\sigma_{rr} - \nu\sigma_{tt}), \quad \varepsilon_{tt} = \frac{1}{E}(\sigma_{tt} - \nu\sigma_{rr}), \\ \varepsilon_{zz} &= -\frac{\nu}{E}(\sigma_{rr} + \sigma_{tt}), \quad \varepsilon_{rt} = \frac{1+\nu}{E}\sigma_{rt}. \end{aligned} \tag{3.3.83}$$

Using this, and with the aid of (3.3.74), the displacements are found.

If, on the other hand, we are dealing with a plane state of strain, then $\varepsilon_{zz} \equiv 0$ is true, and hence $\sigma_{zz} = \nu(\sigma_{rr} + \sigma_{tt})$, and we obtain from (3.3.75)

$$\varepsilon_{rr} = \frac{1-\nu^2}{E}\left(\sigma_{rr} - \frac{\nu}{1-\nu}\sigma_{tt}\right), \quad \varepsilon_{tt} = \frac{1-\nu^2}{E}\left(\sigma_{tt} - \frac{\nu}{1-\nu}\sigma_{rr}\right),$$

$$\varepsilon_{rt} = \frac{1+\nu}{E}\sigma_{rt}. \tag{3.3.84}$$

Thus, the strains are known, and the displacements are calculated by using (3.3.74).

Also, with respect to complex representation, we can easily change over to polar coordinates by notating

$$z = r\,\mathrm{e}^{i\alpha}. \tag{3.3.85}$$

From the transformation formula $u_i' = a_{ik}u_k$ for cartesian vectors it follows, using $u_1' = u_r$, $u_2' = u_t$ and the a_{ik}' from (3.3.70),

$$u_r = \cos \alpha u_1 + \sin \alpha u_2 , \; u_t = -\sin \alpha u_1 + \cos \alpha u_2 ,$$

so that

$$u_r + \mathrm{i} u_t = (u_1 + \mathrm{i} u_2)(\cos \alpha - \mathrm{i} \sin \alpha) = (u_1 + \mathrm{i} u_2)\mathrm{e}^{-\mathrm{i}\alpha}. \tag{3.3.86}$$

Using this relation in (3.3.64), (3.3.65), respectively, we obtain for the plane state of strain

$$u_r + \mathrm{i} u_t = \frac{\mathrm{e}^{-\mathrm{i}\alpha}}{2\mu} \left[\frac{3\mu+\lambda}{\mu+\lambda} F(z) - z\overline{F'(z)} - \overline{G'(z)} \right], \; z = r\mathrm{e}^{\mathrm{i}\alpha}; \tag{3.3.87}$$

and for the plane state of stress

$$u_r + \mathrm{i} u_t = \frac{\mathrm{e}^{-\mathrm{i}\alpha}}{2\mu} \left[\frac{5\lambda+6\mu}{3\lambda+2\mu} F(z) - z\overline{F'(z)} - \overline{G'(z)} \right], \; z = r\,\mathrm{e}^{\mathrm{i}\alpha}, \tag{3.3.88}$$

which makes it possible to calculate components u_r, u_t of the displacement vector in relation to the polar coordinates.

From (3.3.63) we find for the stresses

$$\sigma_{11} + \sigma_{22} = 2[F'(z) + \overline{F'(z)}] = 4\, \mathscr{Re}\, [F'(z)] \tag{3.3.89}$$

when both equations are added.

Subtracting the first equation (3.3.63) from the second,

$$\sigma_{22} - \sigma_{11} - 2\mathrm{i}\sigma_{12} = 2[z\overline{F''(z)} + \overline{G''(z)}]$$

is obtained. Substituting i by $-\mathrm{i}$,

$$\sigma_{22} - \sigma_{11} + 2\mathrm{i}\sigma_{12} = 2[\bar{z}F''(z) + G''(z)] \tag{3.3.90}$$

is obtained. Equations (3.3.89) and (3.3.90) will help us find the stress components σ_{rr}, σ_{tt} and σ_{rt}.

Because of the invariance of $\sigma_{11} + \sigma_{22} = \sigma_{rr} + \sigma_{tt}$ it follows from (3.3.89)

$$\sigma_{rr} + \sigma_{tt} = 4\, \mathscr{Re}\; [F'(z)]. \tag{3.3.91}$$

We find from (3.3.72)

$$\sigma_{tt}-\sigma_{rr}+2\mathrm{i}\sigma_{rt} = (\sigma_{22}-\sigma_{11}+2\mathrm{i}\sigma_{12})(\cos 2\alpha+\mathrm{i}\sin 2\alpha)$$
$$= (\sigma_{22}-\sigma_{11}+2\mathrm{i}\sigma_{12})\mathrm{e}^{2\mathrm{i}\alpha}. \tag{3.3.92}$$

Substituting (3.3.90) in (3.3.92), we obtain

$$\sigma_{tt}-\sigma_{rr}+2\mathrm{i}\sigma_{rt} = 2\,\mathrm{e}^{2\mathrm{i}\alpha}[\bar{z}F''(z)+G''(z)],\ z = r\,\mathrm{e}^{\mathrm{i}\alpha}. \tag{3.3.93}$$

(3.3.91) and (3.3.93) suffice in determining σ_{rr}, σ_{tt} and σ_{rt} when complex functions $F(z)$ and $G'(z)$ are known.

A particularly simple case is given in case of axial symmetry. Here, the quantities involved do not depend on α, and the biharmonic equation (3.3.77) is reduced to Euler's ordinary differential equation

$$\left(\frac{\mathrm{d}^2}{\mathrm{d}r^2}+\frac{1}{r}\frac{\mathrm{d}}{\mathrm{d}r}\right)^2\Phi = 0. \tag{3.3.94}$$

This equation admits the general solution

$$\Phi = C_1+C_2\ln r+C_3r^2\ln r \tag{3.3.95}$$

containing the constants of integration C_1 to C_4.

For the stresses we obtain from (3.3.82), when all derivations with respect to α are equalled to nought,

$$\sigma_{rr} = \frac{1}{r}\frac{\partial\Phi}{\partial r},\ \sigma_{tt} = \frac{\partial^2\Phi}{\partial r^2},\ \sigma_{rt} = 0. \tag{3.3.96}$$

For the strain components, (3.3.83) remains valid, while, however, this time ε_{rt} changes to nought. Relations (3.3.74), from which the displacements are obtained once the strain components are known, change to

$$\varepsilon_{rr} = \frac{\partial u_r}{\partial r},\ \varepsilon_{tt} = \frac{u_r}{r},\ \varepsilon_{zz} = \frac{\partial u_z}{\partial z} \tag{3.3.97}$$

because of the independence of all quantities of α. Through (3.3.95), the boundary conditions for determining the constants of integration, as well as through (3.3.96), (3.3.83), (3.3.84) and (3.3.97), respectively, we are able to establish completely the solution of planar axial symmetrical problems.

3.3.5 *Examples*

1. *Example.* Let us apply the semi-inverse method to the planar problem of a beam which is supported at both sides, and has a uniformly distributed load q (Figure 41). Let h be the beam's height, '1' the breadth, and l the length. Volume forces are omitted.

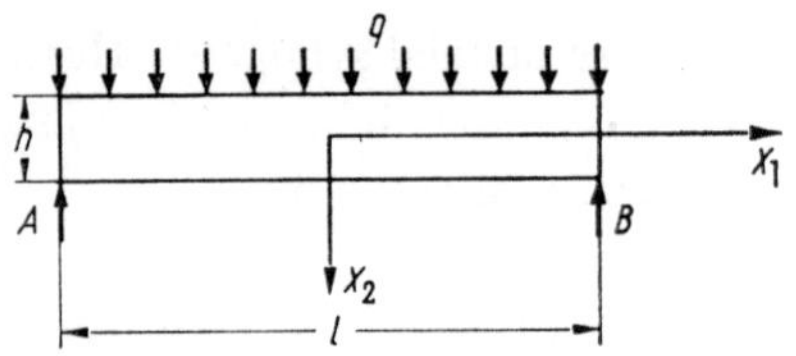

Figure 41

For the stress function Φ let us assume the polynomial

$$\Phi = ax_2^5 + bx_1^2x_2^3 + cx_1^4x_2 + dx_2^3 + ex_1^2x_2 + f(x_1). \tag{3.3.98}$$

Inserting (3.3.98) in the biharmonic equation (3.3.16), we find that

$$a = -\frac{b+c}{5}, \quad f^{\text{IV}}(x_1) = 0$$

must hold if we want to satisfy this equation. Hence, in particular, let us write

$$\begin{aligned} \Phi &= -\frac{b+c}{5}x_2^5 + bx_1^2x_2^3 + cx_1^4x_2 + dx_2^3 + ex_1^2x_2 + f(x_1), \\ f^{\text{IV}}&(x_1) = 0. \end{aligned} \tag{3.3.99}$$

From (3.3.15), for the stresses, we have

$$\begin{aligned} \sigma_{11} &= -4(b+c)x_2^3 + 6bx_1^2x_2 + 6dx_2, \\ \sigma_{22} &= 2bx_2^3 + 12cx_1^2x_2 + 2ex_2 + f''(x_1), \\ \sigma_{12} &= -(6bx_1x_2^2 + 4cx_1^3 + 2ex_1). \end{aligned} \tag{3.3.100}$$

The hitherto unknown constants, and function $f(x_1)$ are to be determined from the boundary conditions. They are

$$\begin{aligned} \sigma_{22} &= -q,\ \sigma_{12} = 0 \text{ for } x_2 = -h/2, \\ \sigma_{22} &= 0,\ \sigma_{12} = 0 \text{ for } x_2 = h/2 \end{aligned} \tag{3.3.101}$$

for the beam's upper and lower surface.

Through $\sigma_{22}(x_2 = h/2) = 0$ it follows from (3.3.100) that

$$\frac{bh^3}{4} + 6cx_1^2 h + eh + f''(x_1) = 0, \tag{3.3.102}$$

which can only be fulfilled for $c = 0$, $f''(x_1) = k = \text{const}$, if the requirement $f^{\text{IV}}(x_1) = 0$ is to be satisfied. Thus, for the stresses

$$\begin{aligned} \sigma_{11} &= 4bx_2^3 + 6bx_1^2 x_2 + 6dx_2, \\ \sigma_{22} &= 2bx_2^3 + 2ex_2 + k, \\ \sigma_{12} &= -6bx_1 x_2^2 - 2ex_1, \end{aligned} \tag{3.3.103}$$

and (3.3.102) changes to

$$\frac{bh^3}{4} + eh = K = 0. \tag{3.3.104}$$

From $\sigma_{22}(x_2 = -h/2) = -q$, and according to (3.3.103), it follows

$$-\frac{bh^3}{4} - eh + k = -q. \tag{3.3.105}$$

Adding (3.3.104) and (3.3.105) results in

$$k = -\frac{q}{2}, \tag{3.3.106}$$

and subtracting equation (3.3.105) from (3.3.104) yields

$$\frac{bh^3}{2} + 2eh = q. \tag{3.3.107}$$

The conditions for the shearing stress σ_{12} result both in

$$-\tfrac{3}{2}bh^2x_1 - 2ex_1 = 0, \; e = -\tfrac{3}{4}bh^2, \text{ respectively,} \tag{3.3.108}$$

by which from (3.3.107) and (3.3.108)

$$b = -\frac{q}{h^3}, \; e = \frac{3}{4}\frac{q}{h} \tag{3.3.109}$$

is obtained. Except for d, all constants have been determined, and we obtain from (3.3.103) for the stresses

$$\begin{aligned} \sigma_{11} &= 6dx_2 - \frac{6q}{h^3}x_1^2x_2 + \frac{4q}{h^3}x_2^3, \\ \sigma_{22} &= -\frac{6q}{h^3}\left(\frac{x_2^3}{3} - \frac{h^2}{4}x_2 + \frac{h^3}{12}\right), \\ \sigma_{12} &= -\frac{6q}{h^3}\left(\frac{h^2}{4} - x_2^2\right)x_1. \end{aligned} \tag{3.3.110}$$

The boundary conditions at both ends of the beam are yet to be fulfilled which are

$$\begin{aligned} &\sigma_{11} = 0 \text{ for } x_1 = \pm\frac{l}{2}, \\ &\int_{-h/2}^{+h/2} \sigma_{12}\, dx_2 = -\frac{ql}{2}. \end{aligned} \tag{3.3.111}$$

The second condition is automatically satisfied if σ_{12} from (3.3.110) is inserted. The first condition cannot be satisfied, since from (3.3.110) it follows that

$$\sigma_{11}(x_1 = \pm l/2) = 6dx_2 - \frac{3ql^2}{2h^3}x_2 + 4\frac{q}{h^3}x_2^3,$$

which is not identically equal to nought. Hence, against the original assumption we have made, there must be normal stresses σ_{11} present in the cross-sections of both ends of the beam. These give rise to the bending moment

$$M = \int_{-h/2}^{+h/2} \sigma_{11} x_2 \, dx_2 = \frac{dh^3}{2} - \frac{ql^2}{8} + \frac{qh^2}{20}. \tag{3.3.112}$$

From (3.3.112) for constant d:

$$d = \frac{2}{h^3}\left(M + \frac{ql^2}{8} - \frac{qh^2}{20}\right). \tag{3.3.113}$$

Thus, for the beam's stresses we obtain finally

$$\begin{aligned}
\sigma_{11} &= \frac{12}{h^3}\left(M + \frac{ql^2}{8} - \frac{qh^2}{20}\right) x_2 - \frac{6q}{h^3} x_1^2 x_2 + \frac{4q}{h^3} x_2^3, \\
\sigma_{22} &= -\frac{6q}{h^3}\left(\frac{x_2^3}{3} - \frac{h^2}{4} x_2 + \frac{h^3}{12}\right), \\
\sigma_{12} &= -\frac{6q}{h^3}\left(\frac{h^2}{4} - x_2^2\right) x_1 .
\end{aligned} \tag{3.3.114}$$

Let us compare (3.3.114) with the stress distribution given by the technical theory of bending for simply supported beams having a uniformly distributed load. It is as follows:

$$\begin{aligned}
\sigma_{11} &= \frac{6q}{h^3}\left(\frac{l^2}{4} - x_1^2\right) x_2, \quad \sigma_{22} = 0, \\
\sigma_{12} &= \frac{6q}{h^3}\left(\frac{h^2}{4} - x_2^2\right) x_1 .
\end{aligned} \tag{3.3.115}$$

Through comparison of (3.3.114) and (3.3.115), the following observation on the usefullness of the technical theory of bending can be made:

1. σ_{12} has already been correctly given.
2. σ_{22} is not included at all.
3. If $M \equiv 0$ is put down additionally in (3.3.114) – since in the case to which the technical theory of bending had been applied, no bending moments occur at the beam's ends – then expression σ_{11} for small x_2 and small h (so that x_2^3 and h^2 can be neglected) change to the expression given through (3.3.115) for σ_{11}. According to (3.3.114) stresses

$$\sigma_{11}^{*} = \frac{12}{h^3}\left(\frac{ql^2}{8} - \frac{qh^2}{20}\right) x_2 - \frac{3ql^2}{2h^3} x_2 + \frac{4q}{h^3} x_2^3$$

occur in the cross-sections at the beam's ends (for $M = 0$).

Since however,

$$\int_{-h/2}^{+h/2} \sigma_{11}^{*}\, \mathrm{d}x_2 = 0$$

is valid, then the distribution σ_{11}^{*} occurring in the cross-sections of the beam's ends will, because it is statically equivalent to a zero-system, hardly have any influence following the principle of Saint-Vénant in a sufficient distance from the beam's supports.

Let us finally say that at a distance sufficiently apart from the supports as well as for beams with a small height, the result given by the technical theory of bending for the normal stresses σ_{11} differs only slightly from that obtained by the exact calculation of the beam's plane stress problem.

2. *Example.* Let us discuss the distribution of stress in an infinite plate with a circular hole, of constant thickness '1', which is under tension. (Figure 42).

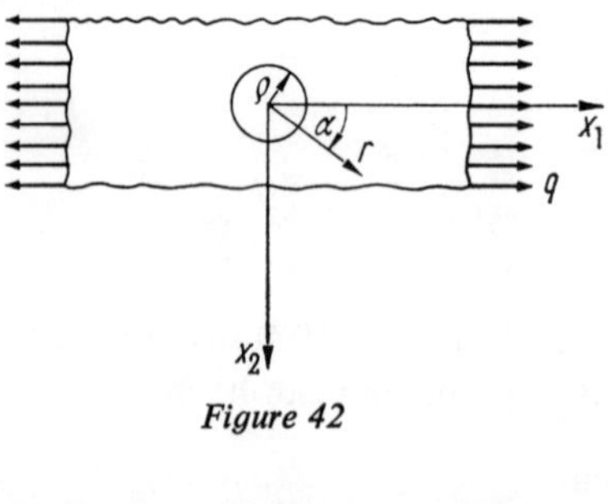

Figure 42

Considering the boundary conditions, let us work with polar coordinates, assuming for the stress function the form

$$\Phi = f_1(r) + f_2(r) \cos 2\alpha. \tag{3.3.116}$$

Substituting in (3.3.77), then for f_1 and f_2 the ordinary differential equations

$$\left(\frac{d^2}{dr^2}+\frac{1}{r}\frac{d}{dr}\right)^2 f_1 = 0, \left(\frac{d^2}{dr^2}+\frac{1}{r}\frac{d}{dr}-\frac{4}{r^2}\right)^2 f_2 = 0$$

are obtained, having the general solutions

$$\left.\begin{aligned} f_1 &= C_1 r^2 \ln r + C_2 r^2 + C_3 \ln r + C_4, \\ f_2 &= C_5 r^2 + C_6 r^4 + \frac{C_7}{r^2} + C_8. \end{aligned}\right\} \text{[see (3.3.94) and (3.3.95)} \tag{3.3.117}$$

Introducing (3.3.116), (3.3.117) in (3.3.82), it yields for the stiess distribution

$$\begin{aligned} \sigma_{rr} &= C_1(1+2\ln r)+2C_2+\frac{C_3}{r^2}-\left(2C_5+\frac{6C_7}{r^4}+\frac{4C_8}{r^2}\right)\cos 2\alpha, \\ \sigma_{tt} &= C_1(3+2\ln r)+2C_2-\frac{C_3}{r^2} \\ &\quad+\left(2C_5+12C_6 r^2+\frac{6C_7}{r^4}\right)\cos 2\alpha, \\ \sigma_{rt} &= \left(2C_5+6C_6 r^2-\frac{6C_7}{r^4}-\frac{2C_8}{r^2}\right)\sin 2\alpha. \end{aligned} \tag{3.3.118}$$

Those constants occurring in (3.3.118) are to be determined by the following boundary conditions:

1. stresses remain finite for $r \to \infty$,
2. for $r = \rho$, $\sigma_{rr} = \sigma_{tt} = 0$,

(3.3.119)

as well as by another condition for $r \to \infty$ yet to be mentioned.

From the first condition of (3.3.119) follows

$$C_1 = C_6 = 0. \tag{3.3.120}$$

From the second condition we obtain to begin with, because of $\sigma_{rr}(r = \rho)=0$,

$$2C_2+\frac{C_3}{\rho^2} = 0, \; 2C_5+\frac{6C_7}{\rho^4}+\frac{4C_8}{\rho^2} = 0, \tag{3.3.121}$$

and therefore, because of $\sigma_{rt}(r = \rho) = 0$,

$$2C_5 - \frac{6C_7}{\rho^4} - \frac{2C_8}{\rho^2} = 0. \tag{3.3.122}$$

Let us discuss the unperforated plate. Notating its stress function by Φ^*, then the following must be true

$$\sigma_{11}^* = \Phi_{,22}^* \equiv q. \tag{3.3.123}$$

From (3.3.123) we obtain through integration

$$\Phi^* = \tfrac{1}{2}qx_2^2 = \tfrac{1}{2}qr^2\sin^2\alpha = \tfrac{1}{4}qr^2 - \tfrac{1}{4}qr^2\cos 2\alpha. \tag{3.3.124}$$

From (3.3.82) for the stress distribution in the unperforated plate

$$\begin{aligned} \sigma_{rr}^* &= \tfrac{1}{2}q(1+\cos 2\alpha),\ \sigma_{tt}^* = \tfrac{1}{2}q(1-\cos 2\alpha), \\ \sigma_{rt}^* &= -\tfrac{1}{2}q\,(\sin 2\alpha) \end{aligned} \tag{3.3.125}$$

is obtained. For $r \to \infty$, the stress distribution (3.3.118) of the perforated plate must change to the distribution (3.3.125) of the unperforated plate in the limit. Then,

$$2C_5 = -\tfrac{1}{2}q,\ 2C_2 = \tfrac{1}{2}q. \tag{3.3.126}$$

From (3.3.121), (3.3.122) and (3.3.126),

$$C_2 = \tfrac{1}{4}q,\ C_3 = -\tfrac{1}{2}q\rho^2,\ C_5 = -\tfrac{1}{4}q,\ C_7 = -\tfrac{1}{4}q\rho^4,\ C_8 = \tfrac{1}{2}q\rho^2 \tag{3.3.127}$$

is finally obtained, and hence, together with (3.3.120), from (3.3.118)

$$\begin{aligned} \sigma_{rr} &= \frac{q}{2}\left\{\left[1-\left(\frac{\rho}{r}\right)^2\right] + \left[1+3\left(\frac{\rho}{r}\right)^4 - 4\left(\frac{\rho}{r}\right)^2\right]\cos 2\alpha\right\} \\ \sigma_{tt} &= \frac{q}{2}\left\{\left[1+\left(\frac{\rho}{r}\right)^2\right] - \left[1+3\left(\frac{\rho}{r}\right)^4\right]\cos 2\alpha\right\}, \\ \sigma_{rt} &= -\frac{q}{2}\left[1-3\left(\frac{\rho}{r}\right)^4 + 2\left(\frac{\rho}{r}\right)^2\right]\sin 2\alpha \end{aligned} \tag{3.3.128}$$

is determined for the stresses. The largest value $3q$ is assumed by σ_{tt} for $r = \rho$, $\alpha = \pm\pi/2$ as can be read off (3.3.128). Thus, $\sigma_{tt,\,max} = 3q$ represents triple value of that stress which, having equal load, would occur in the *unperforated* plate.

3. *Example.* Let us consider the stress distribution in an infinite plate of constant thickness '1' under tension, having an elliptic hole. (Figure 43). We shall deal this problem using complex variables and conformal mapping in order to demonstrate the importance of this method for this and similar problems.

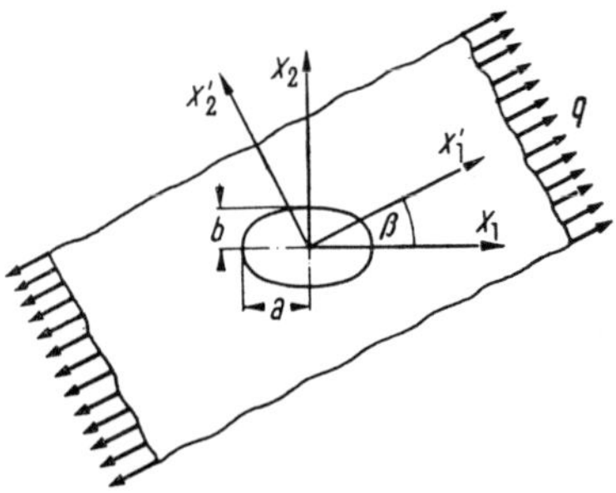

Figure 43

As shown in Figure 43, let the direction of the tension q together with the x_1-axis, which coincides with the major axis a of the elliptic hole, form angle β. Let us introduce the new axes system x_1', x_2' which has been turned about angle β. From the invariance of J_{1s} and fıom (3.3.92), we deduce relations

$$\begin{aligned} &\sigma_{11}'+\sigma_{22}' = \sigma_{11}+\sigma_{22}, \\ &\sigma_{22}'-\sigma_{11}'+2i\sigma_{12}' = (\sigma_{22}-\sigma_{11}+2i\sigma_{12})e^{i\beta} \end{aligned} \tag{3.3.129}$$

if we identify x_1' with r, x_2' with t and β with α. As a boundary condition it has to be demanded that

$$\sigma_{11}' = q,\ \sigma_{22}' = \sigma_{12}' = 0 \text{ at infinity}$$

is valid. From this, and because of (3.3.129), we obtain

$$\sigma_{11}+\sigma_{22} = q,\ \sigma_{22}-\sigma_{11}+2i\sigma_{12} = -q\,e^{-2i\beta}, \text{ at infinity}$$

and hence, because of (3.3.89) and (3.3.90),

$$4\,\mathscr{Re}\,[F'(z)] = q,\ 2[\bar{z}F''(z)+G''(z)] = -q\,\mathrm{e}^{-2i\beta},\ \text{at infinity.} \quad (3.3.130)$$

Let us map the ζ-plane, using conformal mapping

$$z = f(\zeta) = c\left(\zeta+\frac{m}{\zeta}\right),\quad c = \frac{a+b}{2},\quad m = \frac{a-b}{a+b}, \quad (3.3.131)$$

into the z-plane, whereby the unit-circle with contour $|\zeta| = 1$ changes to the elliptic hole. Because of the mapping (3.3.131),

$$\begin{aligned} &F(z) = F_1(\zeta),\ F'(z) = F_1'(\zeta)/f'(\zeta),\\ &F''(z) = [F_1''(\zeta)f'(\zeta)-F_1'(\zeta)f''(\zeta)][f'(\zeta)]^{-3},\\ &G(z) = G_1(\zeta),\ G'(z) = G_1'(\zeta)/f'(\zeta),\\ &G''(z) = [G_1''(\zeta)f'(\zeta)-G_1'(\zeta)f''(\zeta)][f'(\zeta)]^{-3} \end{aligned} \quad (3.3.132)$$

is obtained. Substituting this relation in (3.3.89) and (3.3.90),

$$\begin{aligned} &\sigma_{11}+\sigma_{22} = 4\,\mathscr{Re}\frac{F_1'(\zeta)}{f'(\zeta)},\\ &\sigma_{22}-\sigma_{11}+2i\sigma_{12} = \frac{2}{[f'(\zeta)]^3}[\overline{f(\zeta)}F_1''(\zeta)f'(\zeta)\\ &-\overline{f(\zeta)}F_1'(\zeta)f''(\zeta)+G_1''(\zeta)f'(\zeta)-G_1'(\zeta)f''(\zeta)] \end{aligned} \quad (3.3.133)$$

is the result.

To determine functions $F_1(\zeta)$, $G_1'(\zeta)$, we can use the boundary condition valid for the hole's contour. Since there is no stress at the hole's boundary, from (3.3.66),

$$F(z)+z\overline{F'(z)}+\overline{G'(z)} = 0,\ \text{for the hole's boundary,}$$

from which, because of the conformal mapping, i.e., thıough (3.3.132),

$$F_1(\zeta)+\frac{f(\zeta)}{\overline{f'(\zeta)}}\overline{F_1'(\zeta)}+\frac{\overline{G_1'(\zeta)}}{\overline{f'(\zeta)}} = 0,\ \text{on } |\zeta| = 1 \quad (3.3.134)$$

is set up. Thus we have arrived at a problem belonging to the *first fundamental problem of biharmonic functions.* The various possibilities of solving this problem are given in [15].

For evaluation purposes, (3.3.134) changes to

$$f'(\zeta)\overline{F_1(\zeta)}+\overline{f(\zeta)}F_1'(\zeta)+G_1'(\zeta) = 0 \text{ on } |\zeta| = 1. \tag{3.3.135}$$

To make the stresses unique, then

$$F_1'(\zeta) = \sum_{n=0}^{\infty} A_n\zeta^{-n}, \quad G_1''(\zeta) = \sum_{n=0}^{\infty} B_n\zeta^{-n} \tag{3.3.136}$$

must be true, so that

$$\begin{aligned} F_1(\zeta) &= A_0\zeta+A_1 \ln \zeta+\sum_{n=2}^{\infty} \frac{A_n\zeta^{-n+1}}{-n+1}+A, \\ G_1'(\zeta) &= B_0\zeta+B_1 \ln \zeta+\sum_{n=2}^{\infty} \frac{B_n\zeta^{-n+1}}{-n+1}+B. \end{aligned} \tag{3.3.137}$$

For the displacements we obtain from (3.3.65) with (3.3.132)

$$2\mu(u_1+\mathrm{i}u_2) = \frac{5\lambda+6\mu}{3\lambda+2\mu}F_1(\zeta)-f(\zeta)\frac{\overline{F_1'(\zeta)}}{\overline{f'(\zeta)}}-\frac{\overline{G_1'(\zeta)}}{\overline{f'(\zeta)}}. \tag{3.3.138}$$

Inserting (3.3.137) in (3.3.138), we realize that the displacements become unique if

$$\frac{5\lambda+6\mu}{3\lambda+2\mu}cA_1+\bar{B}_1 = 0 \tag{3.3.139}$$

is true. Constants A, B can be equalled to nought, since they determine a rigid body motion, which is of little interest to us.

Let be, $|\zeta| = 1$ then

$$\begin{aligned} &f'(\zeta) = c(1-m\,\mathrm{e}^{-2\mathrm{i}\alpha}), \quad \overline{f(\zeta)} = c(\mathrm{e}^{-\mathrm{i}\alpha}+m\,\mathrm{e}^{\mathrm{i}\alpha}), \\ &\overline{F_1(\zeta)} = \bar{A}_0\,\mathrm{e}^{-\mathrm{i}\alpha}-\mathrm{i}\bar{A}_1\,\alpha-\sum_{n=2}^{\infty}\frac{\bar{A}_n\,\mathrm{e}^{\mathrm{i}(n-1)\alpha}}{n-1}, \quad F_1'(\zeta) = \sum_{n=0}^{\infty} A_n\,\mathrm{e}^{-\mathrm{i}n\alpha}, \\ &G_1'(\zeta) = B_0\,\mathrm{e}^{\mathrm{i}\alpha}+\mathrm{i}B_1\,\alpha-\sum_{n=2}^{\infty}\frac{B_n\,\mathrm{e}^{-\mathrm{i}(n-1)\alpha}}{n-1}. \end{aligned} \tag{3.3.140}$$

Through (3.3.140) we get from (3.3.135)

$$c(1-m\,\mathrm{e}^{-2\mathrm{i}\alpha})\left[\bar{A}_0\,\mathrm{e}^{-\mathrm{i}\alpha}-\mathrm{i}\bar{A}_1\,\alpha-\sum_{n=2}^{\infty}\frac{\bar{A}_n\,\mathrm{e}^{\mathrm{i}(n-1)\alpha}}{n-1}\right]$$

$$+c(\mathrm{e}^{-\mathrm{i}\alpha}+m\,\mathrm{e}^{\mathrm{i}\alpha})[\sum_{n=0}^{\infty}A_n\,\mathrm{e}^{-\mathrm{i}n\alpha}]$$

$$+B_0\,\mathrm{e}^{\mathrm{i}\alpha}+\mathrm{i}B_1\,\alpha-\sum_{n=2}^{\infty}\frac{B_n\,\mathrm{e}^{-\mathrm{i}(n-1)\alpha}}{n-1}=0. \qquad (3.3.141)$$

When the coefficients of α and $\mathrm{e}^{\mathrm{i}n\alpha}$, occurring in (3.3.141), are made to equal nought, then the equations

$$\begin{aligned}
&-c\bar{A}_1+B_1=0,\ A_n=0 \text{ for } n\geqq 3,\\
&-c\bar{A}_2+\frac{cm}{3}\bar{A}_4+cmA_0+B_0=0,\\
&c\bar{A}_0+cm\bar{A}_2+cA_0+cmA_2-B_2=0, \qquad (3.3.142)\\
&cA_1+cmA_3-\frac{B_3}{2}=0,\\
&-cm\bar{A}_0+cA_2+cmA_4-\frac{B_4}{3}=0,\\
&B_n=0 \text{ for } n\geqq 5
\end{aligned}$$

for the determining of constants are obtained. Let $F_1'(\infty)=A_0$, $F_1''(\infty)=0$, $G_1''(\infty)=B_0$, $f'(\infty)=c$, $f''(\infty)=0$. Hence, from the boundary condition (3.3.130) and using (3.3.132),

$$4\frac{A_0}{c}=q,\ 2\frac{B_0}{c^2}=-q\,\mathrm{e}^{-2\mathrm{i}\beta}$$

is obtained, so that

$$A_0=\frac{cq}{4},\ B_0=-\frac{c^2q\,\mathrm{e}^{-2\mathrm{i}\beta}}{2} \qquad (3.3.143)$$

is determined.

The first equation of (3.3.142) is to be valid as well as (3.3.139). But this is only possible if

$$A_1 = B_1 = 0$$

is true. The remaining equations of (3.3.142) yield

$$A_2 = \frac{cq}{4}(m = 2\mathrm{e}^{2\mathrm{i}\beta}),\ B_2 = \frac{c^2q}{2}(1+m^2-2m\cos 2\beta),$$

$$B_3 = 0,\ B_4 = -\frac{3c^2q}{2}\mathrm{e}^{2\mathrm{i}\beta}$$

Accordingly, (3.3.136) changes to

$$F_1'(\zeta) = \frac{cq}{4} + \frac{cq}{4}(m-2\mathrm{e}^{2\mathrm{i}\beta})\zeta^{-2},$$

$$G_1''(\zeta) = -\frac{c^2q}{2}\mathrm{e}^{-2\mathrm{i}\beta} + \frac{c^2q}{2}(1+m^2-2m\cos 2\beta)\zeta^{-2}$$
$$-\frac{3c^2q}{2}\mathrm{e}^{2\mathrm{i}\beta}\zeta^{-4}.$$

Thus, from (3.3.133) for $|\zeta| = 1$

$$\begin{aligned}\sigma_{11}+\sigma_{22} &= 4\,\mathscr{Re}\frac{F_1'(\zeta)}{f'(\zeta)}\\ &= q[1-m^2-2\cos 2(\beta-\alpha)+2m\cos 2\alpha\cos 2(\beta-\alpha)\\ &\quad -2m\sin 2\alpha\sin 2(\beta-\alpha)]/(1+m^2-2m\cos 2\alpha).\end{aligned} \tag{3.3.144}$$

The greatest stress occurs for $\alpha = 0$, i.e., at the end of the major axis, when $\beta = \pi/2$. Since $\sigma_{11} = 0$ is true at this point, we obtain from (3.3.144)

$$\sigma_{22,\max} = \frac{3+m}{1-m}q = \left(1+2\frac{a}{b}\right)q. \tag{3.3.145}$$

For the circle $a = b$ holds true, so that in this case (3.3.145) refers back to the calculated value $3q$ for the peak stress in the second example. For the ellipse, the peak of stress may increase considerably from that of the circle, if $b \ll a$.

4. *Example.* Let us investigate the state of plane stress in a circular pipe under internal and external pressure. We are then dealing with a planar axial-symmetric problem.

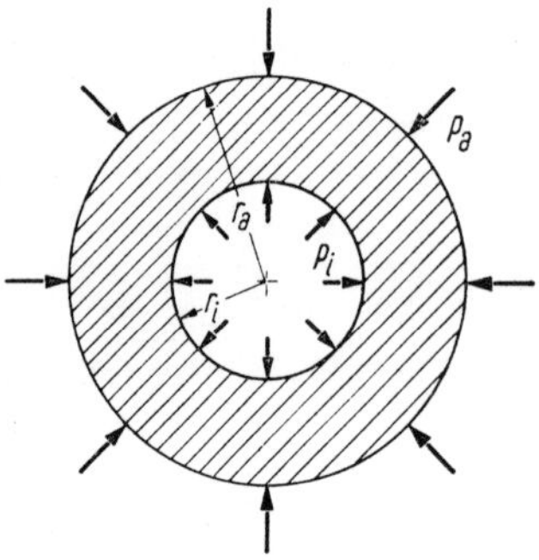

Figure 44

Let us refer back to (3.3.95) and (3.3.96), and obtain for the stresses

$$\sigma_{rr} = \frac{C_2}{r^2} + 2C_3 + C_4(1+2\ln r),$$

$$\sigma_{tt} = -\frac{C_2}{r^2} + 2C_3 + C_4(3+2\ln r), \quad \sigma_{rt} = 0. \tag{3.3.146}$$

As it is supposed to be a state of plane stress, therefore $\sigma_{zz} = 0$ also. We read off Figure 44, as a boundary condition,

$$\sigma_{rr}(r = r_i) = -p_i, \; \sigma_{rr}(r = r_a) = -p_a, \tag{3.3.147}$$

when p_i and p_a are to be taken as the magnitudes of the inside and outside pressure.

From (3.3.146) and (3.3.147), a possible solution for constants C_2 to C_4 is given by

$$\frac{C_2}{r_i^2} + 2C_3 = -p_i, \; \frac{C_2}{r_a^2} + 2C_3 = -p_a, \; C_4 = 0,$$

$$C_2 = \frac{r_a^2 r_i^2}{r_a^2 - r_i^2}(p_a - p_i), \; C_3 = \frac{p_i r_i^2 - p_a r_a^2}{2(r_a^2 - r_i^2)}, \; C_4 = 0, \tag{3.3.148}$$

respectively.

Thus from (3.3.146), the formulae known for thick-walled pipes are obtained:

$$\begin{aligned} \sigma_{rr} &= \frac{r_a^2 r_i^2}{r^2(r_a^2 - r_i^2)}(p_a - p_i) + \frac{p_i r_i^2 - p_a r_a^2}{r_a^2 - r_i^2}, \\ \sigma_{tt} &= -\frac{r_a^2 r_i^2}{r^2(r_a^2 - r_i^2)}(p_a - p_i) + \frac{p_i r_i^2 - p_a r_a^2}{r_a^2 - r_i^2}, \\ \sigma_{rt} &= \sigma_{zz} = 0. \end{aligned} \tag{3.3.149}$$

In the preceding equations we have notated $C_4 = 0$. The justification for doing so follows from a consideration of the displacements, since according to (3.3.83) and (3.3.146),

$$\varepsilon_{zz} = -\frac{\nu}{E}(\sigma_{rr} + \sigma_{tt}) = -\frac{4\nu}{E}[C_3 + C_4(1 + \ln r)]. \tag{3.3.150}$$

Then from $\varepsilon_{zz} = \partial u_z/\partial z$ by means of (3.3.150)

$$u_z = -\frac{4\nu}{E}[C_3 + C_4(1 + \ln r)]z + f(r) \tag{3.3.151}$$

is calculated, when $f(r)$ represents a function of integration. If no warping of the tube's cross-sections which were plane before the application of stress is to occur, then the displacement u_z may only be dependent on z but not on r. Warping does actually not occur so that $f(r) \equiv 0$ is required as well as $C_4 \equiv 0$. This had been done in the calculation of the constants in (3.3.146). Thus, the assumption with respect to C_4 was correct.

The remaining strains are also obtained from (3.3.83) and (3.3.146), and the displacements from (3.3.97). (3.3.151) is valid for u_z, which, together with $f(r) \equiv 0$, $C_4 \equiv 0$, and C_3 according to (3.3.148), yields

$$u_z = 2\frac{p_i r_i^2 - p_a r_a^2}{r_a^2 - r_i^2} z.$$

For $\varepsilon_{rr} = \partial u_r/\partial r$ and $\varepsilon_{tt} = u_r/r$ it follows from (3.3.83) and (3.3.146) that

$$\varepsilon_{rr} = \frac{1}{E}\left[(1+\nu)\frac{C_2}{r^2} + 2(1-\nu)C_3\right],$$

$$\varepsilon_{tt} = \frac{1}{E}\left[-(1+\nu)\frac{C_2}{r^2} + 2(1-\nu)C_3\right], \tag{3.3.152}$$

and from this, through integration of ε_{rr} or by multiplying ε_{tt} with r, accordingly

$$u_r = \frac{1}{E}\left[-(1+\nu)\frac{C_2}{r} + 2(1-\nu)C_3 r\right]. \tag{3.3.153}$$

In (3.3.152) and (3.3.153), the constants C_2 and C_3 are yet to be substituted by the expressions in (3.3.148).

3.4 Three-dimensional problems

3.4.1 *Stress functions*

Let the given problem be formulated in terms of stresses, and assume that no body forces are present. Then,

$$\nabla^2\sigma_{ij} + \frac{1}{1+\nu}\sigma_{kk,ij} = 0, \tag{3.4.1}$$

$$(\sigma_{ij} n_j)_S = p_i \tag{3.4.2}$$

has to be solved. (3.4.1) is the set of Beltrami's equations. (3.4.2) are the boundary conditions. The subscript S in (3.4.2) indicates that the quantity in brackets has to be taken on the surface S of the given body.

Moreover, the stresses have to satisfy the equilibrium conditions

$$\sigma_{ij,j} = 0. \tag{3.4.3}$$

Maxwell Operator

The Maxwell operator

$$M\Phi_{ij} = M\Phi_{ji} \tag{3.4.4}$$

is defined in the following way:

$$\begin{aligned} M\Phi_{11} &= \Phi_{22,33}+\Phi_{33,22}, & M\Phi_{12} &= -\Phi_{33,12}, \\ M\Phi_{22} &= \Phi_{33,11}+\Phi_{11,33}, & M\Phi_{23} &= -\Phi_{11,23}, \\ M\Phi_{33} &= \Phi_{11,22}+\Phi_{22,11}, & M\Phi_{31} &= -\Phi_{22,13}. \end{aligned} \tag{3.4.5}$$

In (3.4.5), the Φ_{11}, Φ_{22}, Φ_{33} are Maxwell's stress functions.

The operator M satisfies two important conditions, namely,

$$(M\Phi_{ij})_{,j} = 0, \tag{3.4.6}$$

and

$$M\Phi_{kk} = \nabla^2\Phi_{kk}-\Phi_{(kk),kk} \tag{3.4.7}$$

(no summation over repeated indices in brackets).

They can be verified by using (3.4.5) in (3.4.6) and (3.4.7).

Solution in terms of Maxwell's stress functions

The conditions (3.4.1), (3.4.2) and (3.4.3) have to be satisfied by any solution. Let

$$\sigma_{ij} = M\Phi_{ij}. \tag{3.4.8}$$

Due to (3.4.6),

$$\sigma_{ij,j} = (M\Phi_{ij})_{,j} = 0. \tag{3.4.9}$$

Hence, (3.4.3) is already satisfied.

Now consider (3.4.1). Let

$$\Phi_{ij} = H_{ij}+\delta_{ij}\Omega, \quad \Phi_{ij} = 0 \text{ for } i \neq j, \tag{3.4.10}$$

where the H_{ij} are harmonic, i.e.,

$$\nabla^2 H_{ij} = 0. \tag{3.4.11}$$

Furthermore, use (3.4.8) in (3.4.1). The result is

$$\nabla^2(M\Phi_{ij}) + \frac{1}{1+\nu}\sigma_{kk,ij} = 0. \tag{3.4.12}$$

First of all, let $i \neq j$. Then, (3.4.5) yields

$$M\Phi_{ij} = -\Phi_{(ll),ij},$$
(no summation with respect to l). (3.4.13)

Hence, (3.4.12) can be transformed to read

$$-\nabla^2\Phi_{(ll),ij} + \frac{1}{1+\nu}\sigma_{kk,ij} = 0.$$

Upon integration, and rearranging terms,

$$\nabla^2\Phi_{(ll)} = \frac{1}{1+\nu}\sigma_{kk}. \tag{3.4.14}$$

Using (3.4.8) for σ_{kk}, also

$$\nabla^2\Phi_{(ll)} = \frac{1}{1+\nu}M\Phi_{kk}.$$

Using (3.4.7), finally,

$$\nabla^2\Phi_{(ll)} = \frac{1}{1+\nu}(\nabla^2\Phi_{kk} - \Phi_{(kk),kk}). \tag{3.4.15}$$

Now (3.4.10), (3.4.11) and

$$\Omega_{,kk} = \nabla^2\Omega \tag{3.4.16}$$

can be used in (3.4.15) to yield

$$\nabla^2\Omega = \frac{1}{1+\nu}[3\nabla^2\Omega - (H_{11,11} + H_{22,22} + H_{33,33} + \nabla^2\Omega)],$$
$$(1-\nu)\nabla^2\Omega = H_{11,11} + H_{22,22} + H_{33,33} = H_{(kk),kk}. \tag{3.4.17}$$

(3.4.17) is the differential equation by means of which Ω can be determined. By virtue of (3.4.11), (3.4.17) yields immediately

$$\nabla^2\nabla^2\Omega = 0, \tag{3.4.18}$$

i.e., Ω is a biharmonic function.

Now, the case $i = j = l$ shall be taken into account in (3.4.12). The result is

$$\nabla^2(M\Phi_{(ll)} + \frac{1}{1+\nu}\sigma_{kk,(ll)} = 0, \tag{3.4.19}$$

(no summation with respect to l).

Since

$$\nabla^2(M\Phi_{(ll)}) \equiv \nabla^2[\nabla^2\Omega - \Omega,_{(ll)}],$$
$$\sigma_{kk,(ll)} \equiv -H_{(kk),kk(ll)} + 2\Omega,_{kk(ll)}$$

according to (3.4.5), (3.4.8), (3.4.10), and (3.4.11), equation (3.4.19) yields

$$\nabla^2[\nabla^2\Omega - \Omega,_{(ll)}] - \frac{1}{1+\nu}H_{(kk),\,kk(ll)} + \frac{2}{1+\nu}\Omega,_{kk(ll)} = 0. \tag{3.4.20}$$

It will be shown that (3.4.20) is identically satisfied under the assumption that Ω satisfies (3.4.17) and (3.4.18). Using (3.4.18) and $\Omega,_{kk} \equiv \nabla^2\Omega$ in (3.4.20),

$$-\nabla^2\Omega,_{(ll)} - \frac{1}{1+\nu}H_{(kk),\,kk(ll)} + \frac{2}{1+\nu}\nabla^2\Omega,_{(ll)} = 0 \tag{3.4.21}$$

results. Integrating, and rearranging terms yields

$$\frac{1-\nu}{1+\nu}\nabla^2\Omega - \frac{1}{1+\nu}H_{(kk),\,kk} = 0.$$

This is in fact an identity by virtue of (3.4.17).

The conclusion is, that (3.4.1) is satisfied by (3.4.10), (3.4.11) and (3.4.17), i.e., the stress functions can be determined using (3.4.10), (3.4.11) and (3.4.17).

The boundary value problem

By means of (3.4.8) and (3.4.10), the boundary conditions (3.4.2) can be rewritten to yield

$$[M(H_{ij}+\delta_{ij}\Omega)n_j]_S = p_i. \tag{3.4.22}$$

Taking (3.4.11) and (3.4.17) into account, the boundary value problem to be solved is given as follows:

$$\begin{aligned} &\nabla^2 H_{ij} = 0,\ (1-\nu)\nabla^2\Omega = H_{(kk),kk}, \\ &[M(H_{ij}+\delta_{ij}\Omega)n_j]_S = p_i. \end{aligned} \tag{3.4.23}$$

i) Let

$$H_{(kk),k} = \psi_k. \tag{3.4.24}$$

Then,

$$\nabla^2\psi_k = \nabla^2[H_{(kk),k}] = [\nabla^2 H_{(kk)}]_k = 0$$

due to (3.4.11), and obviously ψ_k is harmonic. For any harmonic function, the identity

$$\psi_{k,k} = \tfrac{1}{2}\nabla^2(x_k\psi_k) \tag{3.4.25}$$

holds. Hence,

$$(1-\nu)\nabla^2\Omega = H_{(kk),kk} = \psi_{k,k} = \tfrac{1}{2}\nabla^2(x_k\psi_k),$$

and upon integration

$$\Omega = H_{OO} + \frac{1}{2(1-\nu)}x_k\psi_k = H_{OO} + \frac{1}{2(1-\nu)}x_k H_{(kk),k}, \tag{3.4.26}$$

where H_{OO} is an arbitrary harmonic function.

Now (3.4.23) is reduced to

$$\begin{aligned} &\nabla^2 H_{ij} = 0, \\ &\left\{M\left[H_{ij}+\delta_{ij}\left(H_{OO}+\frac{1}{2(1-\nu)}x_k H_{(kk),k}\right)\right]n_j\right\}_S = p_i. \end{aligned} \tag{3.4.27}$$

If H_{11}, H_{22}, H_{33} and H_{OO} are obtained from (3.4.27), the stress functions $\Phi_{11}, \Phi_{22}, \Phi_{33}$ can be calculated, using (3.4.10) and (3.4.26), and the stresses can be evaluated by means of (3.4.8) and (3.4.5).

ii) Let

$$\begin{aligned} H_{11} &= H_{11}(x_2, x_3), \\ H_{22} &= H_{22}(x_1, x_3), \\ H_{33} &= H_{33}(x_1, x_2). \end{aligned} \tag{3.4.28}$$

Then, $H_{(kk),\,kk} = 0$.

Moreover, let

$$[M(H_{ij})n_j]_S = p_i. \tag{3.4.29}$$

Then, from (3.4.23) it follows that

$$[M(\delta_{ij}\Omega)n_j]_S = 0 \tag{3.4.30}$$

has to be satisfied. Thus, (3.4.23) can be replaced by

$$\begin{aligned} &\nabla^2 H_{ij} = 0, \;\; [M(H_{ij})n_j]_S = p_i, \\ &\nabla^2 \Omega = 0, \;\; [M(\delta_{ij}\Omega)n_j]_S = 0. \end{aligned} \tag{3.4.31}$$

In (3.4.31), the equations are decoupled with respect to $H_{(kk)}$ and Ω.

(3.4.30) is obviously satisfied if $(\Omega)_S = \Omega_O = \text{const}$. Hence, (3.4.31) can be replaced by

$$\begin{aligned} &\nabla^2 H_{ij} = 0, \;\; [M(H_{ij})n_j]_S = p_i, \\ &\nabla^2 \Omega = 0, \;\; (\Omega)_S = \Omega_O = \text{const.}, \end{aligned} \tag{3.4.32}$$

which has to be solved for $H_{(kk)}$ and for Ω. In (3.4.32), $\Omega_O \equiv 0$ is included as a special case.

Displacements

Consider (3.4.14). It can be rewritten to yield

$$\sigma_{kk} = (1+\nu)\nabla^2 \Phi_{(ll)}.$$

Assume $l = 1$. Then,

$$\sigma_{kk} = (1+\nu)\nabla^2\Phi_{11} = (1+\nu)(\Phi_{11,11}+\Phi_{11,22}+\Phi_{11,33}). \tag{3.4.33}$$

From (3.4.5) and (3.4.8),

$$\Phi_{11,22} = \sigma_{33}-\Phi_{22,11}, \; \Phi_{11,33} = \sigma_{22}-\Phi_{33,11} \tag{3.4.34}$$

is obtained. Using (3.4.34) in (3.4.33),

$$\sigma_{kk} = (1+\nu)[\Phi_{11,11}+\sigma_{33}-\Phi_{22,11}+\sigma_{22}-\Phi_{33,11}] \tag{3.4.35}$$

results. Rearranging (3.4.35) and using Hooke's Law,

$$Eu_{1,1} = \sigma_{kk}-(1+\nu)(\sigma_{33}+\sigma_{22}) = (1+\nu)[\Phi_{11}-\Phi_{22}-\Phi_{33}]_{,11} \tag{3.4.36}$$

results. Upon integration,

$$u_1 = \frac{1+\nu}{E}[\Phi_{11}-\Phi_{22}-\Phi_{33}]_{,1}. \tag{3.4.37}$$

Using (3.4.10) in (3.4.37), the solution

$$u_1 = \frac{1+\nu}{E}[H_{11}-H_{22}-H_{33}-\Omega]_{,1} \tag{3.4.38}$$

of the displacement u_1 is obtained.

Solution of u_2, u_3 follow from (3.4.38) through cyclic replacement of indices. Thus:

$$\begin{aligned} u_2 &= \frac{1+\nu}{E}[H_{22}-H_{33}-H_{11}-\Omega]_{,2}, \\ u_3 &= \frac{1+\nu}{E}[H_{33}-H_{11}-H_{22}-\Omega]_{,3}. \end{aligned} \tag{3.4.39}$$

In case of (3.4.26), relation (3.4.38) reads

$$u_1 = \frac{1+\nu}{E}\left[H_{11}-H_{22}-H_{33}-H_{oo}-\frac{1}{2(1-\nu)}x_k H_{(kk),k}\right]_{,1}. \tag{3.4.40}$$

Corresponding expressions are obtained for u_2, u_3 using (3.4.26) in (3.4.39).

3.4.2 *Betti's method*

When using the method given by Betti, we start out with the Lamé-Navier equations (3.1.9*). Because of grad div $\boldsymbol{u}$ = curl curl $\boldsymbol{u}+\nabla^2\boldsymbol{u}$, they are changed to

$$\frac{\lambda+2\mu}{\mu}\nabla^2\boldsymbol{u}+\frac{\lambda+\mu}{\mu}\,\mathrm{curl\ curl}\ \boldsymbol{u} = 0, \tag{3.4.41}$$

when for the volume forces $\boldsymbol{K} \equiv 0$ is being assumed. Let us set up the divergence of (3.4.41). Because of div curl curl $\boldsymbol{u} = 0$, this yields

$$\frac{\lambda+2\mu}{\mu}\nabla^2\,\mathrm{div}\ \boldsymbol{u} = 0,\ \ \nabla^2\,\mathrm{div}\ \boldsymbol{u} = 0,\ \text{respectively,} \tag{3.4.42}$$

when the order of differentiation has been changed. But $\Theta = u_{i,\,i} = \mathrm{div}\ \boldsymbol{u}$, so that

$$\nabla^2\Theta = 0 \tag{3.4.43}$$

is obtained, from which follows that Θ represents a harmonic function. According to (3.3.56),

$$\nabla^2(x_i\Theta) = 2\Theta_{,i},\ \Theta_{,i} = \tfrac{1}{2}\nabla^2(x_i\Theta) \tag{3.4.44}$$

is valid. According to (3.1.9), the Lamé-Navier equations change for the vanishing volume forces into

$$\nabla^2 u_i+\frac{\lambda+\mu}{\mu}\Theta_{,i} = 0.$$

This also means that because of (3.4.44)

$$\nabla^2\left(u_i+\frac{\lambda+\mu}{2\mu}x_i\Theta\right) = 0. \tag{3.4.45}$$

Equation (3.4.45) is satisfied, when

$$u_i + \frac{\lambda+\mu}{2\mu} x_i \Theta = h_i \tag{3.4.46}$$

represents a harmonic function.

Let us assume that on the surface O of the body, displacements $u_i(O)$ are specified, and that it would be possible to calculate from these, $\Theta(O)$ for the surface. Then, following (3.4.46), when $u_i(O)$ is known, $h_i(O)$ is a known function for the surface of the body. Through (3.4.45), and (3.4.46), and the boundary conditions, we arrive at Dirichlet's problem

$$\nabla^2 h_i = 0, \; h_i(O) = h_{io} \tag{3.4.47}$$

of the *potential theory*. Once this problem has been solved, and h_i has been determined, we obtain from (3.4.46)

$$u_i = h_i - \frac{\lambda+\mu}{2\mu} x_i \Theta. \tag{3.4.48}$$

Knowing the displacements u_i according to (3.4.48) would be possible, were it also possible to calculate in addition to the known h_i, the calculation of which was based on (3.4.47), function $\Theta = u_{i,i}$ from the given boundary values $u_i(O)$, not only for surface O, but even for all points inside the body. Let us now discuss this extra problem.

For this purpose, *Betti's theorem* has to be used, which states: If an elastic body having vanishing volume forces is exposed to the effects of surface forces systems p_i and p_i^*, which produce the displacements u_i and u_1^*, then for its surface O^*,

$$\int_{O^*} (p_i u_i^* - p_i^* u_i) \mathrm{d}O^* = 0 \tag{3.4.49}$$

is valid. This statement will be proven later on. Let us assume specially

$$u_i^* = (r^{-1})_{,i}, \; r = (x_i x_i)^{\frac{1}{2}}. \tag{3.4.50}$$

This solution of the Lamé-Navier equation shows a singularity at the origin of the coordinate system. The surface forces

$$p_i^* = 2\mu n_j (r^{-1})_{,ij} \tag{3.4.51}$$

correspond to it. Surrounding the origin of the system with a small sphere, having surface Ω and radius tending towards nought, we notate

$$\int_{\Omega}(p_i u_i^* - p_i^* u_i)\mathrm{d}\Omega = \int_{O}(p_i^* u_i - p_i u_i^*)\mathrm{d}O \tag{3.4.52}$$

instead of (3.4.49), when we take $O^* = \Omega + O$, whereby the outer body surface is O. On Ω, $n_j = -x_j/r$ is valid, so that with $p_i = \sigma_{ij}n_j$ and σ_{ij} according to (3.1.3a),

$$\int p_i u_i^* \mathrm{d}\Omega = -\int_{\Omega} \frac{x_j}{r}(\lambda\delta_{ij}\varepsilon_{kk} + 2\mu\varepsilon_{ij})\frac{\partial}{\partial x_i}\left(\frac{1}{r}\right)\mathrm{d}\Omega.$$

But $\varepsilon_{kk} = \Theta$, which yields

$$\int_{\Omega} p_i u_i^* \mathrm{d}\Omega = \int_{\Omega}(\lambda\delta_{ij}\Theta + 2\mu\varepsilon_{ij})\frac{x_i x_j}{r^4}\mathrm{d}\Omega,$$

after completing the operation $\partial(r^{-1})\partial x_i$. Admitting $r \to 0$ in the process of integration, and considering that

$$\int_{\Omega} x_i x_j \mathrm{d}\Omega = \tfrac{4}{3}\pi R^4 \delta_{ij}, \int_{\Omega}\mathrm{d}\Omega = 4\pi R^2,$$

then,

$$\int_{\Omega} p_i u_i^* \mathrm{d}\Omega = 4\pi\lambda\Theta_0 + 2\mu\varepsilon_{ij0}\tfrac{4}{3}\pi\delta_{ij} = 4\pi(\lambda + \tfrac{2}{3}\mu)\Theta_0 \tag{3.4.53}$$

is obtained. In (3.4.53), Θ_0 *represents the value of* Θ *at the origin of the chosen axis system.* Because of (3.4.51), we obtain for

$$\int_{\Omega} p_i^* u_i \mathrm{d}\Omega = -4\mu\int_{\Omega}\frac{x_i u_i}{r^4}\mathrm{d}\Omega.$$

Inserting $u_i = u_{i0} + x_j(u_{i,j})_0$ into this integral, then

$$\int_{\Omega} p_i^* u_i \mathrm{d}\Omega = -4\mu\int_{\Omega}\frac{x_i u_{i0} + x_i x_j (u_{i,j})_0}{r^4}\mathrm{d}\Omega,$$

and correspondingly, for $r \to 0$,

$$\int_{\Omega} p_i^* u_i \, d\Omega = -\frac{16}{3}\pi\mu(u_{i,i})_0 = -\frac{16}{3}\pi\mu\Theta_0 . \tag{3.4.54}$$

Through (3.4.54) and (3.4.53) from (3.4.52)

$$\Theta_0 = \frac{1}{4\pi(\lambda+2\mu)}\int_O (p_i^* u_i - p_i u_i^*) \, dO \tag{3.4.55}$$

is established.

Let us eliminate in (3.4.55), the p_i, since on O the displacements and not the surface forces were supposed to be given. Let us find functions G_i which analogous to Green's functions satisfy the following conditions:

a) The G_i satisfy the Lamé-Navier equations everywhere except at the origin of the axis systems.

b) Let $G_i = 0$ be valid on O.

c) At the origin, the G_i become infinite in the same manner as (3.4.50). When the G_i have been determined, the $u_i^{**} = u_i^* - G_i$ are set up. Because of b), $u_i^{**} \equiv u_i^*$ is on O. For this reason, as well as for Betti's theorem,

$$\int_O p_i^{**} u_i \, dO = \int_O p_i u_i^{**} dO = \int_O p_i u_i^* dO \tag{3.4.56}$$

is valid, whereby p_i^{**} are the surface forces corresponding to u_i^{**}. They are calculable if the u_i^{**} are given and known.

Through (3.4.56) and (3.4.55) we finally arrive at

$$\Theta_0 = \frac{1}{4\pi(\lambda+2\mu)}\int_O (p_i^* - p_i^{**}) u_i \, dO. \tag{3.4.57}$$

The known quantities p_i^*, p_i^{**} and u_i, which are given by the boundary conditions on O, are on the right side. Hence, Θ_0 can be calculated. Since the positioning of the origin in the body was arbitrary, then its position can be assumed at any point, and using (3.4.57), Θ is calculated from $u_i(O)$ for any given point in the body. This was the aim of the auxiliary calculation. Since Θ has been determined from the boundary values $u_i(O)$ for any point of the body, then the displacements u_i can be found from (3.4.48).

In this case we have limited the problem to the fact that the displacements have been given for surface O. When other boundary conditions are present, and surface forces are given on O, then the problem is solved by a calculation of a similar kind.

3.4.3 *Solutions using special functions*

Let us discuss Lamé-Navier equations once again in this paragraph. We shall try to find solutions for displacements u_i which consist of certain *arbitrary* functions, and which will satisfy the system of equations despite all arbitrariness involved. Solutions of this kind are called 'general solutions', because they can conform to the respective boundary conditions owing to the arbitrary functions involved. In reality, however, they represent only particular solutions of the equation system.

Fundamental observations dealing with 'general solutions' were first made by Betti. Let us follow these, at the same time assuming the volume forces to vanish. Then, because of (3.1.9*), there is

$$\text{grad div}\,\boldsymbol{u} + \frac{\mu}{\lambda+\mu}\nabla^2\boldsymbol{u} = 0.$$

Taking the curl of this expression, then because of curl grad div $\boldsymbol{u} \equiv 0$, we arrive at curl $\nabla^2 u = 0$. This is followed by

$$\nabla^2\boldsymbol{u} = \text{grad}\,h,\; u_{i,kk} = h_{,i}, \text{ respectively,} \tag{3.4.58}$$

whereby h must be a harmonic function, because of (3.1.13).

From (3.1.8), i.e., from $\nabla^2\nabla^2 u_i = 0$, it follows that u_i must be biharmonic. We may assume, for example, for u_i the third type of solution belonging to the biharmonic equation, then

$$u_i = \varphi_i + x_i\psi. \tag{3.4.59}$$

Functions φ_i and ψ are assumed to be harmonic, then (3.4.59) should be made to agree with (3.4.58). We find $u_{i,kk}$ from (3.4.59), equating it with (3.4.58) yields

$$u_{i,kk} = \varphi_{i,kk}+(x_i\psi)_{,kk} = h_{,i}. \tag{3.4.60}$$

By assumption, $\varphi_{i,kk} \equiv 0$, because φ_i are harmonic, and according to (3.3.56), $(x_i\psi)_{,kk} = 2\psi_{,i}$ is true. Therefore, (3.4.60) yields $2\psi_{,i} = h_{,i}$, $2\psi = h$, respectively. These are in agreement, since both ψ and h should be arbitrary and harmonic. We can choose the functions in such a way that $2\psi = h$ and, hence, (3.4.60) can be satisfied.

Let us check whether (3.1.13) is being satisfied. Accordingly, $u_{i,kk} = 0$ should be true. This will, in fact, be fulfilled by our assumption (3.4.59), since we have just found $u_{i,kk} = 2\psi_{,i} = h_{,i}$. Hence, $u_{i,ikk} = 2\psi_{,ii} = h_{,ii}$ is true. But since ψ and h have been assumed to be harmonic, then in actual fact $u_{i,ikk} = 2\psi_{,ii} = h_{,ii} = 0$, is also true. This satisfies condition (3.1.43).

The four functions in (3.4.59) are required to fulfill an additional condition with regard to (3.1.10). If (3.4.59) is inserted in (3.1.10), then we obtain

$$\varphi_{j,ii}+(x_j\psi)_{,ii}+\frac{\lambda+\mu}{\mu}[\varphi_{k,kj}+(x_k\psi)_{,kj}] = 0. \tag{3.4.61}$$

By assumption, $\varphi_{j,ii} = 0$ is true, and following (3.3.56), $(x_j\psi)_{,ii} = 2\psi_{,j}$ is true. We obtain for $(x_k\psi)_{,kj}$

$$(x_k\psi)_{,kj} = (\delta_{kk}\psi+x_k\psi_{,k})_{,j} = (3\psi+x_k\psi_{,k})_{,j} = 3\psi_{,j}+(x_k\psi_{,k})_{,j}.$$

Inserting all this in (3.4.61), sorting and integrating it, we then arrive at

$$\frac{5\mu+3\lambda}{\lambda+\mu}\psi+\varphi_{k,k}+x_k\psi_{,k} = \text{const.}, \tag{3.4.62}$$

which is the condition we mentioned above.

Let us finally say the following: 'General solutions' of the Lamé-Navier equation are set up through harmonic functions φ_i ($i = 1, 2, 3$) and ψ in the form of (3.4.59), when the chosen functions satisfy the condition (3.4.62).

From Betti's solution another 'general solution' by Kelvin is derived: Let F be harmonic. Then $F_{,1}$ is harmonic, which can easily be proven. Thus, let us choose foı (3.4.59), that $\psi = F_{,1}$. It will be shown now that $x_i\psi - x_1 F_{,i}$ is harmonic. For this purpose,

$$(x_1 F_{,i})_{,kk} = (\delta_{1k} F_{,i} + x_1 F_{,ik})_{,k} = \delta_{1k} F_{,ik} + \delta_{1k} F_{,ik}$$
$$+ x_1 F_{,ikk} = 2F_{,i1} \tag{3.4.63}$$

is calculated, since $F_{,ikk}$ is equal to nought by definition. Equally, following (3.3.56)

$$(x_i F_{,1})_{,kk} = 2F_{,1i} \tag{3.4.64}$$

is true. Because of (3.4.63) and (3.4.64),

$$(x_i \psi - x_1 F_{,i})_{,kk} = 2F_{,1i} - 2F_{,i1} = 0$$

is notated, which indicates that, as we have stated, $x_i \psi - x_1 F_{,i}$, is harmonic. If H_i are also chosen to be harmonic, then for (3.4.59), $\varphi_i = H_i - (x_i \psi - x_1 F_{,i})$ is put specially. This yields the *Kelvin solution of the first kind*:

$$u_i = H_i + x_1 F_{,i}. \tag{3.4.65}$$

A condition resulting from (3.1.10) connects the four harmonic functions H_i and F. Inserting (3.4.65) into this equation, then, because of $H_{j,ii} \equiv 0$ and (3.4.63),

$$2F_{,j1} + \frac{\lambda + \mu}{\mu} [H_{k,kj} + (x_1 F_{,k})_{,kj}] = 0$$

is obtained. But $(x_1 F_{,k})_{,kj} = (\delta_{1k} F_{,k} + x_1 F_{,kk})_{,j} = F_{,1j}$ holds true, because by definition, $F_{,kk} \equiv 0$ is true. Then, the condition is changed to

$$\frac{\lambda + 3\mu}{\lambda + \mu} F_{,j1} + H_{k,kj} = 0,$$

and finally, through integration, to

$$\frac{\lambda + 3\mu}{\lambda + \mu} F_{,1} + H_{k,k} = \text{const.}. \tag{3.4.66}$$

Thus, the Kelvin solution of the first kind is determined when, as specified in (3.4.65), we set up u_i using the four harmonic functions H_i and F, which satisfy condition (3.4.66).

One of the solutions found by Boussinesq is obtained through specialization of Kelvin's solution (3.4.65). We notate

$$H_i = f_{i,1}, \tag{3.4.67}$$

when f_i are supposed to be harmonic functions so that $f_{i,1}$ and H_i are harmonic functions too.

Also (3.4.66) has to be considered. Inserting (3.4.67) it yields

$$\frac{\lambda+3\mu}{\lambda+\mu} F_{,1} + f_{k,k1} = 0,$$

when the arbitrary constant contained in $F_{,i}$ is properly chosen.

This is followed by

$$\frac{\lambda+3\mu}{\lambda+\mu} F + f_{k,k} = 0, \quad F = -\frac{\lambda+\mu}{\lambda+3\mu} f_{k,k}, \text{ respectively,} \tag{3.4.68}$$

which we obtain through integration, and by determining the unknown functions of integration properly.

The *Boussinesq solution of the first kind*

$$u_i = f_{i,1} - \frac{\lambda+\mu}{\lambda+3\mu} x_1 f_{k,ki} \tag{3.4.69}$$

is obtained from (3.4.65), through (3.4.67) and (3.4.68).
Let us discuss a solution of the homogeneous equation of Lamé-Navier, which was first found by Papkovich and Neuber. Let us represent (3.1.10), using terms $u_{k,k} = \Theta$, $u_{j,ii} = \nabla^2 \boldsymbol{u}$, as

$$\mu \nabla^2 \boldsymbol{u} + (\lambda+\mu) \operatorname{grad} \Theta = 0. \tag{3.4.70}$$

For the solution we start by using a scalar function ψ and a vector $\boldsymbol{A}$ which yields

$$\boldsymbol{u} = \frac{1}{\lambda+\mu} \operatorname{grad} \psi + \frac{1}{\mu} \operatorname{curl} \boldsymbol{A},$$

which, in turn, yields

$$\operatorname{div} \boldsymbol{u} = \Theta = \frac{1}{\lambda+\mu} \operatorname{div} \operatorname{grad} \psi = \frac{1}{\lambda+\mu} \nabla^2\psi. \tag{3.4.71}$$

Inserting (3.4.71) in (3.4.70),

$$\nabla^2(\mu\boldsymbol{u}+\operatorname{grad}\psi) = 0,\ \nabla^2(\mu u_1+\psi_{,i}) = 0, \text{ respectively,}$$

is obtained, which is followed by:

$$\mu u_i+\psi_{,i} = h_i, \tag{3.4.72}$$

when h_i must be harmonic. Through further differentiation, from (3.4.72)

$$\mu u_{i,i}+\psi_{,ii} = h_{i,i},\ \mu\Theta+\nabla^2\ \psi = \operatorname{div} h_i, \text{ respectively,} \tag{3.4.73}$$

is obtained. Let us use (3.4.71) for (3.4.73), which yields

$$\left(\frac{\mu}{\lambda+\mu}+1\right)\nabla^2\psi = \operatorname{div} h_i,\ \nabla^2\psi = \frac{\lambda+\mu}{\lambda+2\mu}\, h_{i,i}, \tag{3.4.74}$$

respectively.

But according to (3.3.56), $h_{i,i} = \nabla^2(x_i h_i)/2$, so that (3.7.74) is changed to

$$\nabla^2\psi = \frac{\lambda+\mu}{2(\lambda+2\mu)} \nabla^2(x_i h_i).$$

We conclude from this relation that

$$\psi = \frac{1}{2}\,\frac{\lambda+\mu}{\lambda+2\mu}\, x_i h_i+h_0 \tag{3.4.75}$$

must be true. Then h_0 is an arbitrary harmonic function.

The 'general solution' by Papkovich-Neuber is easily obtained from (3.4.72) and (3.4.75). It states

$$\mu u_i = h_i-h_{0,i}-\frac{1}{2}\,\frac{\lambda+\mu}{\lambda+2\mu}(x_j h_j)_{,i}, \tag{3.4.76}$$

and contains the four arbitrary harmonic functions h_i and h_0. Studies by Eubanks and Sternberg [23] have been made concerning their completeness.

In (3.4.76), as a special case, we find a solution of the second kind as given by Boussinesq: Let us choose specially $h_i \equiv 0$. Then,

$$\mu u_i = -h_{0,i} \tag{3.4.77}$$

represents this special solution. From it follows $\mu u_{i,i} = -h_{0,ii} \equiv 0$, i.e., $u_{i,i} = \Theta \equiv 0$ holds true. This arises from the fact that $h_{0,ii} = \nabla^2 h_0 \equiv 0$, since h_0 had been assumed as harmonic. For the stresses let us calculate, using (3.1.3a), $\varepsilon_{kk} \equiv u_{k,k} = 0$ and $2\varepsilon_{ij} = u_{,ij} + u_{j,i}$, from (3.4.77)

$$\sigma_{ij} = -2h_{0,ij}. \tag{3.4.78}$$

Another special case is derived from (3.4.76), and Mindlin was first in pointing out this particular possibility. Let us notate by using an arbitrary function F_i for the arbitrary harmonic function h_i:

$$h_i = \frac{\lambda+2\mu}{\lambda+\mu} \nabla^2 F_i. \tag{3.4.79}$$

Because of $\nabla^2 h_i = 0$ we can read off

$$\nabla^2 h_i = \frac{\lambda+2\mu}{\lambda+\mu} \nabla^2 \nabla^2 F_i = 0$$

from (3.4.79), i.e., F_i must be harmonic. Inserting (3.4.79) in (3.4.74),

$$\nabla^2 \psi = \nabla^2 F_{i,i}$$

is obtained, which is satisfied by

$$F_{i,i} = \psi. \tag{3.4.80}$$

Through (3.4.79) and (3.4.80), (3.4.72) changes to

$$\mu u_i = \frac{\lambda+2\mu}{\lambda+\mu} \nabla^2 F_i - F_{j,ji} \tag{3.4.81}$$

following slight changes. This represents a solution which was found by Galerkin, independently of Papkovich-Neuber.

When seeking a solution for the inhomogeneous Lamé-Navier equation, then the previously mentioned 'general solutions' of the homogeneous equation and a particular integral of the inhomogeneous equation have to be added to yield that solution. For this reason we are interested in finding such a particular solution, for example following Kelvin's approach. We shall start from (3.1.9*), assuming, however, that the volume forces are of the form

$$\boldsymbol{K} = \operatorname{grad} \Phi + \operatorname{curl} \boldsymbol{F}. \tag{3.4.82}$$

In (3.4.82), Φ is a scalar and $\boldsymbol{F}$ a vector. Then we start by notating for the solution

$$\boldsymbol{u} = \operatorname{grad} \varphi + \operatorname{curl} \boldsymbol{v}, \tag{3.4.83}$$

which contains scalar φ and vector $\boldsymbol{v}$. It is no difficulty changing (3.1.9*) into

$$(\lambda+2\mu) \operatorname{grad} \operatorname{div} \boldsymbol{u} - \mu \operatorname{curl} \operatorname{curl} \boldsymbol{u} + \boldsymbol{K} = 0$$

because of $\nabla^2 \boldsymbol{u} = \operatorname{grad} \operatorname{div} \boldsymbol{u} - \operatorname{curl} \operatorname{curl} \boldsymbol{u}$. Once (3.4.82) and (3.4.83) have been inserted, then, because of

$$\operatorname{grad} \operatorname{div} \boldsymbol{u} = \operatorname{grad} \nabla^2 \varphi, \ \operatorname{curl} \operatorname{curl} \boldsymbol{u} = -\operatorname{curl} \nabla^2 \boldsymbol{v},$$

we obtain

$$\operatorname{grad} [(\lambda+2\mu)\nabla^2\varphi + \Phi) + \operatorname{curl} [\mu\nabla^2\boldsymbol{v} + \boldsymbol{F}] = 0.$$

Thus, a particular solution of the Lamé-Navier equation is obtained when particular solutions of

$$(\lambda+2\mu)\nabla^2\varphi + \Phi = 0, \ \mu\nabla^2\boldsymbol{v} + \boldsymbol{F} = 0 \tag{3.4.84}$$

are sought. These solutions are

$$\varphi(\boldsymbol{r}) = \frac{1}{4\pi(\lambda+2\mu)} \int \frac{\Phi(\boldsymbol{r}^*)\,\mathrm{d}V^*}{|\boldsymbol{r}-\boldsymbol{r}^*|}, \ \boldsymbol{v}(\boldsymbol{r}) = \frac{1}{4\pi\mu} \int \frac{\boldsymbol{F}(\boldsymbol{r}^*)\,\mathrm{d}V^*}{|\boldsymbol{r}-\boldsymbol{r}^*|}$$

containing

$$\begin{aligned} \Phi(\boldsymbol{r}) &= \frac{1}{4\pi}\int_{V^{\mathrm{T}}} \frac{\boldsymbol{K}(\boldsymbol{r}^*)(\boldsymbol{r}-\boldsymbol{r}^*)}{|\boldsymbol{r}-\boldsymbol{r}^*|^3}\,\mathrm{d}V^*, \\ \boldsymbol{F}(\boldsymbol{r}) &= \frac{1}{4\pi}\int_{V^{\mathrm{T}}} \frac{\boldsymbol{K}(\boldsymbol{r}^*)\times(\boldsymbol{r}-\boldsymbol{r}^*)}{|\boldsymbol{r}-\boldsymbol{r}^*|^3}\,\mathrm{d}V^*. \end{aligned} \tag{3.4.85}$$

During integration for the calculation of $\Phi(\boldsymbol{r})$ and $\boldsymbol{F}(\boldsymbol{r})$ we should integrate over volume V^{T}, beyond which the volume forces vanish. V^{T} may represent the total or partial volume of the body, depending on the special case.

Let us notate in particular $\boldsymbol{r}^* \equiv 0$, $\boldsymbol{K} = (K_1, 0, 0)$, and assume that

$$\int_{V^{\mathrm{T}}} \boldsymbol{K}_1(\boldsymbol{r}^*)\mathrm{d}V^* = \boldsymbol{K}_1^0$$

remains finite, when $V^{\mathrm{T}} \to 0$. From (3.4.85), we obtain for $V^{\mathrm{T}} \to 0$

$$\Phi = \frac{K_1^0}{4\pi}\frac{x_1}{r^3}, \quad F = \frac{K_1^0}{4\pi}\left(0, -\frac{x_3}{r^3}, \frac{x_2}{r^3}\right). \tag{3.4.86}$$

Inserting (3.4.86) in (3.4.84) yields

$$\nabla^2\varphi + \frac{K_1^0}{4\pi(\lambda+2\mu)}\frac{x_1}{r^3} = 0, \quad \nabla^2\boldsymbol{v} + \frac{K_1^0}{4\pi\mu}\left(0, -\frac{x_3}{r^3}, \frac{x_2}{r^3}\right) = 0. \tag{3.4.87}$$

Because of $\nabla^2(x_i r_{,i}^{-1}) = 2r_{,i}^{-1} = -2x_i r^{-3}$, $x_i r^{-3} = -\frac{1}{2}[\nabla^2(x_i r^{-1})]$ is true, through which (3.4.87) is transformed into

$$\nabla^2\left(\varphi - \frac{K_1^0}{8\pi(\lambda+2\mu)}\frac{x_1}{r}\right) = 0, \quad \nabla^2\left[\boldsymbol{v} + \frac{K_1^0}{8\pi\mu}\left(0, \frac{x_3}{r}, -\frac{x_2}{r}\right)\right] = 0.$$

It follows that if the volume force had been a singular force in the direction of x_1 at the origin, then

$$\varphi = \frac{K_1^0}{8\pi(\lambda+2\mu)}\frac{x_1}{r}, \quad \boldsymbol{v} = \frac{K_1^0}{8\pi\mu}\left(0, -\frac{x_3}{r}, \frac{x_2}{r}\right).$$

Also,

$$\operatorname{grad} \varphi = \frac{K_1^0}{8\pi(\lambda+2\mu)} \left(\frac{1}{r} - \frac{x_1^2}{r^3}, \ -\frac{x_1 x_2}{r^3}, \ -\frac{x_1 x_3}{r^3}\right),$$

$$\operatorname{curl} \boldsymbol{v} = \frac{K_1^0}{8\pi\mu} \left(\frac{1}{r} + \frac{x_1^2}{r^3}, \ \frac{x_1 x_2}{r^3}, \ \frac{x_1 x_3}{r^3}\right),$$

so that (3.4.83) changes to

$$\boldsymbol{u} = \frac{K_1^0(\lambda+\mu)}{8\pi\mu(\lambda+2\mu)} \left(\frac{x_1^2}{r^3} + \frac{\lambda+3\mu}{\lambda+\mu}\,\frac{1}{r}, \ \frac{x_1 x_2}{r^3}, \ \frac{x_1 x_3}{r^3}\right).$$

This result is then generalized as follows: For a single force $\boldsymbol{K}$ applied at a point with the radius vector $\boldsymbol{r}^*$,

$$\boldsymbol{u} = \frac{\lambda+\mu}{8\pi\mu(\lambda+2\mu)} \left[(\boldsymbol{r}-\boldsymbol{r}^*)\frac{(\boldsymbol{r}-\boldsymbol{r}^*)K}{|\boldsymbol{r}-\boldsymbol{r}^*|^3} + \frac{\lambda+3\mu}{\lambda+\mu}\,\frac{\boldsymbol{K}}{|\boldsymbol{r}-\boldsymbol{r}^*|}\right], \tag{3.4.88}$$

and, therefore, for distributed forces $\boldsymbol{K}(\boldsymbol{r})$ acting inside of a finite volume V,

$$\boldsymbol{u} = \frac{\lambda+\mu}{8\pi\mu(\lambda+2\mu)} \int_V \left[(\boldsymbol{r}-\boldsymbol{r}^*)\frac{(\boldsymbol{r}-\boldsymbol{r}^*)\boldsymbol{K}(\boldsymbol{r}^*)}{|\boldsymbol{r}-\boldsymbol{r}^*|^3} + \frac{\lambda+3\mu}{\lambda+\mu}\,\frac{\boldsymbol{K}(\boldsymbol{r}^*)}{|\boldsymbol{r}-\boldsymbol{r}^*|}\right] \mathrm{d}V^* \tag{3.4.89}$$

holds true in accordance with the superposition law.

3.4.4 *Further methods*

In addition to the above methods, variational methods can be used. These are based on extremal principles which are themselves derived from energy principles. Variational methods are particularly profitable as far as the practical calculation is concerned, since these serve as a foundation for a series of important approximation methods of numerical mathematics. They will be explained in detail at a later stage. In the meantime reference is made to [16]. Let us also mention other methods where supplementary means of functional analysis are used. We may refer in this context to the method of function spaces or to the method of hyper-circles. An outline is given on page 130 of [18], and reference is made in [24].

3.4.5 *Axial symmetric problems*

An important special case of three-dimensional elasticity is given, once axial symmetry is present. For calculation, it is best to use the cylindrical coordinates r, φ, z, when the z-axis is the symmetric axis. In the case of axial symmetry, all quantities involved are independent of φ, and the stresses $\sigma_{r\varphi}$, $\sigma_{z\varphi}$ as well as the displacements disappear. It is quite possible to use in principle the general equations and methods applying to three-dimensional problems. However, since special conditions do exist, special possibilities present themselves in connection with axial symmetry. For example, it is possible to use only *one* displacement function or *one* stress function, respectively, similar to planar problems. Let us discuss these possibilities in the following. To start with, let us work with the displacements, and let no volume forces be present. We shall, therefore, go back to (3.1.99), assuming $K_r \equiv 0$, and, in addition, because of axial symmetry, we shall equal to nought all derivatives with respect to φ. Following short calculation, we obtain

$$\frac{\lambda+2\mu}{\mu}\left(\frac{1}{r}\frac{\partial u_r}{\partial r}-\frac{u_r}{r^2}+\frac{\partial^2 u_r}{\partial r^2}+\frac{\partial^2 u_z}{\partial r\,\partial z}\right)+\frac{\partial^2 u_r}{\partial z^2}-\frac{\partial^2 u_z}{\partial r\,\partial z}=0. \tag{3.4.90}$$

By anticyclic exchange of indices (1 in 3, 3 in 2, 2 in 1), and by considering (3.1.77) and (3.1.80), as well as omitting $K_3 \equiv K_z$, and by equalling to nought all derivatives with respect to $q_2 = \varphi$, we obtain

$$\frac{\lambda+2\mu}{\mu}\left(\frac{\partial^2 u_z}{\partial z^2}+\frac{1}{r}\frac{\partial u_r}{\partial z}+\frac{\partial^2 u_r}{\partial r\,\partial z}\right)-\frac{\partial^2 u_r}{\partial r\,\partial z}$$
$$-\frac{1}{r}\frac{\partial u_r}{\partial z}+\frac{1}{r}\frac{\partial u_z}{\partial r}+\frac{\partial^2 u_z}{\partial r^2}=0 \tag{3.4.91}$$

from (3.1.98). Equations (3.4.90) and (3.4.91) are fundamental in the calculation of displacements of axial symmetry. (3.4.90) is easily changed to

$$\frac{\partial^2 u_r}{\partial r^2}+\frac{1}{r}\frac{\partial u_r}{\partial r}+\frac{\partial^2 u_r}{\partial z^2}-\frac{u_r}{r^2}+\frac{\lambda+\mu}{\mu}\frac{\partial}{\partial r}\left[\frac{\partial u_r}{\partial r}+\frac{u_r}{r}+\frac{\partial u_z}{\partial z}\right]=0. \tag{3.4.92}$$

From (3.1.107) it follows that

$$\nabla^2 = \frac{\partial^2}{\partial r^2} + \frac{1}{r}\frac{\partial}{\partial r} + \frac{\partial^2}{\partial z^2},$$

and, hence,

$$\frac{\partial^2 u_r}{\partial r^2} + \frac{1}{r}\frac{\partial u_r}{\partial r} + \frac{\partial^2 u_r}{\partial z^2} = \nabla^2 u_r$$

are true. From (3.1.82) we obtain

$$\frac{\partial u_r}{\partial r} + \frac{u_r}{r} + \frac{\partial u_z}{\partial z} = \varepsilon_{rr} + \varepsilon_{\varphi\varphi} + \varepsilon_{zz} = \Theta,$$

and, finally, $(\lambda+\mu)/\mu = 1/(1-2\nu)$. Then (3.4.92) can be changed further to

$$\nabla^2 u_r - \frac{u_r}{r^2} + \frac{1}{1-2\nu}\frac{\partial \Theta}{\partial r} = 0, \tag{3.4.93}$$

and correspondingly (3.4.91) is changed to

$$\nabla^2 u_z + \frac{1}{1-2\nu}\frac{\partial \Theta}{\partial z} = 0. \tag{3.4.94}$$

It is possible to check that identities

$$\begin{aligned} \nabla^2 \frac{\partial^2}{\partial r\,\partial z} &= \frac{\partial^2}{\partial r\,\partial z}\nabla^2 + \frac{1}{r^2}\frac{\partial^2}{\partial r\,\partial z}, \\ \nabla^2 \frac{\partial^2}{\partial z^2} &= \frac{\partial^2}{\partial z^2}\nabla^2 \end{aligned} \tag{3.4.95}$$

are true.

Let us define the displacement function of Love by determining that

$$u_r = -\frac{1}{1-2\nu}\frac{\partial^2 L}{\partial r\,\partial z}, \quad u_z = \frac{1}{1-2\nu}\left[2(1-\nu)\nabla^2 L - \frac{\partial^2 L}{\partial z^2}\right] \tag{3.4.96}$$

is valid.

Through (3.4.95) we find from (3.4.96)

$$\nabla^2 u_r = -\frac{1}{1-2\nu}\left(\frac{\partial^2}{\partial r\,\partial z}\nabla^2 L+\frac{1}{r^2}\frac{\partial^2 L}{\partial r\,\partial z}\right),$$

$$\nabla^2 u_z = \frac{1}{1-2\nu}\left[2(1-\nu)\nabla^2\nabla^2 L-\frac{\partial^2}{\partial z^2}\nabla^2 L\right], \tag{3.4.97}$$

$$\Theta = \frac{\partial}{\partial z}\nabla^2 L.$$

Thus, (3.4.93) is satisfied identically by (3.4.96). From (3.4.94) and (3.4.97),

$$\nabla^2\nabla^2 L = 0 \tag{3.4.98}$$

is yielded, representing the differential equation for the displacement function L. Again we are dealing with a biharmonic equation.

When dealing with a boundary value problem of the first kind, the calculation proceeds to determine displacements u_r and u_z from a solution of (3.4.98), and to adapt these to the boundary conditions. The displacement function L, which has been determined in this way, helps us obtain the strain components

$$\varepsilon_{rr} = -\frac{1}{1-2\nu}\frac{\partial^3 L}{\partial r^2\partial z},\quad \varepsilon_{\varphi\varphi} = -\frac{1}{1-2\nu}\frac{1}{r}\frac{\partial^2 L}{\partial r\,\partial z},$$

$$\varepsilon_{zz} = \frac{1}{1-2\nu}\frac{\partial}{\partial z}\left[2(1-\nu)\nabla^2 L-\frac{\partial^2 L}{\partial z^2}\right], \tag{3.4.99}$$

$$\varepsilon_{rz} = \frac{1}{1-2\nu}\frac{\partial}{\partial r}\left[(1-\nu)\nabla^2 L-\frac{\partial^2 L}{\partial z^2}\right]$$

using the geometric equations (3.1.82) and relation (3.4.96).

From the physical equations (3.1.84), the stresses $\sigma_{rr} = \lambda\Theta+2\mu\varepsilon_{rr}$, $\sigma_{\varphi\varphi} = \lambda\Theta + 2\mu\varepsilon_{\varphi\varphi}$, $\sigma_{zz} = \lambda\Theta+2\mu\varepsilon_{zz}$, $\sigma_{rz} = 2\mu\varepsilon_{rz}$ are obtained. Since $\lambda = 2\mu\nu/(1-2\nu)$ is true, this relation can be changed into

$$\sigma_{rr} = 2\mu\left(\varepsilon_{rr}+\frac{\nu}{1-2\nu}\Theta\right) \text{ etc..}$$

Thus, from the third equation of (3.4.97) and from (3.4.99), we obtain finally

$$
\begin{aligned}
\sigma_{rr} &= \frac{2\mu}{1-2\nu}\frac{\partial}{\partial z}\left(\nu\nabla^2 L - \frac{\partial^2 L}{\partial r^2}\right),\\
\sigma_{\varphi\varphi} &= \frac{2\mu}{1-2\nu}\frac{\partial}{\partial z}\left(\nu\nabla^2 L - \frac{1}{r}\frac{\partial L}{\partial r}\right),\\
\sigma_{zz} &= \frac{2\mu}{1-2\nu}\frac{\partial}{\partial z}\left[(2-\nu)\nabla^2 L - \frac{\partial^2 L}{\partial z^2}\right],\\
\sigma_{rz} &= \frac{2\mu}{1-2\nu}\frac{\partial}{\partial r}\left[(1-\nu)L\nabla^2 - \frac{\partial^2 L}{\partial z^2}\right].
\end{aligned}
\tag{3.4.100}
$$

Let us show how, in the case of axial symmetry, boundary value problems of the second kind are treated when Beltrami's equations for cylindrical coordinates are known (see 3.1.120 and 121). It has already been pointed out that $\sigma_{r\varphi} = \sigma_{\varphi z} = 0$ is true, and that all derivatives with respect to φ disappear. Thus, we obtain from (3.1.120) and (3.1.121)

$$
\begin{aligned}
&\nabla^2\sigma_{rr} + \frac{2}{r^2}(\sigma_{\varphi\varphi}-\sigma_{rr}) + \frac{1}{1+\nu}\frac{\partial^2 J}{\partial r^2} = 0,\\
&\nabla^2\sigma_{\varphi\varphi} - \frac{2(\sigma_{\varphi\varphi}-\sigma_{rr})}{r^2} + \frac{1}{1+\nu}\left(\frac{1}{r}\frac{\partial J}{\partial r}\right) = 0,\\
&\nabla^2\sigma_{zz} + \frac{1}{1+\nu}\frac{\partial^2 J}{\partial z^2} = 0,\\
&\nabla^2\sigma_{rz} - \frac{\sigma_{rz}}{r^2} + \frac{1}{1+\nu}\frac{\partial^2 J}{\partial r\,\partial z} = 0.
\end{aligned}
\tag{3.4.101}
$$

The equation system (3.4.101) is satisfied by stresses which in turn satisfy conditions

$$
\begin{aligned}
\sigma_{rr} &= \frac{\partial}{\partial z}\left(\nu\nabla^2\Phi - \frac{\partial^2\Phi}{\partial r^2}\right), \quad \sigma_{\varphi\varphi} = \frac{\partial}{\partial z}\left(\nu\nabla^2\Phi - \frac{1}{r}\frac{\partial\Phi}{\partial r}\right),\\
\sigma_{zz} &= \frac{\partial}{\partial z}\left[(2-\nu)\nabla^2\Phi - \frac{\partial^2\Phi}{\partial z^2}\right], \quad \sigma_{rz} = \frac{\partial}{\partial r}\left[(1-\nu)\nabla^2\Phi - \frac{\partial^2\Phi}{\partial z^2}\right],
\end{aligned}
\tag{3.4.102}
$$

then a solution of the biharmonic differential equation

$$\left(\frac{\partial^2}{\partial r^2} + \frac{1}{r}\frac{\partial}{\partial r} + \frac{\partial^2}{\partial z^2}\right)^2 \Phi = 0 \tag{3.4.103}$$

is assumed for stress-function Φ. This statement may be checked by substitution in (3.4.101). Again, all can be traced back determining a stress function Φ, the latter adapting to the boundary conditions. Once it has been determined, then from known Φ, using (3.4.102), the stresses are obtained, which in turn are followed by the strains according to the physical equations (3.1.85). The displacements are obtained through integration of the geometric equations (3.1.82), which are in the axial symmetric case

$$\varepsilon_{rr} = \frac{\partial u_r}{\partial r}, \; \varepsilon_{\varphi\varphi} = \frac{u_r}{r}, \; \varepsilon_{zz} = \frac{\partial u_z}{\partial z}, \; \varepsilon_{rz} = \frac{1}{2}\left(\frac{\partial u_r}{\partial z} + \frac{\partial u_z}{\partial r}\right).$$

3.4.6 *Examples*

1. *Example.* Let us use Betti's method for calculating the displacements u_i of a body which fills the half space $x_3 > 0$, and which is limited by plane $x_3 = 0$, in which its displacements are specified.

Let ξ_i be a point P inside the body, and let $(\xi_1, \xi_2, -\xi_3)$ be its mirror image P' with respect to plane $x_3 = 0$. Then,

$$r = [(x_i - \xi_i)(x_i - \xi_i)]^{\frac{1}{2}}, \; R = [(x_1 - \xi_1)^2 + (x_2 - \xi_2)^2 + (x_3 + \xi_3)^2]^{\frac{1}{2}}$$

are the distances of any point in the body x_i from the points P, P', respectively. It follows from (3.4.51)

$$p_i^* = -2\mu(r^{-1}_{,3})_{,i} \tag{3.4.104}$$

because on the body's surface $\boldsymbol{n} = (0, 0, -1)$ is true.

Displacements u_i^{**}, whose significance has been explained in 3.4.2, are given by

$$u_i^{**} = \begin{pmatrix} \dfrac{\partial}{\partial x_1} + \dfrac{2(\lambda+\mu)}{\lambda+3\mu} x_3 \dfrac{\partial^2}{\partial x_1 \partial x_3} \\ \dfrac{\partial}{\partial x_2} + \dfrac{2(\lambda+\mu)}{\lambda+3\mu} x_3 \dfrac{2}{\partial x_2 \partial x_3} \\ -\dfrac{\partial}{\partial x_3} + \dfrac{2(\lambda+\mu)}{\lambda+3\mu} x_3 \dfrac{\partial^2}{\partial x_3^2} \end{pmatrix} \frac{1}{R}. \tag{3.4.105}$$

On checking this, it becomes apparent that for $x_3 = 0$ the requirement that u_{ii}^* is equal to $u_i^* = (r^{-1})$ is being satisfied, since

$$\frac{\partial R^{-1}}{\partial x_1} = \frac{\partial r^{-1}}{\partial x_1}, \quad \frac{\partial R^{-1}}{\partial x_2} = \frac{\partial r^{-1}}{\partial x_2}, \quad \frac{\partial R^{-1}}{\partial x_3} = -\frac{\partial r^{-1}}{\partial x_3} \tag{3.4.106}$$

is true. For the surface-forces p_i^{**} corresponding to u_i^{**}, and resulting from $p_i^{**} = n_j\sigma_{ij}$, because of $\boldsymbol{n} = (0, 0, -1)$, $p_i^{**} = -\sigma_{i3}$, we find through physical equations (3.1.8), and the equations (3.4.105)

$$p_1^{**} = -\sigma_{13} = -\mu\left(\frac{\partial u_1^{**}}{\partial x_3} + \frac{\partial u_3^{**}}{\partial x_1}\right) = -2\mu\frac{\lambda+\mu}{\lambda+3\mu}\frac{\partial^2 R^{-1}}{\partial x_1\,\partial x_3},$$

$$p_2^{**} = -\sigma_{23} = -\mu\left(\frac{\partial u_2^{**}}{\partial x_3} + \frac{\partial u_3^{**}}{\partial x_2}\right) = -2\mu\frac{\lambda+\mu}{\lambda+3\mu}\frac{\partial^2 R^{-1}}{\partial x_2\,\partial x_3},$$

$$p_3^{**} = -\left[\lambda\left(\frac{\partial u_1^{**}}{\partial x_1} + \frac{\partial u_2^{**}}{\partial x_2} + \frac{\partial u_3^{**}}{\partial x_3}\right) + 2\mu\frac{\partial u_3^{*}}{\partial x_3}\right] = 2\mu\frac{\lambda+\mu}{\lambda+3\mu}\frac{\partial^2 R^{-1}}{\partial x_3^2},$$

which because of (3.4.106) is summed up as

$$p_i^{**} = -2\mu\frac{\lambda+\mu}{\lambda+3\mu}(r_{,3}^{-1})_{,i}.$$

Therefore,

$$p_i^* - p_i^{**} = -4\mu\frac{\lambda+2\mu}{\lambda+3\mu}(r_{,3}^{-1})_{,i}$$

results, and from (3.4.57) we obtain

$$\Theta_0 = -\frac{\mu}{\pi(\lambda+3\mu)}\int_0 (r_{,3}^{-1})_{,i}\,u_i\,\mathrm{d}O,$$

whereby for Θ_0 all differentiations must be carried out with respect to ξ_i. Hence, we notate

$$\Theta_0 = -\frac{\mu}{\pi(\lambda+3\mu)}\frac{\partial}{\partial\xi_3}\left[\frac{\partial}{\partial\xi_i}\int_O \frac{u_i}{r}\,\mathrm{d}O\right]. \tag{3.4.107}$$

Using for the sake of brevity

$$\int_O \frac{u_i}{r}\,\mathrm{d}O = \varphi_i,$$

then, (3.4.107) changes to

$$\Theta_0 = \frac{\mu}{\pi(\lambda+3\mu)}\varphi_{i,i3}, \tag{3.4.108}$$

and differentiation is again taken up with respect to ξ_i. The functions φ_i are harmonic on both sides of $\xi_3 = 0$, and

$$u_i = -\frac{1}{2\pi}\lim_{\xi_3\to 0}\varphi_{i,3} \text{ for } \xi_3 = 0, \tag{3.4.109}$$

(differentiation with respect to ξ_i)
is valid. Rewriting (3.4.45) in terms of the variable ξ_i we obtain

$$\nabla^2\left(u_i - \frac{\lambda+\mu}{2\pi(\lambda+3\mu)}\xi_i\varphi_{j,j3}\right) = 0, \tag{3.4.110}$$

(differentiation with respect to ξ_i),

by using (3.4.108).

But it is $\nabla^2(\xi_i\varphi_{j,j3}) = \nabla^2(\xi_3\varphi_{j,ji})$, then (3.4.110) is changed to

$$\nabla^2\left(u_i - \frac{\lambda+\mu}{2\pi(\lambda+3\mu)}\xi_3\varphi_{j,ji}\right) = 0. \tag{3.4.111}$$

The three functions $\partial\varphi_i/\partial\xi_3$ are harmonic in the domain under consideration, and the values of u_i for $\xi_3 \to +0$ have been given by equation (3.4.109). Then, the wanted solution of (3.4.110) must be

$$u_i = -\frac{1}{2\pi}\varphi_{i,3} + \frac{\lambda+\mu}{2\pi(\lambda+3\mu)}\xi_3\varphi_{j,ji}, \tag{3.4.112}$$

whereby all differentiation must be carried out with respect to ξ_i.

2. *Example.* Let there be a single force on the surface of a half space. Let the location and orientation of the axis system be such that $x_3 = 0$ is

valid for the surface of the half space, and that the point of application of the force coincides with the origin, while its line of action is the x_3 axis (Figure 45).

Let us solve the problem by using the 'general solutions' for the displacements through superposition of two particular solutions. To begin with, let us use (3.4.88), notating in accordance with the assumptions $\boldsymbol{r}^* \equiv 0$, $\boldsymbol{K} \equiv (0, 0, K)$, thus obtaining

$$\boldsymbol{u} = \frac{\lambda+\mu}{8\pi\mu(\lambda+2\mu)}\left(\frac{Kx_3}{r^3}\boldsymbol{r}+\frac{\lambda+3\mu}{\lambda+\mu}\frac{\boldsymbol{K}}{r}\right), \quad |\boldsymbol{r}| = r. \tag{3.4.113}$$

In component notation this is

$$u_\alpha = \frac{(\lambda+\mu)K}{8\pi\mu(\lambda+2\mu)}\frac{x_\alpha x_3}{r^3}, \quad \alpha = 1, 2, \quad u_3 = \frac{(\lambda+\mu)K}{8\pi\mu(\lambda+2\mu)}\left(\frac{x_3^2}{r^3}+\frac{\lambda+3\mu}{\lambda+\mu}\frac{1}{r}\right).$$

Because of

$$\sigma_{ij} = \lambda\delta_{ij}u_{k,k}+\mu(u_{i,j}+u_{j,i})$$

it yields for the stresses

$$\begin{aligned}
\sigma_{11} &= -\frac{2\mu Cx_3}{r^3}\left[3\left(\frac{x_1}{r}\right)^2-\frac{\mu}{\lambda+\mu}\right],\\
\sigma_{23} &= -\frac{2\mu Cx_2}{r^3}\left[3\left(\frac{x_3}{r}\right)^2+\frac{\mu}{\lambda+\mu}\right],\\
\sigma_{22} &= -\frac{2\mu Cx_3}{r^3}\left[3\left(\frac{x_2}{r}\right)^2-\frac{\mu}{\lambda+\mu}\right],\\
\sigma_{13} &= -\frac{2\mu Cx_1}{r^2}\left[3\left(\frac{x_3}{r}\right)^2+\frac{\mu}{\lambda+\mu}\right],\\
\sigma_{33} &= -\frac{2\mu Cx_3}{r^3}\left[3\left(\frac{x_3}{r}\right)^2+\frac{\mu}{\lambda+\mu}\right],\\
\sigma_{12} &= -\frac{6\mu Cx_1x_2x_3}{r^5}
\end{aligned}$$

$$C = \frac{(\lambda+\mu)K}{8\pi\mu(\lambda+2\mu)}. \tag{3.4.114}$$

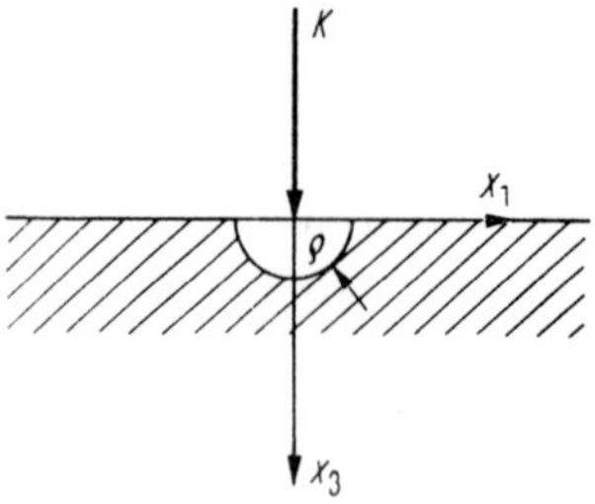

Figure 45

Solutions (3.4.113) and (3.4.114) are valid throughout, except for $r = 0$. Hence, the origin is excluded by a small hemisphere. By (3.4.114), the stresses operating in the half space, which is limited by plane $x_3 = 0$ and a small hemisphere at the origin, are being given when the half space is loaded by a single force in the x_3 direction, and whereby the point of application, coinciding with the origin, is excluded by the hemisphere from the half space.

Let the hemisphere be of radius ρ. Then, for its surface forces $p_i = n_j \sigma_{ij}$ follows from (3.4.114), and because of $n_j = x_j/\rho$,

$$p_\alpha = -\frac{6\mu C x_\alpha x_3}{\rho^4}, \quad p_3 = -\frac{6\mu C x_3^2}{\rho^4} - \frac{2\mu_2^2 C}{(\lambda+\mu)\rho^2}, \quad \alpha = 1, 2.$$

Through integration over the hemisphere we obtain the resultants

$$\int_O p_\alpha \, \mathrm{d}O = 0, \int_O p_3 \, \mathrm{d}O = -\frac{4\pi\mu C(\lambda+2\mu)}{\lambda+\mu}. \tag{3.4.115}$$

As a second particular solution let us set up a Boussinesq solution of the second kind. For this purpose, $h_0 = -D\mu \ln(x_3+r)$ is inserted in (3.4.78), and we obtain for the displacements

$$\boldsymbol{u} = D\left(\frac{x_1}{r(x_3+r)}, \frac{x_2}{r(x_3+r)}, \frac{1}{r}\right), \quad r = (x_i x_i)^{\frac{1}{2}}. \tag{3.4.116}$$

The corresponding stresses are calculated from (3.4.78) using the above given definition for h_0. We arrive at

$$
\begin{aligned}
\sigma_{11} &= 2\mu D\left[\frac{x_2^2+x_3^2}{r^3(r+x_3)}-\frac{x_1^2}{r^2(r+x_3)^2}\right],\\
\sigma_{22} &= 2\mu D\left[\frac{x_1^2+x_3^2}{r^3(r+x_3)}-\frac{x_2^2}{r^2(r+x_3)^2}\right],\\
\sigma_{33} &= -2\mu D\frac{x_3}{r^3},\ \sigma_{13} = -2\mu D\frac{x_1}{r^3},\ \sigma_{23} = -2\mu D\frac{x_2}{r^3},\\
\sigma_{12} &= -2\mu D\frac{x_1 x_2(x_3+2r)}{r^3(r+x_3)^2}.
\end{aligned}
\tag{3.4.117}
$$

Again we put a hemisphere about the origin, and we calculate its surface forces

$$p_\alpha = -2\mu D\frac{x_\alpha}{r^2(r+x_3)},\ \alpha = 1, 2,\ p_3 = -2\mu D\frac{1}{r^2},\ r = \rho.$$

The resultants of the surface forces belonging to the hemisphere are

$$\int_O p_\alpha \mathrm{d}O = 0,\ \int_O p_3\,\mathrm{d}O = -4\pi\mu D. \tag{3.4.118}$$

Let us set up a final solution for the problem which satisfies the following conditions. The half space satisfies $x_3 \geqq 0$. A force $\boldsymbol{K}$ is applied in the x_3 direction at its origin. The singularity brought about by the single force $\boldsymbol{K}$ is removed by exclusion of the origin through a hemisphere, having radius ρ. Only those parts of the half space are of interest to us which lie outside the hemisphere. It is required that the resultants of the surface forces on the hemisphere disappear in the x_1 and x_2 direction, but also counter-balance force $\boldsymbol{K}$ in the x_3 direction. Also, the stresses $\sigma_{13} = \sigma_{23} = \sigma_{33} \equiv 0$ should occur in plane $x_3 = 0$, the latter together with the hemisphere, representing the surface of the half-space.

To reach our goal, let us superpose particular solution (3.4.113) with (3.4.116), (3.4.114) with (3.4.117) and finally (3.4.115) with (3.4.118). From the compound of both the last groups of equations, and the required counter-balancing of the resultant with the concentrated force K, we obtain

$$K = \frac{4\pi\mu C(\lambda+2\mu)}{\lambda+\mu}+4\pi\mu D. \tag{3.4.119}$$

From the combination of (3.4.114) with (3.4.117)

$$\sigma_{\alpha 3} = -\frac{2\mu C x_\alpha}{r^3}\frac{\mu}{\lambda+\mu} - 2\mu D \frac{x_\alpha}{r^3} \quad \alpha = 1, 2, \tag{3.4.120}$$

$$\sigma_{33} = 0$$

is obtained for $x_3 = 0$.

Because of boundary condition $\sigma_{\alpha 3} \equiv 0$ for $x_3 = 0$, condition

$$\frac{2\mu^2 C}{\lambda+\mu} + 2\mu D = 0 \tag{3.4.121}$$

follows from (3.4.120). Both the condition equations (3.4.119) and (3.4.121) supply expression

$$C = \frac{K}{4\pi\mu}, \; D = -\frac{K}{4\pi(\lambda+\mu)} \tag{3.4.122}$$

for the hitherto unknown constants. Superimposing the particular solutions (3.4.113) and (3.4.116) for the displacements, we obtain through (3.4.122) instead of (3.4.114)

$$\begin{aligned} u_\alpha &= \frac{K}{4\pi\mu}\frac{x_\alpha x_3}{r^3} - \frac{K}{4\pi(\lambda+\mu)}\frac{x_\alpha}{r(x_3+r)}, \quad \alpha = 1, 2, \\ u_3 &= \frac{K}{4\pi\mu}\frac{x_3^2}{r^3} + \frac{K(\lambda+2\mu)}{4\pi\mu(\lambda+\mu)}\frac{1}{r}. \end{aligned} \tag{3.4.123}$$

The latter represents the wanted solution.

3. *Example.* Let us deal with the problem of the second example, by using the 'general solutions' (3.4.76) by Papkovich-Neuber. We notate

$$h_0 = -\frac{\mu^2}{\lambda+\mu} g, \; h_i = -\frac{2\mu(\lambda+2\mu)}{\lambda+\mu}\delta_{i3}\frac{\partial g}{\partial x_3}, \tag{3.4.124}$$

and obtain from

$$u_i = \frac{1}{\mu}\left[h_i - h_{0,i} - \frac{1}{2}\frac{\lambda+\mu}{\lambda+2\mu}(x_j h_j)_{,i}\right], \quad i = 1, 2, 3, \tag{3.4.76}$$

through (3.4.124) the expressions

$$\begin{aligned}
u_1 &= \frac{\mu}{\lambda+\mu} g_{,1} + x_3 g_{,13},\\
u_2 &= \frac{\mu}{\lambda+\mu} g_{,2} + x_3 g_{,23},\\
u_3 &= -\frac{\lambda+2\mu}{\lambda+\mu} g_{,3} + x_3 g_{,33}
\end{aligned} \tag{3.4.125}$$

for the stresses.

Conversion to polar coordinates r, φ, z, considering axial symmetry, i.e., using $\partial/\partial\varphi = 0$, yields

$$\begin{aligned}
u_r &= \frac{\mu}{\lambda+\mu} g_{,r} + z g_{,rz},\\
u_z &= -\frac{\lambda+2\mu}{\lambda+\mu} g_{,z} + z g_{,zz}.
\end{aligned} \tag{3.4.126}$$

The corresponding stresses are

$$\begin{aligned}
\sigma_{rr} &= 2\mu\left[z g_{,rrz} + g_{,rr} - \frac{\lambda}{\lambda+\mu}(g_{,rr} + g_{,zz})\right],\\
\sigma_{zz} &= 2\mu(z g_{,zzz} - g_{,zz}),\\
\sigma_{\varphi\varphi} &= -2\mu\left[z g_{,rrz} + z g_{,zzz} + g_{,zz} + \frac{\mu}{\lambda+\mu} g_{,rr}\right]\\
\sigma_z &= 2\mu z g_{,rzz}.
\end{aligned} \tag{3.4.127}$$

Let us notate specially

$$g = -\frac{K}{4\pi\mu}\ln(R_1 + z + \rho), \tag{3.4.128}$$

whereby

$$R_1 = [r^2+(z+\rho)^2]^{\frac{1}{2}}. \tag{3.4.129}$$

Thus, (3.4.126) changes to

$$\begin{aligned} u_r &= \frac{K}{4\pi\mu}\left[\frac{zr}{R_1^3} - \frac{\mu}{\lambda+\mu}\,\frac{r}{R_1(R_1+z+\rho)}\right], \\ u_z &= \frac{K}{4\pi\mu}\left[\frac{z(z+\rho)}{R_1^3} + \frac{\lambda+2\mu}{\lambda+\mu}\cdot\frac{1}{R_1}\right]. \end{aligned} \tag{3.4.130}$$

From (3.4.127) using (3.4.128),

$$\sigma_{zz} = -\frac{K}{2\pi}\left\{\frac{z+\rho}{R_1^3} - z\left[\frac{1}{R_1^3} - \frac{3(z+\rho)^2}{R_1^5}\right]\right\} \tag{3.4.131}$$

is determined.

Vertical normal stresses

$$p_r = -\sigma_{zz}(\text{for } z = 0) = \frac{P\rho}{2\pi(r^2+\rho^2)^{\frac{3}{2}}}$$

occur in the $z = 0$ plane.

For these

$$2\pi\int_0^\infty rp_r\,\mathrm{d}r = P\int_0^\infty \frac{r\rho\,\mathrm{d}r}{(r^2+\rho^2)^{\frac{3}{2}}} = P \tag{3.4.132}$$

is valid. This signifies that the resultant of the normal stresses has the magnitude P. For $\rho \to 0$, we arrive at a single force in the origin which has the magnitude P (Figure 46). Then,

$$R_1 \to R = [r^2+z^2]^{\frac{1}{2}},$$

and from (3.4.130) we obtain

$$\begin{aligned} u_r &= \frac{K}{4\pi\mu}\left[\frac{zr}{R^3} - \frac{\mu}{\lambda+\mu}\,\frac{r}{R(R+z)}\right], \\ u_z &= \frac{K}{4\pi\mu}\left[\frac{z^2}{R^3} + \frac{\lambda+2\mu}{\lambda+\mu}\,\frac{1}{R}\right]. \end{aligned} \tag{3.4.133}$$

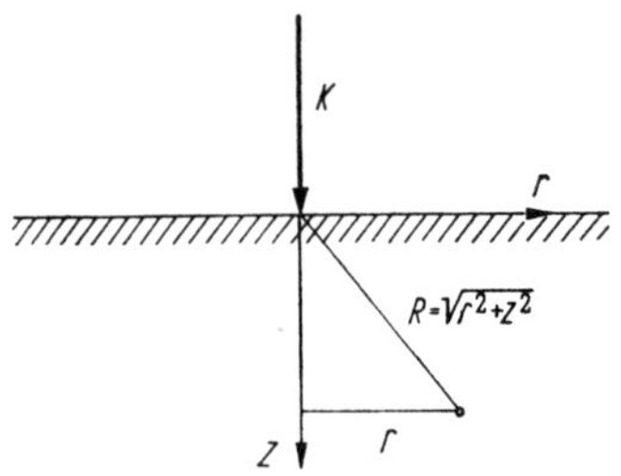

Figure 46

Agreement of (3.4.133) with (3.4.123) becomes obvious, once the quantities z, r, R from (3.4.133) are identified with the corresponding quantities x_3, x_α, r from (3.4.123).

4. *Example.* Let us deal once more with a single force P (Figure 46). This time it is an axial symmetric problem in which we shall use displacement function L. One solution satisfying equation (3.4.98) is

$$L(r, z) = c_1 R + c_2 z \ln(z+R), \quad R = \sqrt{r^2+z^2}. \tag{3.4.134}$$

It contains the yet undetermined constants c_1, c_2. Thus, we obtain from (3.4.96) for the displacements

$$\begin{aligned} u_r &= \frac{1}{1-2\nu}\left[(c_1+c_2)\frac{rz}{R^3} - c_2\frac{r}{R(z+R)}\right], \\ u_z &= \frac{1}{1-2\nu}\left\{[(3-4\nu)c_1 + 2(1-2\nu)c_2]\frac{1}{R} + (c_1+c_2)\frac{z^2}{R^3}\right\}, \end{aligned} \tag{3.4.135}$$

and from (3.4.100) we obtain for example, for the stresses σ_{rz} and σ_{zz}, which are needed for establishing the boundary conditions,

$$\begin{aligned} \sigma_{rz} &= \frac{2\mu}{1-2\nu}\left\{[2\nu c_2 - (1-2\nu)c_1]\frac{r}{R^3} - 3(c_1+c_2)\frac{rz^2}{R^5}\right\}, \\ \sigma_{zz} &= \frac{2\mu}{1-2\nu}\left\{[2\nu c_2 - (1-2\nu)c_1]\frac{z}{R^3} - 3(c_1+c_2)\frac{z^3}{R^5}\right\}. \end{aligned} \tag{3.4.136}$$

On plane $z = 0$, which represents the surface of the half space, let $\sigma_{rz} = \sigma_{zz} = 0$. Because of (3.4.136), condition $\sigma_{rz} = 0$ for $z = 0$ requires that

$$2\nu c_2-(1-2\nu)c_1 = 0,\ c_1+c_2 = \frac{c_1}{2\nu},\ \text{respectively.} \tag{3.4.137}$$

Substituting (3.4.137) in expression σ_{zz} yields

$$\sigma_{zz} = -\frac{3\mu}{\nu(1-2\nu)}c_1\frac{z^3}{R^5}\,. \tag{3.4.138}$$

Thus, the condition requiring that σ_{zz} for $z = 0$ vanishes has already been fulfilled. For reasons of equilibrium,

$$P = -2\pi\int_0^\infty \sigma_{zz}\, r\,\mathrm{d}r$$

is required. According to (3.4.138) it yields

$$P = \frac{6\pi\mu c_1 z^3}{\nu(1-2\nu)}\int_0^\infty \frac{r\,\mathrm{d}r}{(z^2+r^2)^{\frac{5}{2}}} = \frac{2\pi\mu c_1}{\nu(1-2\nu)}\,.$$

This is followed by

$$c_1 = \frac{\nu(1-2\nu)}{2\pi\mu}P, \tag{3.4.139}$$

and thus from (3.4.137)

$$c_2 = \frac{(1-2\nu)^2}{4\pi\mu}P\,. \tag{3.4.140}$$

Through (3.4.137), (3.4.139) and (3.4.140), we obtain from (3.4.135)

$$\begin{aligned} u_r &= \frac{P}{4\pi\mu}\left[\frac{rz}{R^3}-(1-2\nu)\frac{r}{R(z+R)}\right], \\ u_z &= \frac{p}{4\pi\mu}\left[2(1-\nu)\frac{1}{R}+\frac{z^2}{R^3}\right], \end{aligned} \tag{3.4.141}$$

which is in complete agreement with (3.4.133), since

$$1-2\nu = \frac{\mu}{\lambda+\mu}, \quad 2(1-\nu) = \frac{\lambda+2\mu}{\lambda+\mu}.$$

We shall omit the detailed calculation of stresses at this point, like in examples 2 and 3. It is left to the reader to revise these if he wishes to do so. Since the displacements are known it should prove an easy task.

3.5 Energy principles

3.5.1 *Variational principles*

Let us assume the one-dimensional state of stress and strain of a Hooke's body being given. The strain energy density is given by:

$$U = \tfrac{1}{2}\sigma\varepsilon. \tag{3.5.1}$$

Based on the material law, let us use $\sigma = f(\varepsilon)$ (in this case $\sigma = E\varepsilon$ in particular), and hence change (3.5.1) into

$$U(\varepsilon) = \tfrac{1}{2}f(\varepsilon)\varepsilon. \tag{3.5.2}$$

In Figure 47, $U(\varepsilon)$ is shown by the vertically shaded area below the stress-strain curve

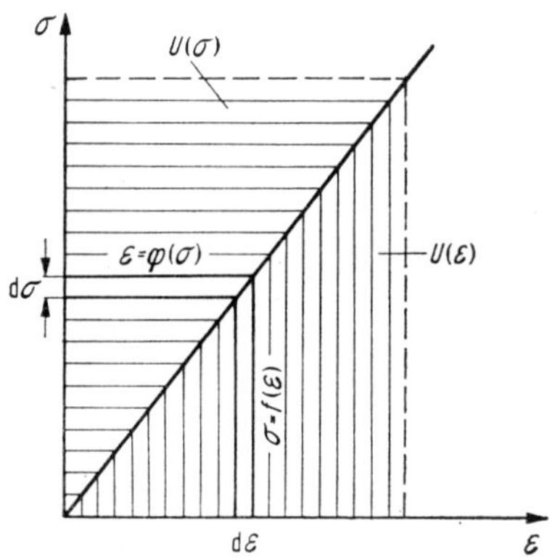

Figure 47

Let us notate expression

$$U(\sigma) = \sigma\varepsilon - U(\varepsilon) \tag{3.5.3}$$

for the horizontally shaded area to the left of the stress-strain curve. Quantity $U(\sigma)$ is also obtained by using the inverse function $\varepsilon = \varphi(\sigma)$,

$$U(\sigma) = \tfrac{1}{2}\varphi(\sigma)\sigma \tag{3.5.4}$$

is the result.

Because of the assumed specific material law, i.e., Hooke's Law, $U(\varepsilon) = U(\sigma)$ holds true in our case. This fact is clearly indicated in Figure 47. The difference between $U(\varepsilon)$ and $U(\sigma)$ is only a formal one for any of the Hooke's bodies: in the first case we are able to express the strain energy density entirely through strain, in the second case entirely through stress.

Let us generalize the concepts obtained by considering a one-dimensional state of stress and strain and apply them to multi-dimensional states. Instead of (3.5.1) there is

$$U = \tfrac{1}{2}\sigma_{ij}\varepsilon_{ij}. \tag{3.5.5}$$

Using Hooke's material law in terms of (2.3.2), we arrive from (3.5.5) at

$$U(\varepsilon) = \tfrac{1}{2}C_{ijkl}\varepsilon_{ij}\varepsilon_{kl}, \tag{3.5.6}$$

representing the generalization of (3.5.2). Also

$$U(\sigma) = \sigma_{ij}\varepsilon_{ij} - U(\varepsilon) \tag{3.5.7}$$

has to be substituted for (3.5.3), which represents nothing more than Legendre's transformation. Through reversal of (2.3.2) we may arrive at the representation

$$\varepsilon_{ij} = C^{*}_{ijkl}\sigma_{kl} \tag{3.5.8}$$

of the material law. Then, analogous to (3.5.4), we arrive at

$$U(\sigma) = \tfrac{1}{2}C^{*}_{ijkl}\sigma_{kl}\sigma_{ij} \tag{3.5.9}$$

using (3.5.8), in (3.5.5), and analogous to the one-dimensional case,

$U(\varepsilon) = U(\sigma)$ applies to the multi-dimensional case also, since we are dealing with Hooke's bodies. Let us term

$$\mathfrak{E}(\varepsilon) = \int_V U(\varepsilon)\,\mathrm{d}V \tag{3.5.10}$$

the *strain energy* and

$$\mathfrak{E}(\sigma) = \int_V U(\sigma)\,\mathrm{d}V \tag{3.5.11}$$

the *complementary energy* or *conjugate strain energy*. This particular difference represents only a formal difference in Hooke's bodies, but develops more than just a formal significance for non-Hookian bodies, since $U(\varepsilon) \neq U(\sigma)$ is valid for the latter. This becomes apparent if in Figure 47 of the one-dimensional case, the linear stress-strain curve of Hooke's law were substituted by a non-linear curve of a more general material law.

As a requirement for the following, the variations of U should be known. From (3.5.6) and (2.3.2) we obtain

$$\delta U(\varepsilon) = \frac{\partial U(\varepsilon)}{\partial \varepsilon_{ij}}\,\delta\varepsilon_{ij} = C_{ijkl}\,\varepsilon_{kl}\,\delta\varepsilon_{ij} = \sigma_{ij}\,\delta\varepsilon_{ij}, \tag{3.5.12}$$

and correspondingly, from (3.5.9) and (3.5.8)

$$\delta U(\sigma) = \frac{\partial U(\sigma)}{\partial \sigma_{ij}}\,\delta\sigma_{ij} = C^*_{ijkl}\,\sigma_{kl}\,\delta\sigma_{ij} = \varepsilon_{ij}\,\delta\sigma_{ij}. \tag{3.5.13}$$

Assuming not just a Hooke's body, but an isotropic Hooke's body, then in (3.5.5), depending on whether $U(\varepsilon)$ or $U(\sigma)$ is wanted, relation (3.1.3a) or (3.1.3b) is to be inserted, thus yielding

$$\begin{aligned} U(\varepsilon) &= \tfrac{1}{2}\lambda J_{1V}^2 + \mu\varepsilon_{ij}\varepsilon_{ij}, \quad J_{1V} = \varepsilon_{kk}, \\ U(\sigma) &= \frac{1}{2}\left(\frac{1+\nu}{E}\,\sigma_{ij}\sigma_{ij} - \frac{\nu}{E}\,J_{1S}^2\right), \quad J_{1S} = \sigma_{kk}, \text{ respectively.} \end{aligned} \tag{3.5.14}$$

Following these preparations, let us start deriving the energy theorems. In order to do this, let us go back to Clapeyron's theorem, which we used in 3.1.2. Changing (3.1.33) by using (3.5.5) yields

$$\int_V \sigma_{ij}\varepsilon_{ij}\mathrm{d}V - \int_V K_i u_i \mathrm{d}V - \int_O p_i u_i \mathrm{d}O = 0. \tag{3.5.15}$$

But (3.5.15) is not only valid for the true stresses, strains and displacements, but also for example for such displacements as u_i^*, which are geometrically permissible, and for such stress conditions as σ_{ij}^*, which are statically permissible. By geometrically permissible displacements we mean those which have differentiability properties as required by the compatibility conditions and which satisfy the boundary conditions

$$u_i^* = u_i^0 \text{ on } O_u \tag{3.5.16}$$

of the true displacements. Then O_u represents that part of the total surface O, where displacements u_i^0 are prescribed, but nothing is specified for the surface forces p_i.

Statically permissible stress conditions are those satisfying the equilibrium and boundary conditions

$$\sigma_{ij,j}^* + K_i = 0,\ \sigma_{ij}^* n_j = p_i^0 \text{ on } O_p \tag{3.5.17}$$

of the true state of stress σ_{ij}. Then O_p is that part of O, for which surface forces p_i^0 have been specified, but no conditions exist for displacements u_i.

Since there is no part of surface O for which not one or the other boundary condition applies, then

$$O - O_u = O_p,\ O - O_p = O_u, \text{ respectively,}$$

must be true; a fact we shall use quite often at a later stage.

Using variation δu_i of the true displacement u_i, the geometric possible displacements u_i^* are represented as $u_i^* = u_i + \delta u_i$. For the true displacements, $u_i = u_i^0$ on O_u is obviously valid. To satisfy (3.5.16), for variations δu_i then

$$\delta u_i = 0 \text{ on } O_u \tag{3.5.18}$$

must be valid, for u_i^* to be permissible. Let us use $u_i^* = u_i + \delta u_i$ in (3.5.15), and consider that $(\delta u_i)_{,j} = \delta(u_{i,j})$, and hence

$$\varepsilon_{ij}^* = \tfrac{1}{2}(u_{i,j}^* + u_{j,i}^*) = \tfrac{1}{2}(u_{i,j} + u_{j,i}) + \tfrac{1}{2}\delta(u_{i,j} + u_{j,i}) =$$
$$= \varepsilon_{ij} + \delta\varepsilon_{ij}$$

is true. Since (3.5.15) is valid for u_i, then several terms are cancelled, leaving

$$\int_V \sigma_{ij}\,\delta\varepsilon_{ij}\,\mathrm{d}V - \int_V K_i\,\delta u_i\,\mathrm{d}V - \int_{O_p} p_i^0\,\delta u_i\,\mathrm{d}O = 0. \tag{3.5.19}$$

In setting up (3.5.19) it has been considered that because of (3.5.18), the integral on the extreme right does not need integration over the total surface O, but only over part $O - O_u = O_p$ on which $p_i = p_i^0$ is specified.

Because of (3.5.12) we find

$$\int_V \sigma_{ij}\,\delta\varepsilon_{ij}\,\mathrm{d}V = \delta\int_V U(\varepsilon)\,\mathrm{d}V.$$

Hence, (3.5.19) is changed to

$$\delta\int_V U(\varepsilon)\,\mathrm{d}V - \int_V K_i\,\delta u_i\,\mathrm{d}V - \int_{O_p} p_i^0\,\delta u_i\,\mathrm{d}O = 0, \tag{3.5.20}$$

and, thus, the *principle of virtual work* is obtained. Considering that the forces in (3.5.20) are not varied, we can change over to

$$\delta\left[\int_V U(\varepsilon)\,\mathrm{d}V - \int_V K_i u_i\,\mathrm{d}V - \int_{O_p} p_i^0 u_i\,\mathrm{d}O\right] = 0. \tag{3.5.21}$$

Let us term

$$\Pi = \int_V U(\varepsilon)\,\mathrm{d}V - \int_V K_i u_i\,\mathrm{d}V - \int_{O_p} p_i^0 u_i\,\mathrm{d}O \tag{3.5.22}$$

the *total potential energy* of the elastic body, thus, notating in short (3.5.21) as

$$\Pi = \text{Extr.}$$
$$\text{for } u_i = u_i^0 \text{ on } O_u. \tag{3.5.23}$$

Thus, we have arrived at the principle of stationary value of the total potential energy. It states that: *the total potential energy Π of an elastic body has a stationary value in the class of geometrically permissible displacements for the true displacements which correspond to the state of equilibrium.* In case the displacements are specified for the whole of the surface O, then $O_u \equiv O$ is true, and thus $O_p = 0$. Hence, integral

$$\int_{O_p} p_i^0 \delta u_i \,\mathrm{d}O$$

vanishes. Let us assume in addition that no volume forces are present, e.g., $K_i \equiv 0$ is valid, then we obtain as a special case of (3.5.20)

$$\delta \int_V U(\varepsilon)\,\mathrm{d}V = \delta \mathfrak{E}(\varepsilon) = 0, \tag{3.5.24}$$

i.e., the *principle of stationary value of the strain energy: strain energy $\mathfrak{E}(\varepsilon)$ assumes a stationary value in the class of geometrically permissible displacements for the true displacements corresponding to the state of equilibrium, if the displacements for the whole surface are specified and no volume forces are present.*

When applied to Hooke's bodies, the statements following from both principles express the fact that the stationary value is especially a minimum. In that case, the principles are called *Green's principle* or *Dirichlet's principle*. Langhaar [25] proves the stationary value for Hooke's bodies to be a minimum. Also, for certain non-linear material laws, the special form of the principles remains valid. Kauderer [12], p. 32, has discussed this.

Until now we have dealt with extremum principles for the displacements. Let us set up now such principles for the stresses.

Let $\sigma_{ij}^* = \sigma_{ij} + \delta\sigma_{ij}$ be a statically permissible state of stress, when $\delta\sigma_{ij}$ are variations of the true stresses σ_{ij}. For the latter, by definition, $\sigma_{ij,j} + K_i = 0$ holds in V and $\sigma_{ij} n_j = p_i^0$ holds on O_p. When σ_{ij}^* are required to be statically permissible, as well as to satisfy (3.5.17), then

$$\begin{aligned} &(\delta\sigma_{ij})_{,j} = 0. \\ &(\delta\sigma_{ij}) n_j = \delta p_i \text{ an } O_u, \\ &\delta\sigma_{ij} = 0 \text{ i.e., } \delta p_i = O \text{ on } O_p \end{aligned} \tag{3.5.25}$$

must be valid for variations $\delta\sigma_{ij}$. As in (3.5.25) it must be noted that the p_i must be varied because of the variation of the stresses, so that in (3.5.15) not only σ_{ij}^* instead of σ_{ij} must be inserted but also $p_i^* = p_i + \delta p_i$ instead of p_i, then several terms will be cancelled and

$$\int_V \delta\sigma_{ij}\varepsilon_{ij}\,\mathrm{d}V - \int_{O_u} \delta p_i u_i^0 \mathrm{d}O = 0 \tag{3.5.26}$$

remains. In the derivation of (3.5.26), it has been considered that the integration of the integral on the right does only have to be carried out over O_u (because of the third line of (3.5.25)), on which $u_i = u_i^0$ is specified additionally. Using (3.5.13), it can be shown that

$$\int_V \delta\sigma_{ij}\varepsilon_{ij}\,\mathrm{d}V = \delta \int_V U(\sigma)\,\mathrm{d}V$$

is valid. Thus, we arrive at

$$\delta \int_V U(\sigma)\,\mathrm{d}V - \int_{O_u} \delta p_i u_i^0 \mathrm{d}O = 0 \tag{3.5.27}$$

from (3.5.26). But the displacements are not varied in this relation, so that for (3.5.27)

$$\delta \left[\int_V U(\sigma)\,\mathrm{d}V - \int_{O_u} p_i u_i^0 \mathrm{d}O \right] = 0 \tag{3.5.28}$$

may also be notated. Let us term

$$\Pi^* = \int_V U(\sigma)\,\mathrm{d}V - \int_{O_u} p_i u_i^0 \mathrm{d}O \tag{3.5.29}$$

the *conjugate total potential energy* and represent (3.5.28) in short as

$$\begin{aligned} &\Pi^* = \text{Extr.} \\ &\text{for } \sigma_{ij} n_j = p_i^0 \text{ on } O_p. \end{aligned} \tag{3.5.30}$$

Thus, we have obtained the *principle of stationary value of the conjugate total potential energy*. It states that: *the conjugate total potential energy in the class of statically permissible states of stress assumes a stationary value for the true state of stress corresponding to the equilibrium.*

A special case is given if the surface forces are prescribed for the whole surface of the body. In that case, $O_p \equiv O$ and $O_u \equiv 0$. Therefore, the second integral from the left in (3.5.27) vanishes, so that this formula simply changes into

$$\delta \int_V U(\sigma)\mathrm{d}V = \delta\mathfrak{C}(\sigma) = 0. \tag{3.5.31}$$

Thus we have obtained the *principle of the stationary value of the conjugate strain energy*. It states that *the conjugate strain energy assumes a stationary value in the class of all statically permissible states of stress for the true state of stress corresponding to the equilibrium under the condition that the p_i are specified on the whole of the body's surface.*

Let us mention again that under suitable conditions, the extremum principles for the stresses may yield a minimum for the stationary value. In that case, let us call them *Castigliano's* and *Menabrea's principle*. However, strictly speaking, the terms *Castigliano's theorem* and *Menabrea's theorem* are actually reserved for more specific cases, which we will now discuss.

Let us suppose that a body in one part O_g of its surface has a fixed support, and that on a very small second part O_b a nearly concentrated load is applied. Since equilibrium of the body is warranted by the reaction forces at O_g, then the variations δp_i outside O_g are arbitrary. Let us arrange that they disappear on $O - O_g - O_b$ but not on O_b. In addition, the displacements u_i on O_g are equal to nought.

Hence, instead of (3.5.27), there is

$$\delta \textstyle\int_V U(\sigma)\mathrm{d}V - \int_{O_b} \delta p_i u_i \mathrm{d}O = 0, \tag{3.5.32}$$

since, because of the conditions we have just set down for δp_i and u_i, in the integral on the right, the integration should be carried out over O_b only. Because of the smallness of O_b, the mean value theorem of integration applies, and, using the mean value u_i^m of the displacements, (3.5.32) may be replaced by

$$\delta \int_V U(\sigma)\,\mathrm{d}V = u_i^m \int_{O_b} \delta p_i\,\mathrm{d}O. \tag{3.5.33}$$

In the limit, i.e., for $O_b \to 0$, (3.5.33) changes to

$$\delta \int_V U(\sigma)\,\mathrm{d}V = u_i^P \delta P_i, \tag{3.5.34}$$

when δP_i represents the variation of a concentrated force, and u_i^P is the displacement vector in the point of application of P. Rearranging the scalar product on the right side yields

$$\delta \int_V U(\sigma)\,\mathrm{d}V = u^P \delta P. \tag{3.5.35}$$

Then u^P is the component of the displacement vector at the point of application of the force P_i *in the direction* of this force. Let this force have the magnitude P and let δP be the variation of its magnitude.

Changing $U(\sigma)$ into $U(P)$ so that U becomes a function of forces P, then

$$\delta \int_V U(\sigma)\,\mathrm{d}V = \frac{\partial}{\partial P}\left[\int_V U(P)\,\mathrm{d}V\right]\delta P$$

is obtained, through which from (3.5.35)

$$\frac{\partial\left[\int_V U(P)\,\mathrm{d}V\right]}{\partial P} = \frac{\partial \mathfrak{S}(P)}{\partial P} = u^P \tag{3.5.36}$$

is obtained. This is the formal representation of the actual *theorem of Castigliano*: *If an elastic body is supported in such a way that any rigid body movements are impossible, and when concentrated forces P affect it, then the displacement component u^P of the point of application of a particular concentrated force P in the direction of this force is obtained from the partial derivative of the conjugate strain energy with respect to the particular force.*

Let us assume that the concentrated force P in question is a reaction A. Then, the support does not allow for displacement of the point of application of A in the direction of the reaction. Therefore, the right side of (3.5.36) vanishes, and changing P into A,

$$\frac{\partial \int_V U(A)\,\mathrm{d}V}{\partial A} = \frac{\partial \mathfrak{E}(A)}{\partial A} = 0 \tag{3.5.37}$$

is obtained, which is the actual statement of *Menabrea's theorem.* It is used in the calculation of reactions of statically indeterminate systems.

The extremum theorems given so far serve to express problems of elastostatics in terms of variational problems. Formulation of this kind, as far as the application of *direct methods of the calculus of variations* is concerned, (it will be dealt with at the end of Section 3.5.3), is of the greatest practical significance. It proves to be a hindrance with respect to practical calculation, however, that geometrically permissible displacements or statically permissible stresses have to be used when operating with the classic theorems. Hence, it has been tried to get rid of these restrictions by supplementing the classic theorems. In particular, D. Rüdiger and E. Reissner have dealt with the generalization of the classic energy principles.

Substituting

$$\int_V U(\varepsilon)\,\mathrm{d}V - \int_V K_i u_i\,\mathrm{d}V - \int_{O_p} p_i^0 u_i\,\mathrm{d}O - \int_{O_u} p_i(u_i - u_i^0)\,\mathrm{d}O = \text{Extr.} \tag{3.5.38}$$

instead of (3.5.23), the principle of the stationary value of the total potential energy has been expanded in such a way that the displacements u_i in (3.5.38) have not to satisfy any boundary conditions. Let us consider that in the integral on the right, quantity p_i has to be taken as

$$p_i = [\lambda \delta_{ij} u_{k,k} + \mu(u_{i,j} + u_{j,i})] n_j,$$

i.e., as a function of the displacements. The proof that (3.5.38) does yield the wanted results is given as follows: According to (3.5.38),

$$\delta \int_V U(\varepsilon)\,\mathrm{d}V - \int_V K_i \delta u_i\,\mathrm{d}V - \int_{O_p} p_i^0 \delta u_i\,\mathrm{d}O - \int_{O_u} \delta p_i u_i\,\mathrm{d}O - \int_{O_u} p_i \delta u_i\,\mathrm{d}O + \int_{O_u} \delta p_i u_i^0\,\mathrm{d}O = 0 \tag{3.5.39}$$

must be true. But, let us calculate

$$\delta\int_V U(\varepsilon)\mathrm{d}V = \int_V \sigma_{ij}\delta\varepsilon_{ij}\mathrm{d}V = \int_V \sigma_{ij}(\delta u_i)_{,j}\mathrm{d}V = \int_V (\sigma_{ij}\delta u_i)_{,j}\mathrm{d}V$$
$$-\int_V \sigma_{ij,j}\delta u_i\mathrm{d}V,$$

and, using Gauss's integral theorem,

$$\delta\int_V U(\varepsilon)\mathrm{d}V = \int_O p_i\delta u_i\mathrm{d}O - \int_V \sigma_{ij,j}\delta u_i\mathrm{d}V \tag{3.5.40}$$

is obtained. Substituting (3.5.40) in (3.5.39), and dividing the first integral on the right of (3.5.40) into

$$\int_O p_i\delta u_i\mathrm{d}O = \int_{O_u} p_i\delta u_i\mathrm{d}O + \int_{O_P} p_i\delta u_i\mathrm{d}O,$$

then, after simple rearranging, we obtain

$$-\int_V (\sigma_{ij,j}+K_i)\delta u_i\mathrm{d}V - \int_{O_P}(p_i^0-p_i)\delta u_i\mathrm{d}O - \int_{O_u}(u_i-u_i^0)\delta p_i\mathrm{d}O = 0,$$

proving that the generalized principle (3.5.38) yields the boundary conditions for the surface forces and for the displacements as *natural boundary conditions*, while no requirements were claimed from the varied quantities. The equilibrium conditions are obtained in addition as the Euler equations of (3.5.38).

Correspondingly, the *principle of the stationary value of the conjugated total potential energy* is expanded and

$$\int_V U(\sigma)\mathrm{d}V + \int_V (\sigma_{ij,j}+K_i)u_i\mathrm{d}V - \int_{O_P}(p_i-p_i^0)u_i\mathrm{d}O - \int_{O_u} p_i u_i^0\mathrm{d}O$$
$$= \text{Extr.} \tag{3.5.41}$$

is substituted for (3.5.30). The quantities contained in (3.5.41) which are to be varied may not have to satisfy any boundary conditions. This is shown as follows: Because of (3.5.41)

$$\delta\int_V U(\sigma)\,dV + \int_V \delta(\sigma_{ij,j})u_i\,dV + \int_V \sigma_{ij,j}\delta u_i\,dV + \int_V K_i\delta u_i\,dV$$
$$-\int_{O_P}\delta p_i u_i\,dO - \int_{O_P} p_i\delta u_i\,dO + \int_{O_P} p_i^0\delta u_i\,dO - \int_{O_u}\delta p_i u_i^0\,dO = 0. \tag{3.5.42}$$

However, the following recalculation can be carried out:

$$\delta\int_V U(\sigma)\,dV = \int_V \varepsilon_{ij}\delta\sigma_{ij}\,dV \tag{3.5.43}$$

holds true, and

$$\int_V \delta(\sigma_{ij,j})u_i\,dV = \int_V (\delta\sigma_{ij}u_i)_{,j}\,dV - \int_V \delta\sigma_{ij}u_{i,j}\,dV.$$

By using Gauss' integral theorem, the latter relation is changed to

$$\int_V \delta(\sigma_{ij,j})u_i\,dV = \int_O \delta p_i u_i\,dO - \int_V \delta\sigma_{ij}u_{i,j}\,dV,$$

and because of $\delta\sigma_{ij}\omega_{ij} \equiv 0$, as well as splitting of the first integral on the right side into two terms, it is again changed to

$$\int_V \delta(\sigma_{ij,j})u_i\,dV = \int_{O_u}\delta p_i u_i\,dO + \int_{O_P}\delta p_i u_i\,dO - \int_V \delta\sigma_{ij}(u_{i,j}-\omega_{ij})\,dV. \tag{3.5.44}$$

Through (3.5.43) and (3.5.44), (3.5.42) changes to

$$\int_V (\varepsilon_{ij}+\omega_{ij}-u_{i,j})\delta\sigma_{ij}\,dV + \int_V (\sigma_{ij,j}+K_i)\delta u_i\,dV$$
$$+\int_{O_P}(p_i^0-p_i)\delta u_i\,dO + \int_{O_u}(u_i-u_i^0)\delta p_i\,dO = 0.$$

We read off from this relation that the generalized principle (3.5.41) yields the boundary conditions for the surface forces and for the displacements in the form of *natural boundary conditions*. It also yields the geometric equations and the equilibrium conditions in the form of Euler equations. In either case, no conditions for the variations are set down.

Inserting known relation

$$\int_V U \, \mathrm{d}V = \frac{1}{2}\int_V K_i u_i \, \mathrm{d}V + \frac{1}{2}\int_O p_i u_i \, \mathrm{d}O$$

in the form of

$$\int_V U \, \mathrm{d}V = -\frac{1}{2}\int_V \sigma_{ij,\,j} u_i \, \mathrm{d}V + \frac{1}{2}\int_O p_i u_i \, \mathrm{d}O$$

into the generalized principles (3.5.38) or (3.5.41), then, in both cases, we are led to a third version of the generalized principles, i.e., to

$$\begin{aligned} &\mp \frac{1}{2}\int_V (\sigma_{ij,\,j} + 2K_i) u_i \, \mathrm{d}V \pm \frac{1}{2}\int_{O_P} (p_i - 2p_i^0) u_i \, \mathrm{d}O \\ &\mp \frac{1}{2}\int_{O_u} p_i (u_i - 2u_i^0) \, \mathrm{d}O = \text{Extr.}. \end{aligned} \qquad (3.4.45)$$

This proves that principles (3.5.38) and (3.5.41) are both equivalent.

A further change of (3.5.41) is possible. Let us rewrite the term

$$\int_V \sigma_{ij,\,j} u_i \, \mathrm{d}V$$

into

$$\int_V \sigma_{ij,\,j} u_i \, \mathrm{d}V = -\int_V \sigma_{ij} u_{i,\,j} \, \mathrm{d}V + \int_{O_P} p_i u_i \, \mathrm{d}O + \int_{O_u} p_i u_i \, \mathrm{d}O, \qquad (3.5.46)$$

which is easily done by using Gauss' integral theorem. Substituting (3.5.46) in (3.5.41), we obtain

$$\begin{aligned} J &= \int_V U(\sigma) \, \mathrm{d}V - \int_V \sigma_{ij} u_{i,\,j} \, \mathrm{d}V + \int_V K_i u_i \, \mathrm{d}V \\ &+ \int_{O_P} p_i^0 u_i \, \mathrm{d}O + \int_{O_u} (u_i - u_i^0) p_i \, \mathrm{d}O = \text{Extr.}, \end{aligned} \qquad (3.5.47)$$

which corresponds to the generalized principle indicated by Reissner. It states that an equilibrium condition is present if δJ becomes nought, when varying u_i and σ_{ij}. It states further that if δJ is identical to nought, the geometric equations, equilibrium conditions and all the boundary conditions are being satisfied.

Reissner's statement need not be proven further because of what has been said about (3.5.41) and because (3.5.47) is equivalent to (3.5.41). Instead, let us emphasize the important fact that according to Reissner, in (3.5.47), as well as in the equivalent version (3.5.41) by Rüdiger, *stresses and displacement may be varied independently of each other*. Thus, the classical principle (3.5.30), in which only the stresses are varied has been considerably expanded not only as far as the freedom of variations of the boundary conditions is concerned, but also with regard to the quantities which are to be varied.

3.5.2 *Reciprocal theorems*

Let there be two states of equilibrium. One may be given by

$$K_i, p_i, \sigma_{ij}, u_i,$$

and the other by

$$K_i^*, p_i^*, \sigma_{ij}^*, u_i^*.$$

Let us use equation (3.5.15) to calculate the work done by forces K_i and p_i of the first state of equilibrium condition along the displacements u_i^* of the second state of equilibrium. Following simple rearrangement of (3.5.15), we obtain

$$\int_O p_i u_i^* \mathrm{d}O + \int_V K_i u_i^* \mathrm{d}V = \int_V \sigma_{ij} \varepsilon_{ij}^* \mathrm{d}V. \tag{3.5.15a}$$

Substituting (2.3.2) on the right side of this equation yields

$$\int_O p_i u_i^* \mathrm{d}O + \int_V K_i u_i^* \mathrm{d}V = \int_V C_{ijkl} \varepsilon_{kl} \varepsilon_{ij}^* \mathrm{d}V. \tag{3.5.48}$$

Secondly, let us calculate the work done by the forces K_i^*, p_i^*, along u_i. We use (3.5.15a) where all starred quantities have been substituted by unstarred quantities, and vice-versa. Thus,

$$\int_O p_i^* u_i \mathrm{d}O + \int_V K_i^* u_i \mathrm{d}V = \int_V \sigma_{ij}^* \varepsilon_{ij} \mathrm{d}V$$

is determined. Inserting (2.3.2), it yields

$$\int_O p_i^* u_i \mathrm{d}O + \int_V K_i^* u_i \mathrm{d}V = \int_V C_{ijkl} \varepsilon_{kl}^* \varepsilon_{ij} \mathrm{d}V. \tag{3.5.49}$$

The important fact that $C_{ijkl} = C_{klij}$ is true is derived from condition (2.3.18), the latter applying to isotropic Hooke's body. This fact is easily checked by using (2.3.18). But then

$$\int_V C_{ijkl} \varepsilon_{kl}^* \varepsilon_{ij} \mathrm{d}V = \int_V C_{ijkl} \varepsilon_{kl} \varepsilon_{ij}^* \mathrm{d}V$$

must be true. This justifies the equating of the left sides of (3.5.48) and (3.5.49) yielding

$$\int_O p_i u_i^* \mathrm{d}O + \int_V K_i u_i^* \mathrm{d}V = \int_O p_i^* u_i \mathrm{d}O + \int_V K_i^* u_i \mathrm{d}V. \tag{3.5.50}$$

On this is based *Betti's reciprocity theorem*: *If a Hooke's body is exposed to two different systems of volume and surface forces, then, the actual work done by the forces of the first system along the displacements of the second system is equal to that work done by the forces of the second system along the displacements belonging to the first system.*

This theorem by Betti has already been used in 3.4.2 in the case of vanishing volume forces. Let us relate it to Maxwell's theorem through further specialization: For $K_i = K_i^* \equiv 0$, (3.5.50) changes to

$$\int_O p_i u_i^* \mathrm{d}O = \int_O p_i^* u_i \mathrm{d}O. \tag{3.5.51}$$

Let there be a beam for which, instead of the distributed surface forces p_i, p_i^*, single forces P_1 and P_2 occur in the points of application x_1 and x_2. Let force P_1 in x_1, x_2 respectively, cause the deflections,

$$y_1 = a_{11} P_1, \; y_2 = a_{21} P_1, \text{ respectively.}$$

In the same manner, let there be a deformation of the beam because of P_2, with deflections in x_1 and x_2, respectively, which amount to

$$y_1' = a_{12}P_2, \; y_2' = a_{22}P_2, \text{ respectively.}$$

The coefficients $a_{ij}(i, j = 1, 2)$ we have just used are termed as *Maxwell's influence factors*. Instead of (3.5.51) let us notate

$$P_1 y_1' = P_2 y_2$$

for the beam, which yields

$$P_1 a_{12} P_2 = P_2 a_{21} P_1$$

and, hence,

$$a_{12} = a_{21},$$

and generally

$$a_{ij} = a_{ji}. \tag{3.5.52}$$

Based on this is *Maxwell's theorem*, making up the symmetry of the influence factors, and which, according to (3.5.52) is true. Maxwell's theorem has proven to be a special case of the more general theorem of reciprocity by Betti.

3.5.3 *Examples*

1. *Example.* Let us apply the principle of stationary value of the total potential energy. Let us discuss a beam clamped on the left side and simply supported on the right side, having a distributed load $q(x_1)$ and a constant rectangular cross-section bh (Figure 48). The state of stress is one-dimensional. Hence, we obtain for $\mathfrak{E}(\varepsilon)$

$$\mathfrak{E}(\varepsilon) = \frac{b}{2}\int_0^l \int_{-h/2}^{+h/2} E\varepsilon^2 \mathrm{d}x_2 \, \mathrm{d}x_1 .$$

But, according to the assumptions of the technical theory of bending,

$$\varepsilon = \frac{x_2}{\rho} \approx -x_2 u_{2,11}$$

is true, whereby ρ represents the radius of curvature of the beam's axis.

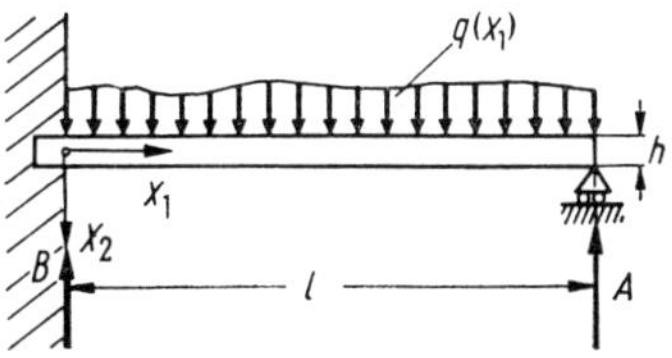

Figure 48

Then,

$$\mathfrak{E}(\varepsilon) = \frac{Eb}{2}\int_0^l \int_{-h/2}^{+h/2} x_2^2 u_{2,11}^2 \, dx_2 \, dx_1 .$$

Using the moment of inertia

$$J = B\int_{-h/2}^{+h/2} x_2^2 \, dx_2 = \frac{bh^3}{12},$$

we obtain

$$\mathfrak{E}(\varepsilon) = \frac{EJ}{2}\int_0^l u_{2,11}^2 \, dx_1 .$$

Omitting the mass forces, let us notate $K_i \equiv 0$. Considering the beam's load as well as the fact that specifications for the displacements only occur at the supports, then the integral

$$\int_{O-O_u} p_i u_i \, dO$$

changes specially into integral

$$\int_0^l q(x_1) \cdot u_2 \, dx_1 ,$$

yielding for the total potential energy the expression

$$\Pi = \frac{EJ}{2}\int_0^l u_{2,11}^2 \, dx_1 - \int_0^l q(x_1) u_2 \, dx_1 . \tag{3.5.53}$$

Requirements

$$u_2(0) = u_{2,1}(0) = u_2(l) = 0 \tag{3.5.54}$$

result as boundary conditions for the displacement from the way the beam is supported. Thus, the principle of stationary value of the total potential energy requires that the functional (3.5.53) in the class of functions $u_2(x_1)$, which satisfy conditions (3.5.54), be made stationary.

2. *Example.* Let us apply the principle of stationary value of the conjugate total potential energy. Again, let us discuss the beam in Figure 48. Since the surface forces along its total length have been specified, then even the special formulation of the principle is obtained which requires the stationary value of the conjugate strain energy $\mathfrak{E}(\sigma)$. Let

$$\mathfrak{E}(\sigma) = \frac{b}{2}\int_0^l \int_{-h/2}^{+h/2} \frac{\sigma 2}{E} \, dx_2 \, dx_1$$

be true. The technical theory of bending yields

$$\sigma = \frac{M(x_1)}{J} x_2$$

through which

$$\mathfrak{E}(\sigma) = \frac{b}{2}\int_0^l \int_{-h/2}^{+h/2} \frac{M^2(x_1)}{EJ^2} x_2^2 \, dx_2 \, dx_1 = \frac{1}{2EJ}\int_0^l M^2(x_1) \, dx_1 \tag{3.5.55}$$

is obtained. For the boundary condition of the stresses, there is $\sigma = 0$ for $x_1 = l$, which together with another auxiliary condition is also notated as

$$M(l) = 0, M_{2,11} = -q(x_1). \tag{3.5.56}$$

The principle of stationary value of the conjugate total potential energy requires that functional (3.5.55) in the class of functions $M(x_1)$, fulfilling condition (3.5.56), be made stationary.

3. *Example.* Let us apply Menabrea's theorem, and this theorem to calculate reaction A at the beam's right support, which is taken to be a statically indeterminate quantity.

Let

$$M(x_1) = A(l-x_1)-\tfrac{1}{2}q(l-x_1)^2$$

be valid. Inserting this expression in (3.5.55) yields

$$\mathfrak{E}(A) = \frac{1}{2EJ}\int_0^l [A(l-x_1)-\tfrac{1}{2}q(l-x_1)^2]^2 \mathrm{d}x_1 .$$

Condition

$$\frac{\partial \mathfrak{E}(A)}{\partial A} = \frac{1}{EJ}\int_0^l [A(l-x_1)-\tfrac{1}{2}q(l-x_1)^2](l-x_1)\,\mathrm{d}x_1 = 0$$

is obtained from Menabrea's theorem. It yields $A = 3ql/8$ for $q = \text{const.}$ for the statically indeterminate reaction of the right support.

4. *Example.* Let us apply the principle of stationary value of the total potential energy. Let us discuss a thin plate in the x_1, x_2 plane, subjected to load p, and having stiffness D and deflecting u_3. Let the plate be clamped along its edges.

The total potential energy of the plate is

$$\Pi = \frac{D}{2}\int_F \left[(\nabla^2 u_3)^2 - 2(1-\nu)(u_{3,11}u_{3,22}-u_{3,12}^2) - \frac{2p}{D}u_3\right] \mathrm{d}x_1\,\mathrm{d}x_2 . \tag{3.5.57}$$

It is easily found that

$$\int_F (u_{3,11}u_{3,22}-u_{3,12}^2)\,\mathrm{d}x_1\,\mathrm{d}x_2 = \frac{1}{2}\int_R \left(\frac{\partial u_3}{\partial x_1}\frac{\partial^2 u_3}{\partial x_2\,\partial s} - \frac{\partial u_3}{\partial x_2}\frac{\partial^2 u_3}{\partial x_1\,\partial s}\right)\mathrm{d}s \tag{3.5.58}$$

is true, when ds represents the line element along the edges. Because the board is clamped at the edge R, then

$$u_3(R) = 0, \frac{\partial u_3}{\partial n}(R) = 0 \tag{3.5.59}$$

is specified for the deflection. $\partial/\partial n$ represents the derivative with respect to normal n. Because of (3.5.59), also $\partial u_3/\partial s$ on R equals nought, so that (3.5.58) becomes nought. Therefore, the principle's requirement reduces itself to the fact that the functional

$$\frac{D}{2}\int_F \left[(\nabla^2 u_3)^2 - \frac{2p}{D} u_3\right] \mathrm{d}x_1 \, \mathrm{d}x_2$$

in the class of functions $u_3(x_1 x_2)$, which fulfill (3.5.59), be stationary.

Finally, let us point out that in examples 1, 2 and 4, we have arrived at problems of the calculus of variations. It is for this reason that working with extremum principles of elastostatics is so successful, since in connection with these, we are able to use the various kinds of direct methods of the calculus of variations for finding solutions, e.g., among others those of Ritz, Galerkin, Kantorowitsch, Trefftz, etc. Operation of these methods will not be discussed here. Let us refer to an appropriate collection of literature instead, and in particular to [14], [15], [16], [19], [25] and [26].

4

Special structures

4.1 Plates

4.1.1 *Differential equation, boundary conditions and internal forces*

Let the term plate mean an elastic body of thickness h, having a *middle-plane* dividing its thickness. Prior to its deformation, let the plate's middle-plane coincide with the x_1, x_2 plane of the reference system. Let load $p = p(x_1, x_2)$ be applied to the plate's surface F in the x_3-direction vertically to the middle-plane (Figure 49). In the most general case, let it be permitted that distributed vertical forces $v(s)$ and distributed moments $m(s)$ are in effect at boundary R of the plate. Let s be the arc length of R. Also, the restricting assumption that thickness h is constant for the plate, shall be used in the following.

In the classic plate theory, which is fundamental to our considerations, and which is also known as *Kirchhoff's plate theory*, the following additional assumptions are made:

a) Let the middle-plane of the plate remain unstrained during bending, so that only displacements u_3 of its points take place.

b) Let the strain ε_{33} be negligibly small, so that u_3 becomes independent of x_3.

c) Let the normals of the middle-plane remain normals of the deformed middle-plane even after deformation of the plate (Kirchhoff's generalization of Bernoulli's hypothesis of the technical bending theory of beams).

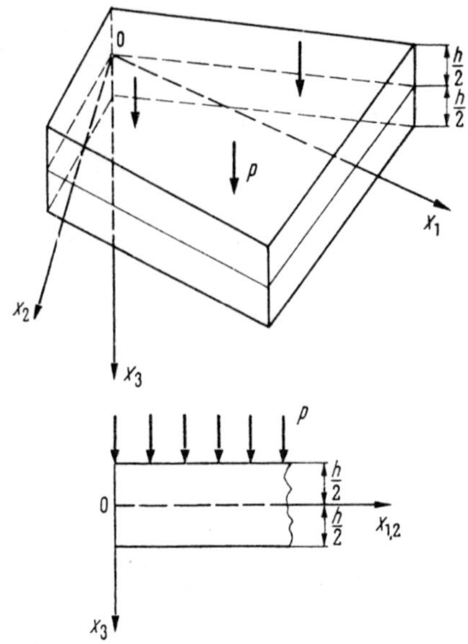

Figure 49

d) Let the stresses σ_{13}, σ_{23} and σ_{33} be so small in comparison to the remaining stresses that they can be omitted in the calculation of the plate's strain energy.

To justify these conditions it is required that the plate's thickness h and the deflection $u_3 = u_3(x_1, x_2)$ of its middle-plane be small. Therefore, we shall be dealing with the *theory of thin plates having small deflections*. Let

$$u_1 = \int_0^{x_3} u_{1,3} \, dx_3$$

be true, which because of the smallness of h is notated as being near enough

$$u_1 = u_{1,3}(x_3 = 0) \cdot x_3 .$$

Because of the above conditions, ε_{13} for $x_3 = 0$ is equal to nought. Therefore, $u_{1,3}(x_3 = 0) = -u_{3,1}(x_3 = 0)$ is valid or more generally, (again because of the plate's little thickness), $u_{1,3}(x_3 = 0) = -u_{3,1}$. From this,

$$u_1 = -u_{3,1} \cdot x_3 \tag{4.1.1}$$

is determined. In the same way we find

$$u_2 = -u_{3,2} \cdot x_3 . \tag{4.1.2}$$

Now we are able to calculate the components of the strain tensor which are needed for further calculation:

$$\begin{aligned} \varepsilon_{11} &= u_{1,1} = -u_{3,11} x_3 , \\ \varepsilon_{22} &= u_{2,2} = -u_{3,22} x_3 , \\ \varepsilon_{12} &= \tfrac{1}{2}(u_{1,2}+u_{2,1}) = -u_{3,12} x_3 . \end{aligned} \tag{4.1.3}$$

On the condition that σ_{33} can be omitted as compared to σ_{11} and σ_{22}, then the latter two normal stresses, by using Hooke's law, have the expressions

$$\sigma_{11} = \frac{E}{1-\nu^2}(\varepsilon_{11}+\nu\varepsilon_{22}), \quad \sigma_{22} = \frac{E}{1-\nu^2}(\varepsilon_{22}+\nu\varepsilon_{11}).$$

The shearing stress σ_{12} is given directly as

$$\sigma_{12} = \frac{E}{1+\nu}\varepsilon_{12} .$$

Through (4.1.3), the foregoing equations change to:

$$\begin{aligned} \sigma_{11} &= -\frac{E}{1-\nu^2} x_3(u_{3,11}+\nu u_{3,22}), \\ \sigma_{22} &= -\frac{E}{1-\nu^2} x_3(u_{3,22}+\nu u_{3,11}), \\ \sigma_{12} &= -\frac{E}{1+\nu} x_3 u_{3,12} . \end{aligned} \tag{4.1.4}$$

Let us use the second equation of (3.5.14). Notating it when it is governed by condition d, then we obtain,

$$U(\sigma) = \frac{1}{2E}[\sigma_{11}^2+\sigma_{22}^2-2\nu\sigma_{11}\sigma_{22}+2(1+\nu)\sigma_{12}^2]. \tag{4.1.5}$$

Inserting (4.1.4), and after simple recalculation,

$$U(\varepsilon) = \frac{Ex_3^2}{2(1-\nu^2)}\left[(u_{3,11}+u_{3,22})^2+2(1-\nu)(u_{3,12}^2-u_{3,11}u_{3,22})\right]$$

is determined. We obtain by integration

$$\mathfrak{E}(\varepsilon) = \int_F\int_{-h/2}^{+h/2} U(\varepsilon)\,\mathrm{d}x_3\,\mathrm{d}F$$

$$= \frac{Eh^3}{24(1-\nu^2)}\int_F\left[(u_{3,11}+u_{3,22})^2+2(1-\nu)(u_{3,12}^2-u_{3,11}u_{3,22})\right]\mathrm{d}F.$$

To abbreviate, let us introduce quantity

$$D = \frac{Eh^3}{12(1-\nu^2)}, \tag{4.1.6}$$

and call it the *plate's flexural rigidity*. Thus, we finally obtain

$$\mathfrak{E}(\varepsilon) = \frac{D}{2}\int_F\left[(u_{3,11}+u_{3,22})^2+2(1-\nu)(u_{3,12}^2-u_{3,11}u_{3,22})\right]\mathrm{d}F \tag{4.1.7}$$

for the plate's strain energy. In the following, let us always omit volume forces K_i. In addition, let R_e be that part of the plate's edge, in which the plate has been clamped. Let R_g be that part in which it is simply supported and let R_f be that part in which it is completely free. Then, from (3.5.22) and (3.5.23) through (4.1.7), the variational problem

$$\Pi = \frac{D}{2}\int_F\left[(u_{3,11}+u_{3,22})^2+2(1-\nu)(u_{3,12}^2-u_{3,11}u_{3,22})-\frac{2p}{D}\,u_3\right]\mathrm{d}F$$

$$-\int_{R_\mathrm{f}} v(s)u_3\,\mathrm{d}s+\int_{R_\mathrm{g}+R_\mathrm{f}} m(s)u_{3,n}\,\mathrm{d}s = \text{Extr.}, \tag{4.1.8}$$

is determined, whereby $u_{3,\mathrm{n}}$ is the derivative of u_3 with respect to the normal of R, and the admissible function u_3 has to satisfy the *geometric conditions*. These are

$$u_3 = u_{3,\mathrm{n}} = 0 \text{ on } R_\mathrm{e}, \tag{4.1.9}$$

where the edge can be neither displaced nor rotated because it has been clamped, as well as

$$u_3 = 0 \text{ on } R_\mathrm{g}, \tag{4.1.10}$$

where the edge, because of the support, cannot be displaced. The requirement (4.1.8) is being satisfied for $\delta\Pi = 0$. Calculating this, we arrive at

$$\int_F \Big[(u_{3,11}+u_{3,22})\cdot\delta(u_{3,11}+u_{3,22})+(1-\nu)\cdot\delta(u_{3,12}^2-u_{3,11}u_{3,22})$$
$$-\frac{p}{D}\delta u_3\Big]\,\mathrm{d}F-\int_{R_\mathrm{f}}\frac{v(s)}{D}\delta u_3\,\mathrm{d}s+\int_{R_\mathrm{g}+R_\mathrm{f}}\frac{m(s)}{D}(\delta u_3)_{,n}\mathrm{d}s = 0. \tag{4.1.11}$$

Now the first and the second term of (4.1.11) may be rearranged. Let

$$\int_F (u_{3,11}+u_{3,22})\cdot\delta(u_{3,11}+u_{3,22})\mathrm{d}F = \int_F u_{3,\alpha\alpha}(\delta u_3)_{,\beta\beta}\,\mathrm{d}F,$$
$$\alpha, \beta = 1, 2,$$

be true. Furthermore,

$$u_{3,\alpha\alpha}(\delta u_3)_{,\beta\beta} = [u_{3,\alpha\alpha}(\delta u_3)_{,\beta}]_{,\beta}-(\delta u_3)_{,\beta}u_{3,\alpha\alpha\beta},$$

so that

$$\int_F u_{3,\alpha\alpha}(\delta u_3)_{,\beta\beta}\,\mathrm{d}F = \int_F [u_{3,\alpha\alpha}(\delta u_3)_{,\beta}]_{,\beta}\,\mathrm{d}F-\int_F (\delta u_3)_{,\beta}u_{3,\alpha\alpha\beta}\,\mathrm{d}F$$

is notated. Using Gauss' integral theorem, this is changed to

$$\int_F u_{3,\alpha\alpha}(\delta u_3)_{,\beta\beta}\,\mathrm{d}F = \int_R u_{3,\alpha\alpha}(\delta u_3)_{,\beta}n_\beta\,\mathrm{d}s-\int_F (\delta u_3)_{,\beta}u_{3,\alpha\alpha\beta}\,\mathrm{d}F.$$

For the integral on the extreme right we calculate in the same way

$$(\delta u_3)_{,\beta}u_{3,\alpha\alpha\beta} = (u_{3,\alpha\alpha\beta}\cdot\delta u_3)_{,\beta}-u_{3,\alpha\alpha\beta\beta}\cdot\delta u_3$$

enabling us to use Gauss' theorem for the integral on the upper right hand side. Finally, it leads to

$$\int_F u_{3,\alpha\alpha}(\delta u_3)_{,\beta\beta}\,\mathrm{d}F = \int_F u_{3,\alpha\alpha\beta\beta}\,\delta u_3\,\mathrm{d}F - \int_R u_{3,\alpha\alpha\beta}\,n_\beta\,\delta u_3\,\mathrm{d}s +$$
$$+ \int_R u_{3,\alpha\alpha}(\delta u_3)_{,\beta}\,n_\beta\,\mathrm{d}s. \tag{4.1.12}$$

Let the symbol ∇^2 be considered two-dimensional. In addition, because of

$$\boldsymbol{n} = (\cos\alpha, \sin\alpha) = \left(\frac{\mathrm{d}x_1}{\mathrm{d}n}, \frac{\mathrm{d}x_2}{\mathrm{d}n}\right),$$

which is read off Figure 50, we notate for any function f,

$$f_{,\beta}n_\beta = f_{,1}n_1 + f_{,2}n_2 = f_{,1}\cos\alpha + f_{,2}\sin\alpha =$$
$$= \frac{\partial f}{\partial x_1}\frac{\mathrm{d}x_1}{\mathrm{d}n} + \frac{\partial f}{\partial x_2}\frac{\mathrm{d}x_2}{\mathrm{d}n} = f_{,n}. \tag{4.1.13}$$

Thus, from (4.1.12),

$$\int_F u_{3,\alpha\alpha}(\delta u_3)_{,\beta\beta}\,\mathrm{d}F = \int_F \nabla^2\nabla^2 u_3\,\delta u_3\,\mathrm{d}F - \int_R (\nabla^2 u_3)_{,n}\,\delta u_3\,\mathrm{d}s +$$
$$+ \int_R \nabla^2 u_3(\delta u_3)_{,n}\,\mathrm{d}s \tag{4.1.14}$$

is obtained. Now, let us change the second term of (4.1.11), and let

$$\int_F \delta(u_{3,12}^2 - u_{3,11}u_{3,22})\,\mathrm{d}F = \int_F [2u_{3,12}(\delta u_3)_{,12} - u_{3,11}(\delta u_3)_{,22} -$$
$$-(\delta u_3)_{,11}u_{3,22}]\,\mathrm{d}F$$

be true. This can also be written as

$$\int_F g_{\alpha,\alpha}\,\mathrm{d}F = \int_R g_\alpha n_\alpha\,\mathrm{d}s = \int_R (g_1\cos\alpha + g_2\sin\alpha)\,\mathrm{d}s \tag{4.1.15}$$

when Gauss' theorem is used. For the sake of brevity,

$$g_1 = (\delta u_3)_{,2} u_{3,12} - (\delta u_3)_{,1} u_{3,22}, \; g_2 = (\delta u_3)_{,1} u_{3,12} - (\delta u_3)_{,2} u_{3,11} \tag{4.1.16}$$

have been used in (4.1.15).

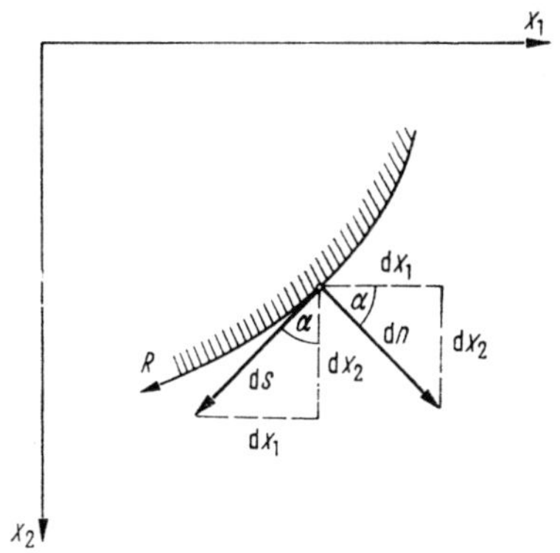

Figure 50

From Figure 50 we read off

$$\frac{dx_1}{ds} = -\sin \alpha, \; \frac{dx_2}{ds} = \cos \alpha,$$

from which for any function f we obtain

$$f_{,s} = \frac{\partial f}{\partial x_1}\frac{dx_1}{ds} + \frac{\partial f}{\partial x_2}\frac{dx_2}{ds} = -f_{,1} \sin \alpha + f_{,2} \cos \alpha. \tag{4.1.17}$$

From (4.1.13) and (4.1.17)

$$\begin{aligned} f_{,1} &= -f_{,s} \sin \alpha + f_{,n} \cos \alpha, \\ f_{,2} &= f_{,s} \cos \alpha + f_{,n} \sin \alpha \end{aligned} \tag{4.1.18}$$

is derived.

Let us apply (4.1.18) to the derivatives of δu_3 in (4.1.16) and insert the thus changed quantities g_1 and g_2 in (4.1.15). Through corresponding summary

$$\int_F \delta(u_{3,12}^2 - u_{3,11} u_{3,22}) \mathrm{d}F = \int_R \{(2 \sin\alpha \cos\alpha u_{3,12} - \sin^2 \alpha u_{3,11} - \\ -\cos^2 \alpha u_{3,22})(\delta u_3)_{,n} + [\sin\alpha \cos\alpha (u_{3,22} - u_{3,11}) + \\ + u_{3,12}(\cos^2\alpha - \sin^2\alpha)](\delta u_3)_{,s}\} \mathrm{d}s \tag{4.1.19}$$

is obtained. The second term on the right side of (4.1.19) is changed again through integration by parts yielding

$$\int_F \delta(u_{3,12}^2 - u_{3,11} u_{3,22}) \mathrm{d}F = \int_R \{(2 \sin\alpha \cos\alpha u_{3,12} - \sin^2 \alpha u_{3,11} - \\ -\cos^2 \alpha u_{3,22})(\delta u_3)_{,n} - [\sin\alpha \cos\alpha (u_{3,22} - u_{3,11}) + \\ + u_{3,12}(\cos^2\alpha - \sin^2\alpha)]_{,s} \delta u_3\} \mathrm{d}s. \tag{4.1.20}$$

These intermediate results, i.e., (4.1.14) and (4.1.20), are inserted in (4.1.11). It yields

$$\int_F \left[\nabla^2\nabla^2 u_3 - \frac{p}{D}\right] \delta u_3 \, \mathrm{d}F \\ + \int_R \{\nabla^2 u_3 + (1-\nu)(2 \sin\alpha \cos\alpha u_{3,12} - \sin^2\alpha u_{3,11} - \cos^2\alpha u_{3,22})\} \\ \times (\delta u_3)_{,n} \mathrm{d}s - \int_R \{(\nabla^2 u_3)_{,n} + (1-\nu)[\sin\alpha \cos\alpha (u_{3,22} - u_{3,11}) \\ + u_{3,12}(\cos^2\alpha - \sin^2\alpha)]_{,s}\} \delta u_3 \, \mathrm{d}s \\ + \int_{R_g + R_f} \frac{m(s)}{D} (\delta u_3)_{,n} \mathrm{d}s - \int_{R_f} \frac{v(s)}{D} \delta u_3 \, \mathrm{d}s = 0. \tag{4.1.21}$$

From the first term of (4.1.21) we read off Euler's differential equation

$$\nabla^2\nabla^2 u_3 = \frac{p}{D}. \tag{4.1.22}$$

This inhomogeneous biharmonic equation is termed *Kirchhoff's plate equation* and represents the differential equation governing Kirchhoff's plate theory.

Through the requirement that the line integral in (4.1.21) disappear, we arrive at the *natural* or *dynamic boundary conditions*, respectively, of the plate theory. Considering the already-mentioned geometric boundary conditions (4.1.9) and (4.1.10), the following conditions for the various kinds of support of the plate are obtained:

a) *Clamped edge* R_e: In this case,

$$u_3 = u_{3,n} = 0 \tag{4.1.9}$$

is valid, implying $\delta u_3 = (\delta u_3)_{,n} = 0$, so that the line integrals of (4.1.21) disappear for this reason on R_e.

b) *Simply supported edge* R_g: In this case,

$$u_3 = 0 \tag{4.1.10}$$

is valid, so that $\delta u_3 = 0$ must be true also. In order that the line integrals of (4.1.21) disappear on R_g completely, then the natural boundary condition

$$D\{\nabla^2 u_3 + (1-\nu)(2 \sin\alpha \cos\alpha u_{3,12} - \sin^2\alpha u_{3,11} - \cos^2\alpha u_{3,22}\} + m(s) = 0 \tag{4.1.23}$$

must be satisfied in addition. If we are specially dealing with a straight edge parallel to the x_2-axis, then $\alpha = 0$ is true, and hence, from (4.1.23)

$$D(u_{3,11} + \nu u_{3,22}) + m(s) = 0 \tag{4.1.24}$$

is obtained. If no boundary moments $m(s)$ have been specified as an external load, then (4.1.24) changes into

$$u_{3,11} + \nu u_{3,22} = 0. \tag{4.1.25}$$

c) *Free edge* R^f: No geometric conditions have been specified for it, so that δu_3 as well as $(\delta u_3)_n$ may be arbitrary. When the line integrals of (4.1.21) on R_f are required to disappear in spite of this, then both the natural boundary conditions

$$D\{\nabla^2 u_3+(1-\nu)(2\sin\alpha\cos\alpha u_{3,12}-\sin^2\alpha u_{3,11}-\cos^2\alpha u_{3,22}\}$$
$$+m(s) = 0, \tag{4.1.26}$$

and

$$D\{(\nabla^2 u_3)_{,n}+(1-\nu)[\sin\alpha\cos\alpha(u_{3,22}-u_{3,11})+u_{3,12}(\cos^2\alpha$$
$$-\sin^2\alpha)]_{,s}\}+v(s) = 0$$

must be satisfied. If we are dealing specially with a straight edge parallel to the x_2-axis, then from (4.1.26), and because of $\alpha = 0$, (4.1.13), and (4.1.17),

$$\begin{aligned} &D(u_{3,11}+\nu u_{3,22})+m(s) = 0,\\ &D[u_{3,111}+(2-\nu)u_{3,122}]+v(s) = 0 \end{aligned} \tag{4.1.27}$$

applies to it. When no boundary moments in $m(s)$ and no boundary forces $v(s)$ have been specified as external loads, then (4.1.27) changes to

$$\begin{aligned} &u_{3,11}+\nu u_{3,22} = 0,\\ &u_{3,111}+(2-\nu)u_{3,122} = 0. \end{aligned} \tag{4.1.28}$$

Since the differential equation (4.1.22), and the accompanying boundary conditions (4.1.9), (4.1.10), and (4.1.23) to (4.1.28) are known, then we could calculate the deflection u_3 of the plate's middle-plane and following this we could find the strains and stresses through (4.1.3) and (4.1.4). However, let us put this problem off until the next section and first determine the *internal forces*. It is quite usual to use these in engineering, especially for illustrating the boundary conditions. The internal forces are defined as forces or moments per unit length (Figure 51).

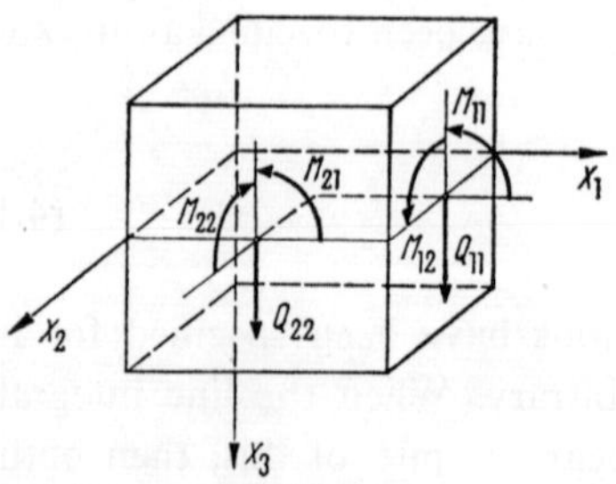

Figure 51

Quantities

$$Q_{11} = \int_{-h/2}^{+h/2} \sigma_{13}\, dx_3\,, Q_{22} = \int_{-h/2}^{+h/2} \sigma_{23}\, dx_3\,, \tag{4.1.29}$$

are termed *shearing forces* and quantities

$$\begin{aligned} M_{11} &= \int_{-h/2}^{+h/2} \sigma_{11}\, x_3\, dx_3\,, M_{22} = \int_{-h/2}^{+h/2} \sigma_{22}\, x_3\, dx_3\,, \\ M_{12} &= -M_{21} = -\int_{-h/2}^{+h/2} \sigma_{12}\, x_3\, dx_3\ ^{1}) \end{aligned} \tag{4.1.30}$$

are termed *moments*. It is at this point that we notice that Kirchhoff's plate theory is not entirely free of contradictions. To begin with, stresses σ_{i3}, $i = 1, 2, 3$ were omitted with respect to the others. However, this point we are using for (4.1.29) shearing stresses σ_{13}, σ_{23}, and we shall be using all results obtained by means of them in the following calculations, regardless of the original assumptions.

Firstly, let us calculate the stresses σ_{i3}, whereby the plate equation (4.1.22) is again obtained as an additional result.

Omitting volume forces K_i, let us notate for the equilibrium conditions simply

$$\sigma_{ij,j} = 0. \tag{4.1.31}$$

From (4.1.31) for $i = 1$,

$$\sigma_{13,3} = -\sigma_{12,2} - \sigma_{11,1}$$

is yielded, which because of (4.1.4) leads to

$$\sigma_{13,3} = \frac{E}{1-\nu^2}\, x_3 (u_{3,111} + u_{3,122}). \tag{4.1.32}$$

Correspondingly, we obtain from (4.1.31) for $i = 2$, and from (4.1.4),

[1] At times, these two moments are defined by having the *same* sign in all cases.

$$\sigma_{13,3} = \frac{E}{1-\nu^2} x_3(u_{3,112}+u_{3,222}). \tag{4.1.33}$$

Because of condition b), which states that $\varepsilon_{33} = u_{3,3} = 0$ is true, it has been deduced that $u_3 = u_3(x_1, x_2)$ is also true, i.e., u_3 remains independent of x_3. Through integration, from (4.1.32) and (4.1.33),

$$\begin{aligned} \sigma_{13} &= \frac{E}{1-\nu^2}(u_{3,111}+u_{3,122})\left[\frac{x_3^2}{2}+c_1(x_1, x_2)\right], \\ \sigma_{23} &= \frac{E}{1-\nu^2}(u_{3,112}+u_{3,222})\left[\frac{x_3^2}{2}+c_2(x_1, x_2)\right] \end{aligned} \tag{4.1.34}$$

is obtained. Using boundary conditions

$$\sigma_{13}\left(x_1, x_2, x_3 = \pm\frac{h}{2}\right) = \sigma_{23}\left(x_1, x_2, x_3 = \pm\frac{h}{2}\right) = 0,$$

which are valid for σ_{13}, σ_{23}, the constants of integration c_1, c_2 in (4.1.34) are determined as

$$c_1 = c_2 = -\frac{h^2}{8},$$

so that

$$\begin{aligned} \sigma_{13} &= \frac{E}{1-\nu^2}(u_{3,111}+u_{3,122})\left(\frac{x_3^2}{2}-\frac{h^2}{8}\right), \\ \sigma_{23} &= \frac{E}{1-\nu^2}(u_{3,112}+u_{3,222})\left(\frac{x_3^2}{2}-\frac{h^2}{8}\right). \end{aligned} \tag{4.1.35}$$

Because of (4.1.31), $\sigma_{ij,ji} = 0$ is also true. Notating this for $i = 3$, we obtain, after rearranging, $\sigma_{33,33} = -\sigma_{31,13}-\sigma_{32,23}$. Those terms which in this relation are on the right side, are obtained from (4.1.32) and (4.1.33) by differentiation. Thus, we obtain

$$\begin{aligned} \sigma_{33,33} &= -\frac{E}{1-\nu^2} x_3(u_{3,1111}+2u_{3,1122}+u_{3,2222}) = \\ &- \frac{E}{1-\nu^2} x_3 \nabla^2\nabla^2 u_3. \end{aligned} \tag{4.1.36}$$

From (4.1.36),

$$\sigma_{33} = -\frac{E}{1-\nu^2}\nabla^2\nabla^2 u_3\left[\frac{x_3^3}{6} + c_3(x_1, x_2)\cdot x_3 + c_4(x_1, x_2)\right] \tag{4.1.37}$$

is determined by double integration with respect to x_3. But for σ_{33}, boundary conditions

$$\sigma_{33}\left(x_1, x_2, x_3 = \frac{h}{2}\right) = 0,\ \sigma_{33}\left(x_1, x_2, x_3 = -\frac{h}{2}\right) = -p$$

are given and, hence, constants of integration c_3 and c_4 are determined as

$$c_3 = -\frac{h^2}{24} - \frac{1-\nu^2}{E\nabla^2\nabla^2 u_3}\frac{p}{h},\ c_4 = \frac{1}{2}\frac{1-\nu^2}{E\nabla^2\nabla^2 u_3}p.$$

Thus, (4.1.37) changes to

$$\sigma_{33} = -\frac{E}{1-\nu^2}\nabla^2\nabla^2 u_3\left(\frac{x_3^2}{6} - \frac{h^2 x_3}{24}\right) - p\left(\frac{1}{2} - \frac{x_3}{h}\right). \tag{4.1.38}$$

Inserting (4.1.35) and (4.1.38) into equilibrium condition $\sigma_{3j,j} = 0$, then we obtain

$$\frac{E}{1-\nu^2}\nabla^2\nabla^2 u_3\left(\frac{x_3^2}{2} - \frac{h^2}{8} - \frac{x_3^2}{2} + \frac{h^2}{24}\right) + \frac{p}{h} = 0,$$

which through (4.1.6) and for $x_3 = \pm h/2$ is changed to

$$\nabla^2\nabla^2 u_3 = \frac{p}{D}, \tag{4.1.22}$$

i.e., the already known Kirchhoff's plate equation. Let us point out in addition that (4.1.38) through (4.1.22) can be changed to

$$\sigma_{33} = -p\left[2\left(\frac{x_3}{h}\right)^3 - \frac{3}{2}\frac{x_3}{h} + \frac{1}{2}\right]. \tag{4.1.39}$$

We shall use this result at a later stage. Let us calculate now the internal forces according to (4.1.29), (4.1.30) through (4.1.35) and (4.1.4). Then,

$$M_{11} = -D(u_{3,11}+\nu u_{3,22}),\ M_{22} = -D(u_{3,22}+\nu u_{3,11}),$$
$$M_{12} = -M_{21} = D(1-\nu)u_{3,12} \tag{4.1.40}$$

and

$$Q_{11} = -D(u_{3,11}+u_{3,22})_{,1},\ Q_{22} = -D(u_{3,11}+u_{3,22})_{,2} \tag{4.1.41}$$

is obtained. Comparison of (4.1.40) with (4.1.41) indicates that

$$Q_{11} = M_{11,1}+M_{21,2} \text{ and } Q_{22} = M_{22,2}-M_{12,1} \tag{4.1.42}$$

is true.

Let us define as vertical supporting forces, which result from the torsional moment M_{12} and from the shearing force Q_{11} at the cut or at the plate's edge, respectively,

$$V_{11} = Q_{11}-M_{12,2}, \tag{4.1.43}$$

which because of (4.1.10) and (4.1.41) yields

$$V_{11} = -D[u_{3,111}+(2-\nu)u_{3,122}]. \tag{4.1.44}$$

The internal forces in a vertical cut at a slope to the coordinate axes x_1, x_2 (Figures 52, 53), are obtained analogous to (4.1.29) and (4.1.30).

Hence, for the moments

$$M_{nn} = \int_{-h/2}^{+h/2} \sigma_{nn} x_3 \, dx_3,\ M_{nt} = -\int_{-h/2}^{+h/2} \sigma_{nt} x_3 \, dx_3 \tag{4.1.45}$$

is valid.

With respect to the axis system $x'_1, x'_2 \equiv n, t$, which is rotated through angle α relative to axis system x_1, x_2, the stresses

$$\sigma'_{11} = \sigma_{nn} = \cos^2\alpha\sigma_{11}+\sin^2\alpha\sigma_{22}+2\sin\alpha\cos\alpha\sigma_{12},$$
$$\sigma'_{12} = \sigma_{nt} = (\sigma_{22}-\sigma_{11})\sin\alpha\cos\alpha+\sigma_{12}(\cos^2\alpha-\sin^2\alpha) \tag{4.1.46}$$

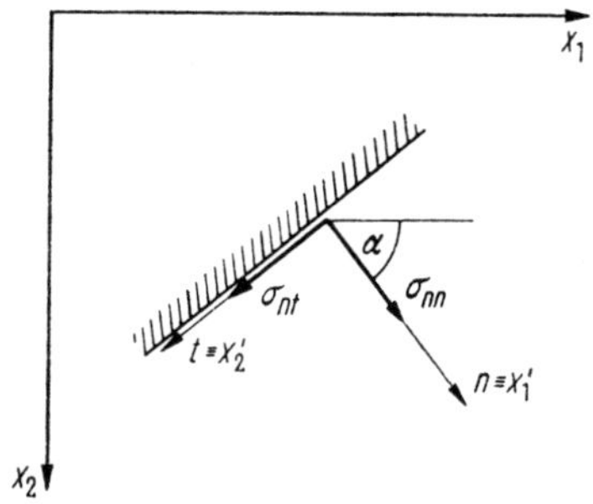

Figure 52

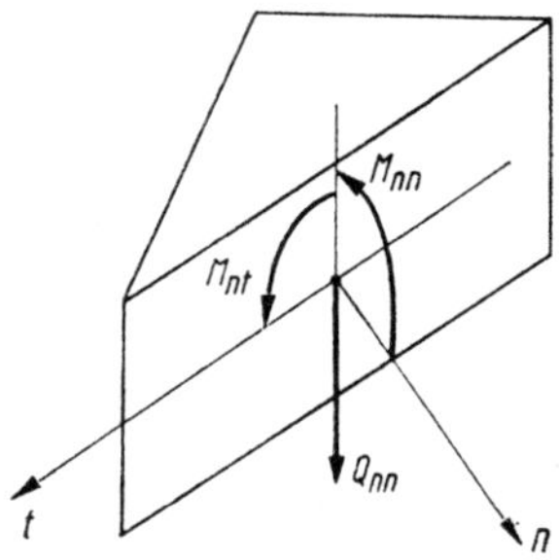

Figure 53

are given by the transformation formula $\sigma'_{ij} = a_{ik}a_{jl}\sigma_{kl}$ containing transformation matrix

$$(a_{ik}) = \begin{pmatrix} \cos\alpha & \sin\alpha & 0 \\ -\sin\alpha & \cos\alpha & 0 \\ 0 & 0 & 1 \end{pmatrix}.$$

Inserting (4.1.46) in (4.1.45), then for the moments we obtain

$$\begin{aligned} M_{nn} &= M_{11}\cos^2\alpha + M_{22}\sin^2\alpha - 2M_{12}\sin\alpha\cos\alpha, \\ M_{nt} &= (M_{11}-M_{22})\sin\alpha\cos\alpha + M_{12}(\cos^2\alpha - \sin^2\alpha), \end{aligned} \tag{4.1.47}$$

when (4.1.30) is considered.

Using (4.1.40), and following simple change, we obtain from (4.1.47)

$$\begin{aligned} M_{nn} &= -D[\nabla^2 u_3 + (1-\nu)(2\sin\alpha\cos\alpha u_{3,12} - \sin^2\alpha u_{3,11} \\ &\quad -\cos^2\alpha u_{3,22})], \\ M_{nt} &= D(1-\nu)[(u_{3,22}-u_{3,11})\sin\alpha\cos\alpha + u_{3,12}(\cos^2\alpha - \sin^2\alpha)]. \end{aligned} \tag{4.1.48}$$

Correspondingly, the expression

$$Q_{nn} = \int_{-h/2}^{+h/2} \sigma_{n3}\, dx_3 \tag{4.1.49}$$

applies to the shearing force in the diagonal cut. Let $\sigma_{n3} = \sigma'_{13} = a_{11}a_{33}\sigma_{13}+a_{12}a_{33}\sigma_{23} = \cos\alpha\sigma_{13}+\sin\alpha\sigma_{23}$ be true, so that (4.1.49) containing the latter, and because of (4.1.29), changes to

$$Q_{nn} = Q_{11}\cos\alpha+Q_{22}\sin\alpha. \tag{4.1.50}$$

When relations (4.1.41) and (4.1.13) are used in (4.1.50), then we obtain

$$Q_{nn} = -D(\nabla^2 u_3)_{,n}. \tag{4.1.51}$$

Finally, analogous to (4.1.43), let us notate

$$V_{nn} = Q_{nn}-M_{nt,s}, \tag{4.1.52}$$

which changes to

$$V_{nn} = -D\{(\nabla^2 u_3)_{,n}+(1-\nu)[\sin\alpha\cos\alpha(u_{3,22}-u_{3,11})+ u_{3,12}(\cos^2\alpha-\sin^2\alpha)]_{,s}\} \tag{4.1.53}$$

because of the second line (4.1.48) and because of (4.1.51).

All internal forces are now known, and they can be used to interpret the previously obtained natural or dynamic boundary conditions: Condition (4.1.23) of the simply supported edge means that $M_{nn} = m(s)$ is valid because of the first line of (4.1.48). Correspondingly, (4.1.40) indicates that conditions (4.1.24), (4.1.25) mean nothing more than $M_{11} = m(s)$ and $M_{11} \equiv 0$, respectively. Conditions (4.1.26) of the free edge are recognizable by the first equation of (4.1.48) and by (4.1.53) as $M_{nn} = m(s)$, and $V_{nn} = Q_{nn}-M_{nt,s} = v(s)$. The comparison with (4.1.40) and (4.1.44) indicates that the conditions of (4.1.27) and (4.1.28) respectively, are equivalent with $M_{11} = m(s)$, $V_{11} = v(s)$ and $M_{11} \equiv 0$. $V_{11} \equiv 0$, respectively.

4.1.2 *Methods of calculation and examples*

Only in the rarest of cases is it possible to solve the boundary value problem as given by the plate equation (4.1.22) and the boundary conditions (4.1.9), (4.1.10), as well as (4.1.23) to (4.1.28), in an exact and direct manner. Therefore, let us go back to a series of methods which yield solutions indirectly or approximately. Let us discuss them in this section.

To begin with, the *semi-inverse method* can be used. Let us apply it in connection with the example of an elliptic plate which has a clamped edge and a uniformly distributed load p (Figure 54). Let the plate's edge R be given by equation

$$\left(\frac{x_1}{a}\right)^2 + \left(\frac{x_1}{b}\right)^2 = 1. \tag{4.1.54}$$

We start the solution of (4.1.22) by

$$u_3 = C\left[\left(\frac{x_1}{a}\right)^2 + \left(\frac{x_2}{b}\right)^2 - 1\right]^2 = CF^2(x_1, x_2), \tag{4.1.55}$$

which obviously satisfies boundary conditions

$$u_3(R) = 0,\ u_{3,n}(R) = u_{3,1}(R)\cos\alpha + u_{3,2}(R)\sin\alpha = 0$$

of the clamped edge, since $F(R) = 0$ is valid, and hence

$$u_3(R) = CF^2(R) = 0,\ u_{3,1}(R) = 2CF(R)F_{,1}(R) = 0,$$
$$u_{3,2}(R) = 2CF(R)F_{,2}(R) = 0.$$

The hitherto unknown constant C is now determined in such a way that relation (4.1.55) also satisfies differential equation (4.1.22). Let

$$u_{3,1111} = \frac{24C}{a^4},\ u_{3,1122} = \frac{8C}{a^2b^2},\ u_{3,2222} = \frac{24C}{b^4}$$

be true, which when inserted in (4.1.22) yields

$$\frac{24C}{a^4} + \frac{16C}{a^2b^2} + \frac{24C}{b^4} = \frac{p}{D}.$$

From this follows that C must have value

$$C = \frac{p}{\left(\dfrac{24}{a^4} + \dfrac{16}{a^2b^2} + \dfrac{24}{b^4}\right)D},$$

so that the solution is

$$u_3 = \frac{pa^4b^4}{8D(3b^4+2a^2b^2+3a^4)}\left[\left(\frac{x_1}{a}\right)^2+\left(\frac{x_2}{b}\right)^2-1\right]^2. \tag{4.1.56}$$

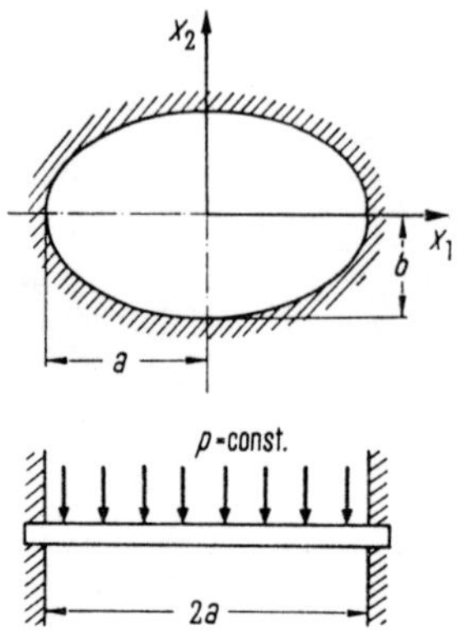

Figure 54

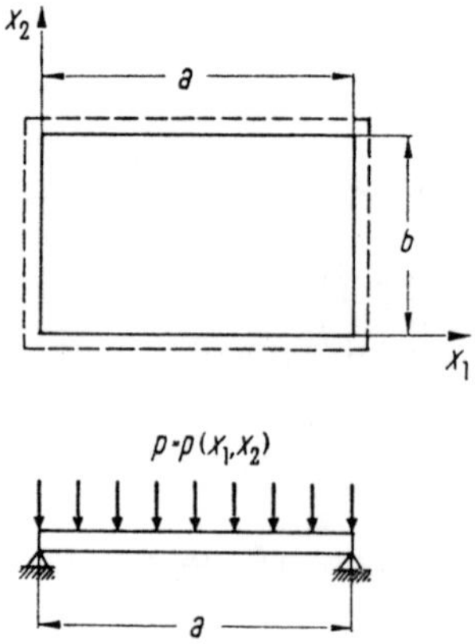

Figure 55

For the circular plate, let $a = b = \rho$ be true, so that its solution from (4.1.56) changes to

$$u_3 = \frac{p}{64D}(x_1^2+x_2^2-\rho^2)^2. \tag{4.1.57}$$

It is important for the practical work of the engineer to calculate additionally to the deflection u_3 the internal forces and the stresses through (4.1.40), (4.1.41) and (4.1.4), and to find their maximal values. This is quite easy since the solutions (4.1.56), (4.1.57) for u_3 are known and can be used for calculating the internal forces and the stresses. We shall leave it to the reader to carry these calculations out.

A second method is using *series expansions*. This method has been used from the very beginning of plate theory, and it will be introduced here by studying its application to two classic solutions of the rectangular plate problem.

One of the solutions is by Navier and is valid for the plate which is simply supported on all edges and has a distributed load p (Figure 55).

Let the boundary conditions (4.1.10) and (4.1.23) be specified at the plate's edges. They are

$$u_3(R) = 0,$$

and furthermore because of $m(s) = 0$ as well as $\alpha = 0°$ at one time and $\alpha = 90°$ at the other,

$$u_{3,11} + \nu u_{3,22} = 0 \text{ for } x_1 = 0,\ x_1 = a,$$
$$u_{3,22} + \nu u_{3,11} = 0 \text{ for } x_2 = 0,\ x_2 = b.$$

They are obviously satisfied by the series

$$u_3 = \sum_{m,n} A_{mn} \sin \frac{m\pi x_1}{a} \sin \frac{n\pi x_2}{b}. \tag{4.1.58}$$

Let us set up the load as the Fourier series

$$\begin{aligned} p &= \sum_{m,n} a_{mn} \sin \frac{m\pi x_1}{a} \sin \frac{n\pi x_2}{b}, \\ a_{jk} &= \frac{4}{ab} \int_0^a \int_0^b p \sin \frac{j\pi x_1}{a} \sin \frac{k\pi x_2}{b} \, \mathrm{d}x_1 \, \mathrm{d}x_2, \end{aligned} \tag{4.1.59}$$

and insert (4.1.58) and (4.1.59) into differential equation (4.1.22). This yields

$$\sum_{m,n} \left\{ A_{mn} \left[\left(\frac{m\pi}{a}\right)^4 + 2\left(\frac{m\pi}{a}\right)^2 \left(\frac{n\pi}{b}\right)^2 + \left(\frac{n\pi}{b}\right)^4 \right] - \frac{a_{mn}}{D} \right\} \sin \frac{m\pi x_1}{a} \sin \frac{n\pi x_2}{b} = 0.$$

Since this relation is to be valid for any value of x_1, x_2, then it becomes necessary that the content of the curled bracket is equal to nought. This yields

$$A_{mn} = \frac{1}{\pi^4 D} \frac{a_{mn}}{(m^2/a^2 + n^2/b^2)^2}, \tag{4.1.60}$$

whereby a_{mn} must be notated as a known quantity according to the second line of (4.1.59). Using the series coefficients A_{mn}, specified by (4.1.60),

$$u_3 = \frac{4}{\pi^4 Dab} \sum_{m,n} \frac{\int_0^a \int_0^b p \sin \frac{m\pi x_1}{a} \sin \frac{n\pi x_2}{b} \mathrm{d}x_1 \mathrm{d}x_2}{(m^2/a^2+n^2/b^2)^2} \sin \frac{m\pi x_1}{a} \sin \frac{n\pi x_2}{b} \tag{4.1.61}$$

is determined according to (4.1.58).

In the special case of constant load, i.e., for $p = p_0 = \text{const}$,

$$a_{mn} = \frac{16p_0}{\pi^2 mn} \text{ for } m, n \text{ odd}$$

is true. Then (4.1.61) changes into

$$u_3 = \frac{16p_0}{\pi^6 D} \sum_{m,n} \frac{\sin(m\pi x_1/a)\sin(n\pi x_2/b)}{mn[(m^2/a^2)+(n^2/b^2)]^2}, \quad m, n \text{ odd}. \tag{4.1.62}$$

The largest deflection is from (4.1.62) for $x_1 = a/2$, $x_2 = b/2$,

$$u_{3,\max} = \frac{16p_0}{\pi^6 D} \sum_{m,n} \frac{(-1)^{m+n/2-1}}{mn(m^2/a^2+n^2/b^2)^2}, \quad m, n \text{ odd}, \tag{4.1.63}$$

occurring in the plate's centre. When the plate is square ($a = b$), and when $\nu = 0, 3$ is assumed, then

$$u_{3,\max} = 0.0443 p_0 \frac{a^4}{Eh^3}$$

is yielded from (4.1.63). The bending moments, serving as an example for the calculation of internal forces, are obtained from (4.1.40) by means of (4.1.62) as

$$M_{11} = \frac{16p_0}{\pi^4} \sum_{m,n} \frac{(m^2/a^2)+\nu(n^2/b^2)}{mn(m^2/a^2+n^2/b^2)^2} \sin \frac{m\pi x_1}{a} \sin \frac{n\pi x_2}{b},$$

$$M_{22} = \frac{16p_0}{\pi^4} \sum_{m,n} \frac{\nu(m^2/a^2)+(n^2/b^2)}{mn(m^2/a^2+n^2/b^2)^2} \sin \frac{m\pi x_1}{a} \sin \frac{n\pi x_2}{b}, \quad m, n \text{ odd}.$$

The greatest bending moment in the plate's centre for the square plate $(a = b)$ and for $\nu = 0.3$ is for example

$$M_{11}\left(x_1 = \frac{a}{2},\ x_2 = \frac{a}{2}\right) = M_{22}\left(x_1 = \frac{a}{2},\ x_2 = \frac{a}{2}\right) = 0.0479 p_0 a^2.$$

The bending stresses σ_{11}, σ_{22}, serving as an example for the calculation of stresses, follow from (4.1.4). Comparison with (4.1.40) indicates that

$$\sigma_{11} = \frac{12}{h^3} M_{11} x_3,\ \sigma_{22} = \frac{12}{h^3} M_{22} x_3$$

is true. The greatest bending stresses of the square plate occur in the centre of the plate at $x_3 = \pm h/2$. Hence, its value (for $\nu = 0.3$) is

$$\sigma_{\max} = \frac{6}{h^2} M_{\max} = \frac{6}{h^2} \cdot 0.0479 p_0 a^2 = 0.287 p_0 \left(\frac{a}{h}\right)^2.$$

If instead of the uniformly distributed load p, a concentrated force P is assumed to be the plate's load in $x_1 = x_1^0, x_2 = x_2^0$, then in the calculation of coefficients a_{jk} we must consider the fact that p in the formula of the second line of (4.1.59) will be equal to nought except for point $x_1 = x_1^0$, $x_2 = x_2^0$ where $p = P/\mathrm{d}x_1\,\mathrm{d}x_2$. The double integral is reduced to

$$\frac{P}{\mathrm{d}x_1\,\mathrm{d}x_2} \sin\frac{m\pi x_1^0}{a} \sin\frac{n\pi x_2^0}{b}\,\mathrm{d}x_1\,\mathrm{d}x_2 = P \sin\frac{m\pi x_1^0}{a} \sin\frac{n\pi x_2^0}{b},$$

and instead of (4.1.61),

$$u_3 = \frac{4P}{\pi^4 Dab} \sum_{m,n} \frac{\sin\dfrac{m\pi x_1^0}{a} \sin\dfrac{n\pi x_2^0}{b}}{(m^2/a^2 + n^2/b^2)^2} \sin\frac{m\pi x_1}{a} \sin\frac{n\pi x_2}{b}$$

is obtained as the solution.

The other solution has been given by M. Levy and is of interest, since it shows that through

$$u_3 = \sum_m f_m(x_2) \sin\frac{m\pi x_1}{a}, \tag{4.1.64}$$

containing unknown function $f_m(x_2)$, much scope is gained as far as the fulfilment of the boundary conditions is concerned. Solutions of this kind are used, e.g., in Kantorowitsch's method [15, p. 282].

Let there be a rectangular plate. Let the two opposite edges be simply supported and the two other edges be either supported arbitrarily or free (Figure 56). Let the plate have a uniformly distributed load p.

The boundary conditions for the simply supported edges, i.e., for $x_1 = 0$, $x_1 = a$, are according to (4.1.10) and (4.1.25), $u_3 = 0$ and $u_{3,11} + \nu u_{3,22} = 0$. Through the particular support of the plate, $u_{3,22} = 0$ applies along the two straight lines $x_1 = 0$, $x_1 = a$, and the condition $u_{3,11} + \nu u_{3,22} = 0$ is reduced for $x_1 = 0$, $x_1 = a$ to $u_{3,11} = 0$. Hence,

$$u_3 = 0,\ u_{3,11} = 0 \text{ for } x_1 = 0,\ x_1 = a$$

must be satisfied, which is carried out indeed by solution (4.1.64).

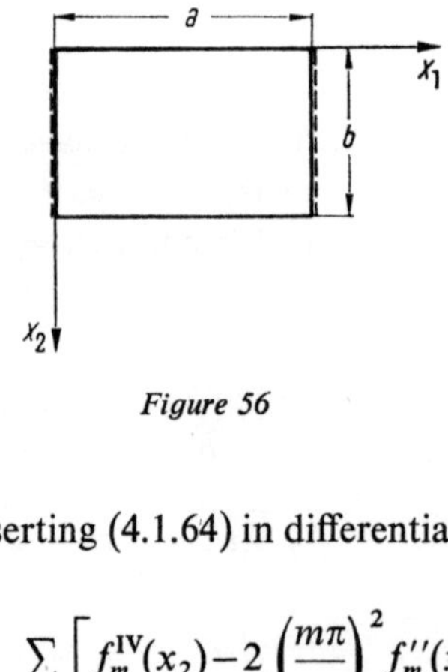

Figure 56

Inserting (4.1.64) in differential equation (4.1.22) it is followed by

$$\sum_m \left[f_m^{\mathrm{IV}}(x_2) - 2\left(\frac{m\pi}{a}\right)^2 f_m''(x_2) + \left(\frac{m\pi}{a}\right)^4 f_m(x_2) \right] \sin \frac{m\pi x_1}{a} = \frac{p}{D}. \quad (4.1.65)$$

Let us use operator

$$L_m(x_2) = f_m^{\mathrm{IV}}(x_2) - 2\left(\frac{m\pi}{a}\right)^2 f_m''(x_2) + \left(\frac{m\pi}{a}\right)^4 f_m(x_2), \quad (4.1.66)$$

and the Fourier series

$$p = \sum_m F_m(x_2) \sin \frac{m\pi x_1}{a},$$
$$F_m(x_2) = \frac{2}{a}\int_0^a p(x_1, x_2) \sin \frac{m\pi x_1}{a}\, dx_1, \tag{4.1.67}$$

for load p. Inserting (4.1.66) and (4.1.67) in (4.1.65), $L_m(x_2) = F_m(x_2)/D$ is obtained. Thus, the plate problem has been reduced to the problem of solving the inhomogeneous differential equation

$$L_m(x_2) = \frac{F_m(x_2)}{D}. \tag{4.1.68}$$

Its general solution is

$$f_m(x_2) = A_m \cosh \frac{m\pi x_2}{a} + B_m \sinh \frac{m\pi x_2}{a} + C_m x_2 \cosh \frac{m\pi x_2}{a}$$
$$+ D_m x_2 \sinh \frac{m\pi x_2}{a} + f_m^*(x_2). \tag{4.1.69}$$

A_m, B_m, C_m, D_m are the constants of integration, and $f_m^*(x_2)$ is a particular solution of (4.1.68).

The constants of integration are determined by using the boundary conditions for the other two boundaries. Should these be also simply supported, then, corresponding to the conditions of the boundary we have dealt with first,

$$u_3 = 0, \; u_{3,22} = 0 \text{ for } x_2 = 0, \; x_2 = b$$

is now valid. According to (4.1.64) it yields

$$f_m(0) = f_m''(0) = f_m(b) = f_m''(b) = 0,$$

leading finally to four equations for the constants of integration. Choosing the particular solution $f_m^*(x_2)$ in such a way that $f_m^*(0) = f_m^{*\prime\prime}(0) = 0$, then

$$A_m = D_m = 0,$$

$$B_m = \left(2\frac{m\pi}{a}\sinh^2\frac{m\pi b}{a}\right)^{-1}\left\{b\cosh\frac{m\pi b}{a}\, f_m^{*\prime\prime}(b)\right.$$

$$\left. - \left[2\frac{m\pi}{a}\sinh\frac{m\pi b}{a} + \left(\frac{m\pi}{a}\right)^2 b\cosh\frac{m\pi b}{a}\right] f^*(b)\right\}, \qquad (4.1.70)$$

$$C_m = \left(2\frac{m\pi}{a}\sinh^2\frac{m\pi b}{a}\right)^{-1}\left[\left(\frac{m\pi}{a}\right)^2 f_m^*(b) - f_m^{*\prime\prime}(b)\right]$$

is obtained. The particular solution of the inhomogeneous equation (4.1.68), having the two previously mentioned initial conditions, is as follows:

$$f_m^*(x_2) = \frac{1}{D}\int_0^{x_2} \psi_m(x_2-\xi)F_m(\xi)\,\mathrm{d}\xi,$$

$$\psi_m(x_2) = \frac{a^2}{2(m\pi)^2}\left[x_2\cosh\frac{m\pi x_2}{a} - \frac{a}{m\pi}\sinh\frac{m\pi x_2}{a}\right], \qquad (4.1.71)$$

$F_m(x_2)$ according to (4.1.67).

Inserting (4.1.69), (4.1.70) and (4.1.71) in (4.1.64), another solution for the simply supported rectangular plate is the result. This solution is remarkable as it contains excellent converging series.

The third method is based on the *extremal principles of elasto-mechanics*, which we have discussed in 3.5.1. It takes us to the application of the approximate methods by Ritz, Galerkin, and others. The principle of the stationary value of the total potential energy leads to the variation problem (4.1.8). For its solution, permissible functions are used which satisfy the *geometric conditions* at the plate's edge R. Given the special case that the plate has also edge portions R_f, which are a *free, yet unloaded*, then

$$\Pi = \frac{D}{2}\int\left[(\nabla^2 u_3)^2 + 2(1-\nu)(u_{3,12}^2 - u_{3,11}u_{3,22}) - \frac{2p}{D}u_3\right]\mathrm{d}F = \text{Extr.},$$

$$u_3 = 0 \text{ on } R-R_\mathrm{f} \qquad (4.1.72)$$

$$(\text{and } u_{3,n} = 0 \text{ on } R_\mathrm{e})$$

has to be used. Since such a case dealing with plates having no edge loads is of the greatest practical importance, let us use it as a basis for all following calculations.

The *Ritz method* consists of using in (4.1.72) finite series for u_3 of the kind

$$u_3 = \sum_{m,n} a_{mn} F_{mn}(x_1, x_2), \tag{4.1.73}$$

and satisfying the geometric boundary conditions. The coefficients of the series are determined from the condition

$$\frac{\partial \Pi}{\partial a_{mn}} = 0. \tag{4.1.74}$$

Inserting the coefficient a_{mn} obtained through (4.1.74) in (4.1.73), then an approximate solution for u_3 is determined. It is also used in the derivation of all other quantities of interest, e.g., the internal forces and stresses.

Many possibilities and proposals exist for choosing functions F_{mn}. In any case, they should represent a *complete*, and as far as facilitation of calculation is concerned, an *orthonormal* system of functions.

If for series (4.1.73), functions F_{mn} have been chosen, all of which do not only satisfy the geometric conditions but all other boundary conditions, then using *Galerkin's equation system*

$$\begin{aligned} &\sum_{r,s} a_{rs} \int_F [\nabla^2\nabla^2 F_{rs}(x_1, x_2)] F_{mn}(x_1, x_2) \, dx_1 \, dx_2 \\ &\quad = \int_F \frac{p(x_1, x_2)}{D} F_{mn}(x_1, x_2) \, dx_1 \, dx_2 , \end{aligned} \tag{4.1.75}$$

the coefficients a_{mn} are calculated. We must remember too, that for plates of polygonal edge which have *no free edge portion*, the integral

$$\int_F (u_{3,12}^2 - u_{3,11} u_{3,22}) \, dF$$

vanishes. Let us prove this as follows. For any two f and g

$$\int_F gh_{,1} \mathrm{d}F = -\int_F hg_{,1} \mathrm{d}F + \int_R hg \cos\alpha \, \mathrm{d}s,$$
$$\int_F gh_{,2} \mathrm{d}F = -\int_F hg_{,2} \mathrm{d}F + \int_R hg \sin\alpha \, \mathrm{d}s \tag{4.1.76}$$

is valid, where angle α is between normal $\boldsymbol{n}$ of the edge and the x_1 axis (Figure 50).

If the first formula (4.1.76) is used in $\int_F u_{3,12}^2 \mathrm{d}F$, and the second formula (4.1.76) in $\int_F u_{3,2} u_{3,112} \mathrm{d}F$, then

$$\int_F u_{3,12}^2 \mathrm{d}F = -\int_F u_{3,2} u_{3,112} \mathrm{d}F + \int_R u_{3,2} u_{3,12} \cos\alpha \, \mathrm{d}s,$$

$$\int_F u_{3,2} u_{3,112} \mathrm{d}F = -\int_F u_{3,11} u_{3,22} \mathrm{d}F + \int_R u_{3,2} u_{3,11} \sin\alpha \, \mathrm{d}s,$$

and

$$\int_F u_{3,12}^2 \mathrm{d}F = \int_F u_{3,22} u_{3,11} \mathrm{d}F + \int_R u_{3,2} [u_{3,12} \cos\alpha - u_{3,11} \sin\alpha] \mathrm{d}s$$

is obtained.

Using the second formula (4.1.76) in $\int_F u_{3,12}^2 \mathrm{d}F$, and then the first formula (4.1.76) in a term of the previous result, then, accordingly,

$$\int_F u_{3,12}^2 \mathrm{d}F = \int_F u_{3,11} u_{3,22} \mathrm{d}F + \int_R u_{3,1} [u_{3,12} \sin\alpha - u_{3,22} \cos\alpha] \mathrm{d}s$$

is obtained. Summarizing the two solutions yields

$$\int_F (u_{3,12}^2 - u_{3,11} u_{3,22}) \mathrm{d}F = \frac{1}{2} \int_R [-u_{3,2} u_{3,11} \sin\alpha + u_{3,2} u_{3,12} \cos\alpha$$
$$+ u_{3,1} u_{3,12} \sin\alpha - u_{3,1} u_{3,22} \cos\alpha] \mathrm{d}s = \frac{1}{2} \int_R J \, \mathrm{d}s.$$

Since there is to be no free edge portion, then because of the support, $u_{3,s} = 0$ on R is true, i.e., because of (4.1.17), $u_{3,1} \sin\alpha = u_{3,2} \cos\alpha$, which yields

$$J = \frac{u_{3,2}}{\sin\alpha}[-u_{3,11}\sin^2\alpha+2u_{3,12}\sin\alpha\cos\alpha-u_{3,22}\cos^2\alpha].$$

Considering this fact, it is realised that the square bracket contains nothing more than $1/r_s$, whereby r_s is the radius of curvature of surface $u_3(x_1, x_2)$ for a vertical cut in the s-direction. For the linear, supported edge portions of R, $r_s = \infty$ applies throughout, so that, as stated,

$$\int_F (u_{3,12}^2 - u_{3,11}u_{3,22})\mathrm{d}F = \frac{1}{2}\int_R J\,\mathrm{d}s = \frac{1}{2}\int_R \frac{u_{3,2}}{r_s \sin\alpha}\,\mathrm{d}s = 0$$

is valid. For this reason, variational problem (4.1.72) is simplified to

$$\Pi = \frac{D}{2}\int_F \left[(\nabla^2 u_3)^2 - \frac{2p}{D}u_3\right]\mathrm{d}F = \text{Extr.},$$

$$u_3 = 0 \text{ on } R, \tag{4.1.77}$$

$$(\text{and } u_{3,n} = 0 \text{ on } R_e),$$

which pertains to plates with supported polygonal edges. Through (4.1.77) let us discuss, for example, Navier's problem in Figure 55, and start by using (4.1.58), which satisfies the geometric boundary conditions. Then,

$$\Pi = \frac{\pi^4 ab}{8} D \sum_{m,n}\left(\frac{m^2}{a^2}+\frac{n^2}{b^2}\right)A_{mn}^2 - \sum_{m,n} A_{mn}\int_F p\sin\frac{m\pi x_1}{a}\sin\frac{n\pi x_2}{b}\,\mathrm{d}F \tag{4.1.78}$$

is obtained. From $\partial\Pi/\partial A_{kj} = 0$,

$$A_{mn} = \frac{4}{\pi^4 abD}\,\frac{\int_F p\sin(m\pi x_1/a)\sin(n\pi x_2/b)\,\mathrm{d}F}{(m^2/a^2+n^2/b^2)^2} \tag{4.1.79}$$

is determined, which equals to (4.1.60). Thus, we arrive at Navier's solution.

Already in Section 3.5.1 it has been tried to extend the class of permissible functions in connection with the extremal principles and to reach total exemption from the boundary conditions. In order to achieve this, extra terms have been added to the classic extremal principles. Thus, for example, from (3.5.23), (3.5.38) has been obtained. Using the variational problem (4.1.72), or (4.1.77) for problems dealing with plates, then principles of stationary value of potential energy are used, in actual fact. It is advisable therefore, that as far as plate problems are concerned, (4.1.72), (4.1.77) respectively should be exempt from boundary conditions. One way of achieving this is to follow U. Wegner's [27] method, e.g., start from special problem (4.1.77) and carry out mathematically the exemption of the boundary conditions in terms of the well-known *method of Lagrange's parameter*. Details are given in the quoted paper. Another way lies in the realization of the fact that the plate problems are related to the general facts of elasticity theory. Hence, the extended principle (3.5.38) may be taken as a starting point and notated for the special case dealing with plates. The following is a discussion of this procedure.

The first three terms of (3.5.38) lead to

$$\frac{D}{2}\int_F \left[(\nabla^2 u_3)^2 - \frac{2p}{D}u_3\right] \mathrm{d}F.$$

We have considered this expression already for plates with supported polygonal edges and in the case of vanishing volume forces K_i. For the sake of simplicity, let us assume that in fact, $K_i = 0$ is true. The fourth additional term of (3.5.38) must still be changed to suit the plate problem. Let us assume, for example, that the plate is clamped along its entire edge. Then R is assumed for O_u, because u_3 and $u_{3,n}$ are specified everywhere on R. All these quantities must equal to nought. Therefore, in the case of plates, nought corresponds identically to part

$$\int_{O_u} p_i u_i^0 \mathrm{d}O.$$

For the remaining part

$$\int_{O_u} p_i u_i \mathrm{d}O$$

it must be considered that p_i must be expressed by u_i and that in the case of plates, the p_i must be understood as 'generalized' forces. This means that they are not always actual forces but also moments, so that parts of the energy expression involved originate from product $V_{nn}u_3$ as well as from product $M_{nn}u_{3,n}$. Hence, in the plate problem, let us notate for

$$-\int_{O_u} p_i u_i \, dO$$

the line integral

$$-\int_R (V_{nn}u_3 - M_{nn}u_{3,n})\, ds.$$

Thus, instead of (4.1.77), we obtain the extended principle

$$\frac{D}{2}\int_F \left[(\nabla^2 u_3)^2 - \frac{2p}{D}u_3\right] dF - \int_R (V_{nn}u_3 - M_{nn}u_{3,n})\, ds = \text{Extr.} \qquad (4.1.80)$$

To use this expression in the calculation, V_{nn} is substituted by (4.1.53), and M_{nn} by (4.1.48). Let

$$\frac{1}{r_{sn}} = u_{3,ns} = \sin\alpha\cos\alpha(u_{3,22} - u_{3,11}) + u_{3,12}(\cos^2\alpha - \sin^2\alpha) \quad (4.1.81)$$

be the twist of middle-plane u_3 with respect to the s and n directions. It had been pointed out earlier that

$$\frac{1}{r_s} = -u_{3,11}\sin^2\alpha + 2u_{3,12}\sin\alpha\cos\alpha - u_{3,22}\cos^2\alpha \qquad (4.1.82)$$

represents the curvature of the middle-plane u_3 relative to the direction s. Considering both of these relations in (4.1.52) and (4.1.48), then we obtain for (4.1.80)

$$\frac{D}{2}\int_F \left[(\nabla^2 u_3)^2 - \frac{2p}{D}u_3\right] dF + D\int_R \left[(\nabla^2 u_3)_{,n} + (1-\nu)\left(\frac{1}{r_{sn}}\right)_{,s}\right] u_3\, ds$$

$$-D\int_R \left(\nabla^2 u_3 + \frac{1-\nu}{r_s}\right) u_{3,n}\, ds = \text{Extr.}$$

Because of the assumption that R is polygonal,

$$\left(\frac{1}{r_{sn}}\right) = 0, \ \frac{1}{r_s} = 0 \text{ on } R$$

may be notated, thus reducing the variational problem after division with D to

$$\frac{1}{2}\int_F \left[(\nabla^2 u_3)^2 - \frac{2p}{D} u_3\right] \mathrm{d}F + \int_R [(\nabla^2 u_3)_{,n} u_3 - \nabla^2 u_3 \cdot u_{3,n}] \mathrm{d}s = \text{Extr.} \tag{4.1.83}$$

This represents exactly the relation as given by U. Wegner in [27] on top of page 215. Another important fact is that *the functions permitted for (4.1.83) are not required to satisfy even the geometric conditions*, and hence *are not required to satisfy any boundary conditions whatsoever.*

In the same manner as the integral

$$\int_F u_{3,\alpha\alpha}(\delta u_3)_{,\beta\beta} \mathrm{d}F$$

has been changed, and (4.1.14) had been obtained, relation

$$\int_F u_{3,\alpha\alpha} \cdot u_{3,\beta\beta} \mathrm{d}F \equiv \int_F (\nabla^2 u_3)^2 \mathrm{d}F$$

$$= \int_F (\nabla^2\nabla^2 u_3) u_3 \, \mathrm{d}F - \int_R [(\nabla^2 u_3)_{,n} u_3 - \nabla^2 u_3 \cdot u_{3,n}] \mathrm{d}s \tag{4.1.84}$$

is set up. The calculation for this procedure does not need repeating at this point. Following U. Wegner's proposal, and allowing permissible functions u_3 which satisfy $\nabla^2\nabla^2 u_3 = p/D$, then (4.1.84) changes to

$$\int_F \frac{p}{D} u_3 \, \mathrm{d}F = \int_F (\nabla^2 u_3)^2 \mathrm{d}F - \int_R [\nabla^2 u_3 \cdot u_{3,n} - u_3(\nabla^2 u_3)_{,n}] \mathrm{d}s. \tag{4.1.85}$$

Inserting (4.1.85) in (4.1.83), the new variational problem

$$\frac{1}{2}\int_F (\nabla^2 u_3)^2 \mathrm{d}F = \text{Extr.},$$

$$\text{secondary condition: } \nabla^2\nabla^2 u_3 = \frac{p}{D}, \tag{4.1.86}$$

is obtained. It can replace (4.1.77). The class of permissible functions has been reduced by the secondary condition (4.1.86). Yet it is easier to set up functions which satisfy this secondary condition than to indicate functions satisfying the geometric boundary conditions which are required when discussing the classic variational problem, for example (4.1.77). To demonstrate the ease with which (4.1.86) is handled, let us outline the calculation as indicated by U. Wegner in [27]. Let us introduce the new quantity

$$\nabla^2 u_3 = M(x_1, x_2) \tag{4.1.87}$$

which through (4.1.40) is

$$M = -\frac{M_{11}+M_{12}}{D(1+\nu)}. \tag{4.1.88}$$

Because of the secondary condition of (4.1.86), only those functions M are permitted which satisfy

$$\nabla^2 M = \frac{p}{D}. \tag{4.1.89}$$

Let M_0 be a particular solution of (4.1.89). Notating $M = M_0 + M^*$, then $\nabla^2 M^* = 0$ is valid, i.e., permissible functions M are obtained by adding to M_0 the solutions M^* of Laplace's differential equation (potential functions), which are known in great numbers. Let us use the series

$$M = M_0 + \sum_\nu c_\nu M_\nu^*, \tag{4.1.90}$$

containing the yet unknown coefficients c_ν and potential functions M_ν^*, inserting it in

$$J = \frac{1}{2}\int_F (M_0 + \sum_\nu c_\nu M_\nu^*)^2 \mathrm{d}F = \text{Extr.} \tag{4.1.91}$$

This is a condition obtained from the first line of (4.1.86) through (4.1.87) and (4.1.90). The calculation of c_ν is carried out in accordance with requirement $\partial J/\partial c_i = 0$, $i = 1, 2, 3, \ldots$, leading to the infinite system of equations

$$\int_F (M_0 + \sum_j c_j M_j^*) M_i^* \mathrm{d}F = 0,\ i, j = 1, 2, 3, \ldots \tag{4.1.92}$$

When the c_ν have been found by using (4.1.92), then by using (4.1.90), quantity M is determined. It is often adequate to derivate the data which interest the engineer. If, for example, we are dealing with a square plate which has its edges clamped, then, in the plate's centre, $M_{11} = M_{22}$ is true, so that

$$M_{11} = M_{22} = -\tfrac{1}{2}D(1+\nu)M, \text{ in plate's centre}$$

is obtained. For the plate's edges $x_1 = 0, x_1 = a$ is $u_{3,22} = 0$, and hence, because of (4.1.40),

$$M_{11} = -Du_{3,11} = -D\nabla^2 u_3$$

is true. Because of (4.1.87) this is followed by $M_{11} = -DM$ on $x_1 = 0$ and on $x_1 = a$. Correspondingly, $M_{22} = -DM$ on $x_2 = 0, x_2 = a$ is obtained. But this indicates that the boundary moments can be determined from the known boundary values of M.

From (4.1.41), $Q_{(\alpha\alpha)} = -D(\nabla^2 u_3)_{,\alpha}$, $\alpha = 1,2$, is obtained and thus, because of (4.1.87),

$$Q_{(\alpha\alpha)} = -DM_{,\alpha},\ \alpha = 1, 2.$$

Hence, the shearing forces are determined from M. The vertical supporting force V_{11} on the edge, which is changed from (4.1.44) to

$$V_{11} = -D[\nabla^2 u_3 + (1-\nu)u_{3,22}]_{,1}, \tag{4.1.93}$$

is on boundary $x_1 = 0, x_1 = a$, where, because it is clamped, $u_{3,1} = 0$ is true,

$$V_{11} = -D(\nabla^2 u_3)_{,1},$$

which according to (4.1.87) is

$$V_{11} = -D \cdot M_{,1} \text{ on } x_1 = 0,\ x_1 = a.$$

In the same way

$$V_{22} = -D \cdot M_{,2} \text{ on } x_2 = 0, \ x_2 = a$$

is determined. Therefore, the vertical forces at the plate's edges are also obtained from M. Thus, it has been shown that in the case we have just discussed, as well as in other cases, the calculation of M, as we have demonstrated, is adequate. How the calculation is carried out in more general cases, when u_3 has to be known, is discussed in U. Wegner's paper [27].

A further possibility for solving plate problems consists in applying *complex variables*. This is obvious, since there is a close connection between the planar problems dealt with in 3.3 and the actual plate problems.

Solution u_3 of the plate's equation (4.1.22) is assumed to have the form

$$u_3 = u_3^0 + u_3^* \tag{4.1.94}$$

where u_3^0 is a particular solution of (4.1.22) so that

$$\nabla^2\nabla^2 u_3^0 = \frac{p}{D}$$

holds. Then,

$$\nabla^2\nabla^2 u_3^* = 0 \tag{4.1.95}$$

remains to be satisfied by u_3^*, which therefore turns out to be a biharmonic function.

Particular solutions of (4.1.22) are for example the following:

$$\text{for } p = \text{const, } u_3^0 = \frac{p}{64D}(x_1^2+x_2^2)^2,$$

$$\text{for } p = 2p_1 x_1, \ p_1 = \text{const, } u_3^0 = \frac{2p_1}{192D} x_1(x_1^2+x_2^2)^2,$$

$$\text{for } p = p_2(x_1^2+x_2^2), \ p_2 = \text{const, } u_3^0 = \frac{p_2}{572D}(x_1^2+x_2^2)^3.$$

Now, let us refer back to 3.3 and identify Φ with u_3^*. Then,

$$u_3^* = \tfrac{1}{2}[\bar{z}F(z) + z\overline{F(z)} + G(z) + \overline{G(z)}] \tag{4.1.96}$$

is a solution of (4.1.95) according to (3.3.60a). As pointed out in 3.3, $F(z)$ and $G(z)$ are analytic functions of the complex variable $z = x_1 + ix_2$. Quantities being overbarred are complex conjugate.

The as yet unknown functions $F(z)$ and $G(z)$ can be determined using the boundary conditions of the respective problem. We shall here deal only with the case of the plate having clamped edges. Other boundary conditions are dealt with by G. N. Sawin [28], and a thorough representation of this method of solution including various examples has been given by H. Wilken [37].

In our case, $u_3 = u_{3,n} = 0$ on R. Due to the kind of support of the plate, also $u_{3,s} = 0$ must hold on R. However, from (4.1.18) follows

$$u_{3,1} + iu_{3,2} = (u_{3,n} + iu_{3,s})e^{i\alpha}.$$

Using the boundary condition,

$$u_{3,1} + iu_{3,2} = 0 \text{ on } R.$$

According to (4.1.94) this also reads

$$u^*_{3,1} + iu^*_{3,2} = -(u^0_{3,1} + iu_{3,2}) \text{ on } R. \tag{4.1.97}$$

Inserting (4.1.96) in (4.1.97) yields

$$F(z) + z\overline{F'(z)} + \overline{G'(z)} = -(u^0_{3,1} + iu^0_{3,2}) \text{ on } R. \tag{4.1.98}$$

This is the condition which may be used for determining $F(z)$ and $G'(z)$. For a numerical calculation, a conformal mapping $z = f(\zeta)$ has to be carried out in addition, which maps the unit circle $|\zeta| < 1$ on the simply connected surface of the plate. According to (3.3.132), relation (4.1.98) then changes into

$$F_1(\zeta) + \frac{f(\zeta)}{f'(\zeta)}\overline{F_1'(\zeta)} + \frac{\overline{G_1'(\zeta)}}{\overline{f'(\zeta)}} = U(\zeta) \text{ on } R, \tag{4.1.99}$$

where $U(\zeta)$ is obtained by expressing $-(u^0_{3,1} + iu^0_{3,2})$ in terms of ζ. Hence, we have arrived at the first fundamental problem of biharmonic functions. Solutions of this problem can be found in [15].

Having solved (4.1.99), then through (4.1.96) and (4.1.94) u_3 is determined. As well known, having this quantity at hand enables us to determine all remaining data such as internal forces and stresses.

However, internal forces might also be determined directly from $F(z)$ and $G'(z)$ without referring back to u_3. Because (4.1.40), (4.1.41) and (4.1.96) yield

$$M_{11}-M_{11}^0 = -D\{1+\nu)[F'(z)+\overline{F'(z)}]+\frac{1-\nu}{2}[\bar{z}F''(z)+z\overline{F''(z)}$$

$$+G''(z)+\overline{G''(z)}]\},$$

$$M_{22}-M_{22}^0 = -D\left\{(1+\nu)[F'(z)+\overline{F'(z)}]-\frac{1-\nu}{2}[G''(z)+\overline{G''(z)}\right.$$

$$\left.+\bar{z}F''(z)+z\overline{F''(z)}]\right\},$$

$$M_{21}-M_{21}^0 = -\mathrm{i}\frac{D(1-\nu)}{2}[\bar{z}F''(z)+G''(z)-z\overline{F''(z)}-\overline{G''(z)}],$$

$$Q_{11}-Q_{11}^0 = -2D[F''(z)+\overline{F''(z)}],$$

$$Q_{22}-Q_{22}^0 = -2\mathrm{i}D[F''(z)-\overline{F''(z)}],$$

where

$$M_{11}^0 = -D(u_{3,11}^0+\nu u_{3,22}^0),\ M_{22}^0 = -D(u_{3,22}^0+\nu u_{3,11}^0),$$

$$M_{21}^0 = D(1-\nu)u_{3,12}^0,$$

$$Q^0 = -D(u_{3,11}^0+u_{3,22}^0)_{,1},\ Q_{22}^0 = -D(u_{3,11}^0+u_{3,22}^0)_{,2}.$$

4.1.3 *Transition to curvilinear coordinates*

It will be suitable in many cases to use planar curvilinear coordinates because of the course of the plate's boundary and the kind of load distribution. This is quite possible because certain facts in 3.1.3 have been made available. Through (3.1.107), for example, the plate equation (4.1.22) is changed to

$$\left\{\frac{1}{h_1 h_2}\left[\frac{\partial}{\partial q_1}\left(\frac{h_2}{h_1}\frac{\partial}{\partial q_1}\right)+\frac{\partial}{\partial q_2}\left(\frac{h_1}{h_2}\frac{\partial}{\partial q_2}\right)\right]\right\}^2 u_3=\frac{p}{D}, \tag{4.1.100}$$

i.e., considering the fact that $q_3 = x_3$ is notated, and that based on previous assumptions no dependence of u_3 on q_3 exists. Then, p also must be given with respect to the particular curvilinear coordinates used.

Let us keep to giving details for only the important case of polar coordinates r, φ (Figure 57). Then, $q_1 = r$, $q_2 = \varphi$ is valid, and according to (3.1.80), $h_1 = 1$, $h_2 = r$. Therefore, (4.1.100) changes to

$$\left(\frac{\partial^2}{\partial r^2}+\frac{1}{r}\frac{\partial}{\partial r}+\frac{1}{r^2}\frac{\partial^2}{\partial \varphi^2}\right)^2 u_3=\frac{p}{D}. \tag{4.1.101}$$

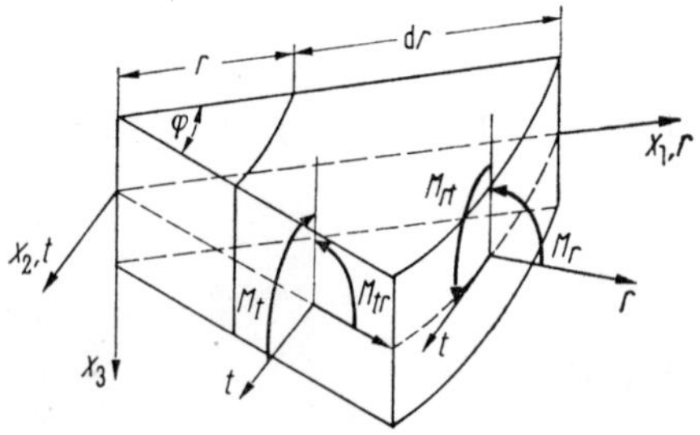

Figure 57

Without limiting generality, let us assume that for the cuts in question, directions r, t coincide with those of x_1, x_2. Then from (4.1.40),

$$M_{rr} = -D(u_{3,rr}+\nu u_{3,tt}) \tag{4.1.102}$$

is obtained by substituting in M_{11} indices 1 and 2 by indices r and t. By changing α into φ, we obtain from (3.3.79) that

$$\frac{\partial}{\partial t}=\frac{1}{r}\frac{\partial}{\partial \varphi} \tag{4.1.103}$$

is true. Moreover, comparing (3.3.78) with (3.3.81) yields

$$\frac{\partial^2}{\partial t^2} = \frac{1}{r}\frac{\partial}{\partial r} + \frac{1}{r^2}\frac{\partial^2}{\partial \varphi^2}, \tag{4.1.104}$$

and finally relation

$$\frac{\partial^2}{\partial r\,\partial t} = \frac{\partial}{\partial r}\left(\frac{1}{r}\frac{\partial}{\partial \varphi}\right) = \frac{1}{r}\frac{\partial^2}{\partial r\,\partial \varphi} - \frac{1}{r^2}\frac{\partial}{\partial \varphi} \tag{4.1.105}$$

is obtained by comparing (3.3.78) with (3.3.80). Through (4.1.104), (4.1.102) is changed to

$$M_{rr} = -D[u_{3,rr} + \nu(r^{-1}u_{3,r} + r^{-2}u_{3,\varphi\varphi})]. \tag{4.1.106}$$

Similarly, we obtain

$$\begin{aligned} M_{tt} &= -D(r^{-1}u_{3,r} + r^{-2}u_{3,\varphi\varphi} + \nu u_{3,rr}), \\ M_{rt} &= -M_{tr} = D(1-\nu)(r^{-1}u_{3,r\varphi} - r^{-2}u_{3,\varphi}), \end{aligned} \tag{4.1.107}$$

starting from (4.1.40) by renaming indices, and through (4.1.104), and (4.1.105). Thus, all moments acting in the cuts have been obtained.

In the same manner, i.e., by changing in (4.1.41) 1,2 into r, t and by using (4.1.104) and (4.1.105),

$$\begin{aligned} Q_{rr} &= -D(u_{3,rr} + r^{-1}u_{3,r} + r^{-2}u_{3,\varphi\varphi})_{,r}, \\ Q_{tt} &= -Dr^{-1}(u_{3,rr} + r^{-1}u_{3,r} + r^{-2}u_{3,\varphi\varphi})_{,\varphi} \end{aligned} \tag{4.1.108}$$

is determined. Assuming edges which are free from m and v, the following boundary conditions are added to the plate equation (4.1.101):

$$\begin{aligned} &1.\ \textit{Clamped edge } R_e\text{:}\quad u_3 = 0,\ u_{3,r} = 0, \\ &2.\ \textit{Simply supported edge } R_g\text{:}\quad u_3 = 0,\ M_{rr} = 0, \\ &3.\ \textit{Free edge } R_f\text{:}\quad M_{rr} = 0,\ Q_{rr} - \frac{1}{r}\frac{\partial M_{rt}}{\partial \varphi} = 0. \end{aligned} \tag{4.1.109}$$

The boundary conditions are obtained from those indicated in 4.1.1, by appropriate change of indices.

Let us show the use of the theory by applying it to the simple example of a circular plate which bears a load symmetric to the x_3 axis. Then, everything is independent of φ. The plate equation (4.1.101) changes to the ordinary differential equation

$$\left(\frac{d^2}{dr^2}+\frac{1}{r}\frac{d}{dr}\right)^2 u_3 = \frac{1}{r}\frac{d}{dr}\left\{r\frac{d}{dr}\left[\frac{1}{r}\frac{d}{dr}\left(r\frac{du_3}{dr}\right)\right]\right\} = \frac{p}{D}. \tag{4.1.110}$$

It is assumed that the plate has radius ρ and that p = const. Both sides of (4.1.110) are multiplied by r, and through integration, as well as division by r,

$$\frac{d}{dr}\left[\frac{1}{r}\frac{d}{dr}\left(r\frac{du_3}{dr}\right)\right] = \frac{pr}{2D}+\frac{C_1}{r} \tag{4.1.111}$$

is obtained. Carrying out the calculation leads from (4.1.111) to solution

$$u_3 = \frac{pr^4}{64D}+\frac{C_1 r^2}{4}(\ln r-1)+\frac{C_2 r^2}{4}+C_3 \ln r+C_4. \tag{4.1.112}$$

Comparison of (4.1.108) with (4.1.111) indicates that

$$Q_{rr} = -D\frac{d}{dr}\left(\frac{d^2u_3}{dr}+\frac{1}{r}\frac{du_3}{dr}\right) = -D\left(\frac{pr}{2D}+\frac{C_1}{r}\right). \tag{4.1.113}$$

Since for $r = 0$, $Q_{rr} = \infty$ is impossible, then $C_1 = 0$ must be valid. Similarly, for $r = 0$, $u_3 = \infty$ is impossible. According to (4.1.112), $C_3 = 0$ must, therefore, be valid. The remaining constants of integration are determined by using the boundary conditions. Let us, for example, assume a clamped edge, then according to (4.1.109)

$$u_3 = 0, \quad \frac{du_3}{dr} = 0 \text{ for } r = \rho$$

must be true. From this through (4.1.112), using $C_1 = C_3 = 0$, equation system

$$\frac{p\rho^4}{64D} + \frac{C_2\rho^2}{4} + C_4 = 0, \quad \frac{p\rho^3}{16D} + \frac{C_2\rho}{2} = 0$$

containing solutions

$$C_2 = -\frac{p\rho^2}{8D}, \quad C_4 = \frac{p\rho^4}{64D} \tag{4.1.114}$$

is obtained. Inserting $C_1 = C_3 = 0$ and (4.1.114) in (4.1.112), then in agreement with (4.1.57)

$$u_3 = \frac{p}{64D}(\rho^2 - r^2)^2, \quad r^2 = x_1^2 + x_2^2$$

is obtained

4.1.4 *Supplements and expansions*

Up to this point, and as far as the plates are concerned, we have discussed the most simple problems under simplifying assumptions, so that Kirchhoff's theory was adequate. In generalizing the class of problems, however, we are forced to go further and supplement Kirchhoff's theory or substitute it by other more rigorous theories.

Instead of discussing isotropic plates, we may change to *orthotropic* or completely *anisotropic* plates. Orthotropic plates are of great technical importance, for example, in using them for bridges. Similarly, plate-like structural elements are used for the construction of ships and aircraft. They consist of a system of metal sheets which are reinforced by stringers and ribs (Figure 58). Particulars on the theory and calculation of orthotropic plates are given by K. Girkmann [29]. E. Giencke [30] has written on the calculation of aircraft constructions, and gives a comprehensive bibliography.

Further supplementation of the theory is demanded when *multiple connected plates*, i.e., plates with holes, are discussed. G. N. Sawin [28] has discussed these as well as anisotropic and stiffened plates. A. S. Wolmir [31] discusses, in the context of plates, problems like *anisotropy, plates of various layers, post-critical behaviour, plastic deformation, web plates*, etc. *Plates of variable thickness* are also of importance, and A. Kacner [32], in his paper, makes reference to these, and gives details on calculating plates with holes.

Let us continue by discussing in more detail two problems. One consists of improving the theory of thin plates in such a way that all contradictions contained in the previously mentioned theory are removed. The second problem occurs if large deflections are allowed, and which then renders Kirchhoff's theory as ineffectual.

E. Reissner has improved Kirchhoff's theory indicating the following: From (4.1.4), (4.1.10) and (4.1.6), relations

$$\sigma_{11} = \frac{M_{11}}{h^2/6}\frac{x_3}{h/2}, \quad \sigma_{22} = \frac{M_{22}}{h^2/6}\frac{x_3}{h/2}, \quad \sigma_{12} = \frac{M_{21}}{h^2/6}\frac{x_3}{h/2} \tag{4.1.115}$$

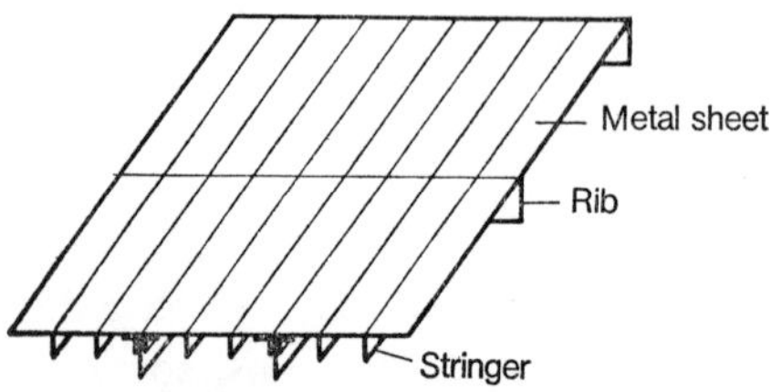

Figure 58

are notated, and from (4.1.35), (4.1.41) and (4.1.6),

$$\sigma_{12} = \frac{Q_{11}}{2h/3}\left[1-\left(\frac{x_3}{h/2}\right)^2\right], \quad \sigma_{23} = \frac{Q_{22}}{2h/3}\left[1-\left(\frac{x_3}{h/2}\right)^2\right] \tag{4.1.116}$$

is obtained. Using (4.1.42) to express $Q_{11,1}+Q_{22,2}$,

$$Q_{11,1}+Q_{22,2} = M_{21,21}+M_{11,11}-M_{22,22}-M_{12,12}$$

is determined, which because of (4.1.40) changes to

$$Q_{11,1}+Q_{22,2} = -D\nabla^2\nabla^2 u_3. \tag{4.1.117}$$

Comparing (4.1.22) with (4.1.117) shows that

$$Q_{11,1}+Q_{22,2} = -p \tag{4.1.118}$$

is valid. Let us use the principle of the stationary value of the conjugate total potential energy. According to (3.5.14) and (3.5.29) we obtain after simple changes

$$
\begin{aligned}
\Pi^* &= \frac{1}{2E}\iiint [\sigma_{11}^2+\sigma_{22}^2+\sigma_{33}^2-2\nu(\sigma_{11}\sigma_{22}+\sigma_{11}\sigma_{33}+\sigma_{22}\sigma_{33}) \\
&\quad +2(1+\nu)(\sigma_{12}^2+\sigma_{13}^2+\sigma_{23}^2)]\,\mathrm{d}x_1\,\mathrm{d}x_2\,\mathrm{d}x_3 \\
&\quad -\iint (\sigma_{nn}u_{nn}+\sigma_{nt}u_s+\sigma_{n3}u_3)\,\mathrm{d}s\,\mathrm{d}x_3
\end{aligned} \tag{4.1.119}
$$

for the conjugate total potential energy.

Inserting equations (4.1.115), (4.1.116), and (4.1.39) in (4.1.119), then following integration with respect to x_3 as far as this is possible, and using the relations

$$
\sigma_{nn} = \frac{M_{nn}}{h^2/6}\frac{x_3}{h/2}, \quad \sigma_{nt} = \frac{M_{tn}}{h^2/6}\frac{x_3}{h/2}, \quad \sigma_{n3} = \frac{Q_{nn}}{2h/3}\left[1-\frac{x_3}{h/2}\right]^2,
$$

which correspond to (4.1.115) and (4.1.116), from (4.1.119)

$$
\begin{aligned}
\Pi^* &= \frac{12}{2Eh^3}\iint \Big[(M_{11}+M_{22})^2+2(1+\nu)(M_{21}^2-M_{11}M_{22}) \\
&\quad +\frac{1+\nu}{5}h^2(Q_{11}^2+Q_{22}^2) \\
&\quad -\frac{\nu}{5}h^2p(M_{11}+M_{22})+\frac{h^3}{12}\int_{-h/2}^{+h/2}\sigma_{33}^2\,\mathrm{d}x_3\Big]\,\mathrm{d}x_1\,\mathrm{d}x_2 \\
&\quad -\int\Big[\frac{M_{nn}}{h^2/6}\int_{-h/2}^{+h/2}u_{nn}\frac{x_3}{h/2}\,\mathrm{d}x_3+\frac{M_{tn}}{h^2/6}\int_{-h/2}^{+h/2}u_s\frac{x_3}{h/2}\,\mathrm{d}x_3 \\
&\quad +\frac{Q_{nn}}{2h/3}\int_{-h/2}^{+h/2}u_3\left[1-\left(\frac{x_3}{h/2}\right)^2\right]\mathrm{d}x_3\Big]\,\mathrm{d}s
\end{aligned} \tag{4.1.120}
$$

is obtained. Similarly as in Kirchhoff's plate theory, let us assume

$$
u_{nn} = u_n^{*\prime}(s)\cdot x_3, \; u_s = u_t^{*\prime}(s)\cdot x_3, \; u_3 = u_3^*(s), \tag{4.1.121}
$$

corresponding to (4.1.1), (4.1.2), and to $u_3 = u_3(x_1, x_2)$. The dash represents the derivative with respect to s. The line integral in (4.1.120) thus becomes

$$\int [M_{nn} u_n^{*\prime} + M_{tn} u_t^{*\prime} + Q_{nn} u_3^*] \, ds.$$

For a plate with clamped edges,

$$u_n^{*\prime} = u_t^{*\prime} = u_3^* = 0,$$

holds true. Nothing is being specified for a plate with free edges as far as these three quantities are concerned. The principle of the stationary value of the conjugate total potential energy requires that $\delta\Pi^* = 0$ is true. At the same time, the equilibrium conditions (4.1.118) are to be satisfied. This secondary condition can be included in the principle using the method of Lagrange's multipliers.

Let λ be the multiplier. Then from (4.1.120) and (4.1.118),

$$\delta \left\{ \frac{6}{Eh^3} \iint \left[(M_{11}+M_{22})^2 + 2(1+\nu)(M_{21}^2 - M_{11} M_{22}) \right. \right.$$
$$\left. + \frac{1+\nu}{5} h^2 (Q_{11}^2 + Q_{22}^2) - \frac{\nu}{5} h^2 p (M_{11}+M_{22}) \right] dx_1 \, dx_2$$
$$- \int [M_{nn} u_n^{*\prime} + M_{tn} u_t^{*\prime} + Q_{nn} u_3^*] \, ds + \iint \lambda(x_1, x_2)(Q_{11,1} + Q_{22,2}$$
$$\left. + p) \, dx_1 \, dx_2 \right\} = 0$$

is obtained. Carrying out the variations,

$$\frac{12}{Eh^3} \iint \left[(M_{11}+M_{22})(\delta M_{11} + \delta M_{22}) + (1+\nu)(2M_{21} \delta M_{21} \right.$$
$$- M_{11} \delta M_{22} - M_{22} \delta M_{11}) + \frac{1+\nu}{5} h^2 (Q_{11} \delta Q_{11} + Q_{22} \delta Q_{22})$$
$$\left. - \frac{\nu}{10} h^2 p (\delta M_{11} + \delta M_{22}) \right] dx_1 \, dx_2 - \int [\delta M_{nn} u_n^{*\prime} + \delta M_{tn} u_t^{*\prime}$$
$$+ \delta Q_{nn} u_3^*] \, ds + \iint \lambda [\delta(Q_{11,1}) + \delta(Q_{22,2})] \, dx_1 \, dx_2 = 0 \qquad (4.1.122)$$

is determined. Through integration by parts, the last integral of (4.1.122) becomes

$$\iint \lambda[\delta(Q_{11,1})+\delta(Q_{22,2})]\,\mathrm{d}x_1\,\mathrm{d}x_2$$
$$= -\iint [\lambda_{,1}\delta Q_{11}+\lambda_{,2}\delta Q_{22}]\,\mathrm{d}x_1\,\mathrm{d}x_2+\int \lambda\delta Q_{nn}\,\mathrm{d}s.$$

Inserting this in (4.1.122), the fact that variation δQ_{nn} is arbitrary, yields

$$u_3^*(s) = \lambda(s). \tag{4.1.123}$$

Since the result is valid for any curve which might occur in the plate's inside, it is possible that λ is identifiable with u_3.

The variations of Q_{11}, Q_{12} are connected with the variations of M_{11}, M_{22} and M_{12} because of (4.1.42). For this reason, and through integration by parts, from (4.1.122)

$$\iint \left[\frac{12}{Eh^3}\left(M_{11}-\nu M_{22}-\frac{1+\nu}{5}h^2Q_{11,1}-\frac{\nu}{10}h^2p\right)+u_{3,11}\right]\delta M_{11}$$
$$+\left[\frac{12}{Eh^3}\left(M_{22}-\nu M_{11}-\frac{1+\nu}{5}h^2Q_{22,2}-\frac{\nu}{10}h^2p\right)+u_{3,22}\right]\delta M_{22}$$
$$+\left[\frac{12}{Eh^3}\left(2(1+\nu)M_{21}-\frac{1+\nu}{5}h^2(Q_{11,2}+Q_{22,1})\right)\right.$$
$$\left.+2u_{3,12}\right]\delta M_{21}\Big\}\,\mathrm{d}x_1\,\mathrm{d}x_2-\int\left\{\left[u_n^{*\prime}+u_{3,n}-\frac{12(1+\nu)}{5hE}Q_{nn}\right]\delta M_{nn}\right.$$
$$\left.+\left[u_t^{*\prime}+u_{3,s}-\frac{12(1+\nu)}{5hE}Q_{tt}\right]\delta M_{tn}-[u_3^*-u_3]\delta Q_{nn}\right\}\mathrm{d}s = 0 \tag{4.1.124}$$

is obtained. From (4.1.124), the differential equations, and the boundary conditions of Reissner's theory are determined. In order to cause the disappearance of the line integral in (4.1.124), either the dynamic conditions

$$M_{nn} = m(s),\ M_{tn} = d(s),\ Q_{nn} = q(s) \tag{4.1.125}$$

or the geometric conditions

$$u_{3,n}-\frac{12(1+\nu)}{5hE}Q_{nn} = -u_n^{*\prime},\ u_{3,s}-\frac{12(1+\nu)}{5hE}Q_{tt} = -u_t^{*\prime},$$

$$u_3 = u_3^* \tag{4.1.126}$$

must be satisfied. The first two relations of (4.1.126) indicate that, because of the influence of the shear, normal and tangential line elements *in the middle plane* do not remain vertical to the line element which used to be vertical to the middle plane before deformation occurred.

With respect to the arbitrary nature of variations $\delta M_{11}, \delta M_{22}, \delta M_{21}$, three equations follow from (4.1.124), which by using (4.1.118), and following a few rearrangements, change into the differential equation system

$$M_{11}-\frac{h^2}{5}Q_{11,1}+\frac{\nu h^2}{10(1-\nu)}p = -D(u_{3,11}+\nu u_{3,22}),$$

$$M_{22}-\frac{h^2}{5}Q_{22,2}+\frac{\nu h^2}{10(1-\nu)}p = -D(u_{3,22}+\nu u_{3,11}), \tag{4.1.127}$$

$$M_{21}-\frac{h^2}{10}(Q_{11,2}+Q_{22,1}) = -(1-\nu)Du_{3,12}.$$

From (4.1.118) and (4.1.127), further differential equations are obtained:

$$Q_{11}-\frac{h^2}{10}\nabla^2 Q_{11}+\frac{h^2}{10(1-\nu)}p_{,1} = -D(\nabla^2 u_3)_{,1},$$

$$Q_{22}-\frac{h^2}{10}\nabla^2 Q_{22}+\frac{h^2}{10(1-\nu)}p_{,2} = -D(\nabla^2 u_3)_{,2}. \tag{4.1.128}$$

These are determined by using (4.1.42), through differentiation, combination, and using symbol $\nabla = \partial^2/\partial x_1^2+\partial^2/\partial x_2^2$. Equations (4.1.117), (4.1.118), and (4.1.128) allow for calculation of quantities Q_{11}, Q_{22}, and u_3. Once this has been carried out, then the still missing internal forces M_{11}, M_{22}, and $M_{21} = -M_{12}$ are obtained from (4.1.127) by using the previously mentioned quantities.

It is remarkable that the equations (4.1.125) easily yield Poisson's three controversial boundary conditions for free edges, instead of the two Kirchhoff boundary conditions $M_{nn} = m(s)$, $V_{nn} = Q_{nn} - M_{nt,s} = v(s)$ of the classic theory.

The development of a theory for plates having large deflections has been carried out by T. von Kármán as a substitute for Kirchhoff's theory, which becomes invalid. Von Kármán retains the assumption that the plate should be thin. When no shearing forces are present on the plate's surface, then ε_{13} and ε_{23} may be neglected as in the classic theory. Since the thickness of the plate is small, then $\varepsilon_{33} = u_{3,3}$ can be equated to nought and $u_3 = u_3(x_1, x_2)$ becomes independent of x_3. This indicates that Kirchhoff's hypothesis should also apply to plates with large deflections. Furthermore, for the assumed thin plate, σ_{33} is omitted, and according to the condition $\varepsilon_{13} = \varepsilon_{23} = 0$, and Hooke's law, $\sigma_{13} = \sigma_{23} = 0$ must be notated. Thus, according to Hooke's law

$$\varepsilon_{11} = \frac{1}{E}(\sigma_{11} - \nu\sigma_{22}),\ \varepsilon_{22} = \frac{1}{E}(\sigma_{22} - \nu\sigma_{11}),\ \varepsilon_{12} = \frac{1+\nu}{E}\sigma_{12}$$

$$\sigma_{11} = \frac{E}{1-\nu^2}(\varepsilon_{11} + \nu\varepsilon_{22}),\ \sigma_{22} = \frac{E}{1-\nu^2}(\varepsilon_{22} + \nu\varepsilon_{11}), \tag{4.1.129}$$

$$\sigma_{12} = \frac{E}{1+\nu}\varepsilon_{12}, \text{ respectively,}$$

is yielded, which are relations already occurring in the classic theory. The difference to that theory is, that with respect to the large deflections, now, strain in the plate's middle plane cannot be excluded, and for the strain, non-linear terms must be at least partially considered.

If $u_i^{(m)}$, $i = 1, 2, 3$, represent the displacements of the middle plane, then, instead of (4.1.1) and (4.1.2), which are valid for Kirchhoff's theory,

$$u_1 = u_1^{(m)} - x_3 u_{3,1},\ u_2 = u_2^{(m)} - x_3 u_{3,2} \tag{4.1.130}$$

is notated. Thus, the influence of deformation of the middle plane has been more adequately considered than in the classic theory. Because of u_3 being independent of x_3, the relation $u_3 = u_3^{(m)}$ remains intact.

The transition to the strains yields, because of (4.1.130), and in difference to (4.1.3),

$$\varepsilon_{11} = \varepsilon_{11}^{(m)} - x_3 u_{3,11}, \; \varepsilon_{22} = \varepsilon_{22}^{(m)} - x_3 u_{3,22}, \; \varepsilon_{12} = \varepsilon_{12}^{(m)} - x_3 u_{3,12}. \tag{4.1.131}$$

Then, in the arrangement of $\varepsilon_{ij}^{(m)}$, the non-linearities will be considered, i.e.,

$$\begin{aligned} \varepsilon_{11}^{(m)} &= u_{1,1}^{(m)} + \tfrac{1}{2} u_{3,1}^2, \; \varepsilon_{22}^{(m)} = u_{2,2}^{(m)} + \tfrac{1}{2} u_{3,2}^2, \\ \varepsilon_{12}^{(m)} &= \tfrac{1}{2}(u_{2,1}^{(m)} + u_{1,2}^{(m)}) + \tfrac{1}{2} u_{3,1} u_{3,2} \end{aligned} \tag{4.1.132}$$

is required. Let us use the principle of the stationary value of the total potential energy. To begin with let

$$\begin{aligned} \mathfrak{E}(\varepsilon) = \frac{E}{1-\nu^2} \int_F \int_{-h/2}^{+h/2} &[(\varepsilon_{11} + \varepsilon_{22})^2 + 2(1-\nu) \\ &\cdot(\varepsilon_{12}^2 - \varepsilon_{11}\varepsilon_{22})] \, \mathrm{d}x_3 \, \mathrm{d}F \end{aligned} \tag{4.1.133}$$

be true, from which, because of (4.1.131),

$$\begin{aligned} \mathfrak{E}(\varepsilon) = \frac{Eh}{1-\nu^2} \int_F &[(\varepsilon_{11}^{(m)} + \varepsilon_{22}^{(m)})^2 + 2(1-\nu)(\varepsilon_{11}^{(m)2} - \varepsilon_{11}^{(m)}\varepsilon_{22}^{(m)})] \, \mathrm{d}F \\ + D \int_F &[(u_{3,11} + u_{3,22})^2 + 2(1-\nu)(u_{3,12}^2 - u_{3,11} u_{3,22})] \, \mathrm{d}F \end{aligned}$$

is yielded.

From the requirement

$$\delta \left[\mathfrak{E}(\varepsilon) - \int_F p u_3 \, \mathrm{d}F \right] = 0$$

of the principle, the three equations

$$\begin{aligned} &(\varepsilon_{11}^{(m)} + \nu\varepsilon_{22}^{(m)})_{,1} + (1-\nu)\varepsilon_{12,2}^{(m)} = 0, \\ &(\varepsilon_{22}^{(m)} + \nu\varepsilon_{11}^{(m)})_{,2} + (1-\nu)\varepsilon_{12,1}^{(m)} = 0, \\ &\nabla^2\nabla^2 u_3 = \frac{p}{D} + \frac{12}{h^2} [(\varepsilon_{11}^{(m)} + \nu\varepsilon_{22}^{(m)}) u_{3,11} + (\varepsilon_{22}^{(m)} + \nu\varepsilon_{11}^{(m)}) u_{3,22} \\ &\qquad + 2(1-\nu)\varepsilon_{12}^{(m)} u_{3,12}] \end{aligned} \tag{4.1.134}$$

are obtained as Euler's equations of the variational problem.

In comparison with (4.1.22), it is clear to see the extent of the expansion as compared to the classic theory. The system (4.1.134) diminishes to (4.1.22), once the strain of the plate's middle plane is omitted, i.e., if all $\varepsilon_{ij}^{(m)}$ are equated to nought.

Let us introduce the internal forces

$$
\begin{aligned}
N_{11} &= \int_{-h/2}^{+h/2} \sigma_{11}\,\mathrm{d}x_3 = \frac{Eh}{1-\nu^2}\left(\varepsilon_{11}^{(m)}+\nu\varepsilon_{22}^{(m)}\right),\\
N_{22} &= \int_{-h/2}^{+h/2} \sigma_{22}\,\mathrm{d}x_3 = \frac{Eh}{1-\nu^2}\left(\varepsilon_{22}^{(m)}+\nu\varepsilon_{11}^{(m)}\right),\\
N_{12} = N_{21} &= \int_{-h/2}^{+h/2} \sigma_{12}\,\mathrm{d}x_3 = \frac{Eh}{1+\nu}\,\varepsilon_{12}^{(m)},
\end{aligned}
\tag{4.1.135}
$$

which are additional with respect to the classic plate theory. They have been calculated using (4.1.129) and (4.1.131). Inserting (4.1.135) in (4.1.134), the differential equation system

$$
\begin{aligned}
&N_{11,1}+N_{12,2} = 0, \quad N_{22,2}+N_{12,1} = 0,\\
&\nabla^2\nabla^2 u_3 = \frac{p}{D}+\frac{1}{D}\left(N_{11}u_{3,11}+N_{22}u_{3,22}+2N_{12}u_{3,12}\right)
\end{aligned}
\tag{4.1.136}
$$

is obtained. The general solutions of the equations in the first line of (4.1.136) are indicated by using stress function Φ, for which

$$
N_{11} = \Phi_{,22},\ N_{22} = \Phi_{,11},\ N_{12} = -\Phi_{,12} \tag{4.1.137}
$$

is valid. Through (4.1.137), the third line of (4.1.136) changes to

$$
\nabla^2\nabla^2 u_3 = \frac{p}{D}+\frac{1}{D}\left(\Phi_{,22}u_{3,11}+\Phi_{,11}u_{3,22}-2\Phi_{,12}u_{3,12}\right). \tag{4.1.138}
$$

Let us set up a relation between Φ and u_3. From (4.1.135) and (4.1.137),

$$\varepsilon_{11}^{(m)} = \frac{1}{Eh}(\Phi_{,22}-\nu\Phi_{,11}),\ \varepsilon_{22}^{(m)} = \frac{1}{Eh}(\Phi_{,11}-\nu\Phi_{,22}),$$

$$\varepsilon_{12}^{(m)} = -\frac{1+\nu}{Eh}\Phi_{,12} \tag{4.1.139}$$

is obtained. The equations (4.1.132) require that

$$\varepsilon_{11,22}^{(m)}+\varepsilon_{22,11}^{(m)}-2\varepsilon_{12,12}^{(m)} = u_{3,12}^{2}-u_{3,11}u_{3,22} \tag{4.1.140}$$

is true. Inserting into this 'compatibility equation' the $\varepsilon_{ij}^{(m)}$ of (4.1.139), then

$$\nabla^2\nabla^2\Phi+Eh(u_{3,11}u_{3,22}-u_{3,12}^{2}) = 0 \tag{4.1.141}$$

is determined. Equations (4.1.138) and (4.1.141) represent the fundamental differential equations of von Kármán's theory. They substitute Kirchhoff's plate equation (4.1.22). Reference to solution methods of the von Kármán equations have been made by Timoshenko [33] and Wang [34].

Through limit process, it is quite easy to arrive at the differential equation of the membrane from the third equation (4.1.136). According to the usual assumptions, the membrane has the property of being stretched only, and of having no bending stiffness. Therefore, those stresses caused by bending as compared to those caused by stretching can be omitted. Then the term on the left side of the equation is ignored and

$$N_{\alpha\beta}u_{3,\alpha\beta} = -p,\ \alpha, \beta = 1, 2, \tag{4.1.142}$$

is determined. If we are dealing with isotropic stretching, that is, if the stresses are equal in all directions of the membrane, then $N_{\alpha\beta} = S\delta_{\alpha\beta}$ is true, whereby S represents the tension per unit length. Hence, (4.1.142) changes to

$$S\delta_{\alpha\beta}u_{3,\alpha\beta} = S(u_{3,11}+u_{3,22}) = -p,$$

$$\nabla^2 u_3 = -\frac{p}{S}, \tag{4.1.143}$$

respectively. Then, the above represents the already known *membrane equation*. Thus, it has been demonstrated in what the nature of the expansion consists, which carries the von Kármán theory further than the Kirchhoff's theory.

4.2 Shells

4.2.1 *Introduction*

In our context, a *shell* is represented by an elastic body having two limiting curved surfaces. The distance of the limiting surfaces is called shell thickness, h. Let us keep to *thin* shells in our discussion. For these, the thickness h is a relatively small quantity as compared with other dimensions and the radii of curvature. Let us introduce the term *middle surface*, which, in a similar way, has already been used in connection with plates. The middle surface is the curved surface dividing the thickness of the shell. The geometry of the shell is completely determined by the middle surface and by the shell's thickness. Since in the following chapter we will refer quite often to the middle surface, it might be useful to repeat the more important facts about the differential geometry of surfaces.

4.2.2 *Elements of the differential geometry of surfaces*

Let the points of a surface be determined relative to a fixed point O by the radius vector

$$\boldsymbol{r} = (x_1, x_2, x_3) \tag{4.2.1}$$

(Figure 59). Using *Gauss' coordinates* q_α, $\alpha = 1, 2$, we can change over to representation

$$x_i = x_i(q_1, q_2), \; i = 1, 2, 3, \tag{4.2.2}$$

and by eliminating q_α from (4.2.2), a third possibility for the surface equation is obtained,

$$F(x_1, x_2, x_3) = 0. \tag{4.2.3}$$

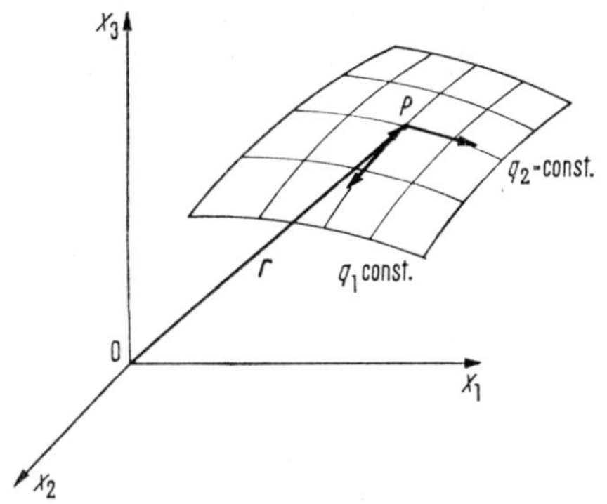

Figure 59

Gauss' coordinates, parameters, q_α, $\alpha = 1, 2$, respectively, which will consistently be used, are curvilinear coordinates on the surface. In connection with them, we will therefore constantly refer back to the results of 3.1.3. For a sphere of radius R are for example $q_1 = \varphi$, $q_2 = \vartheta$, and the equations (4.2.2) are

$$x_1 = R \sin \varphi \cos \vartheta, x_2 = R \sin \varphi \sin \vartheta, x_3 = R \cos \varphi.$$

From point P of the surface to the adjacent point Q, vector $\mathrm{d}\boldsymbol{r}$ is drawn (Figure 60), for which

$$\mathrm{d}\boldsymbol{r} = \boldsymbol{r}_{,\alpha} \mathrm{d}q_\alpha, \alpha = 1, 2, \tag{4.2.4}$$

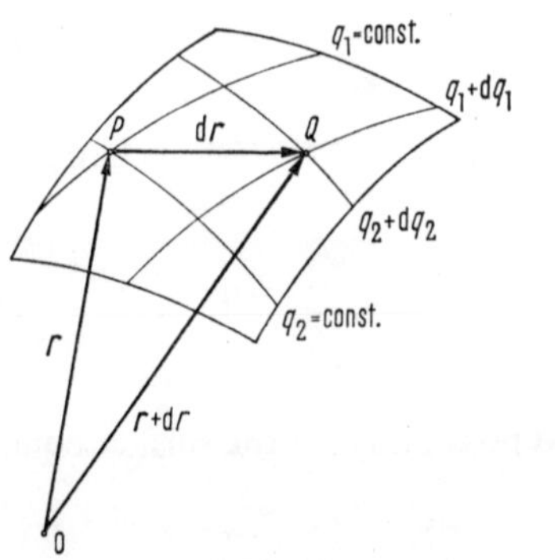

Figure 60

and because of (4.2.1),

$$\mathrm{d}x_i = x_{i,\alpha}\mathrm{d}q_\alpha,\ i = 1, 2, 3,\ \alpha = 1, 2, \tag{4.2.5}$$

is true, respectively. Through (4.2.5) we obtain for the square of the line element of the surface, which is $\mathrm{d}s^2 = \mathrm{d}x_i\mathrm{d}x_i$,

$$\mathrm{d}s^2 = x_{i,\alpha}x_{i,\beta}\mathrm{d}q_\alpha\mathrm{d}q_\beta,\ \alpha, \beta = 1, 2. \tag{4.2.6}$$

But

$$\boldsymbol{r}_{,\alpha}\boldsymbol{r}_{,\beta} = x_{i,\alpha}x_{i,\beta} = g_{\alpha\beta} \tag{4.2.7}$$

is true, whereby $g_{\alpha\beta}$ represents the metric-tensor of the particular surface. Hence, the relation

$$\mathrm{d}s^2 = g_{\alpha\beta}\mathrm{d}q_\alpha\mathrm{d}q_\beta \tag{4.2.8}$$

is obtained from (4.2.6). It is called the *first fundamental form* of the surface. Let us discuss two surface curves C and C^*, intersecting at point P (Figure 61), and forming angle ϑ, and having arch lengths s and s^*. Angle ϑ is evidently obtained from

$$\cos\vartheta = \frac{\mathrm{d}\boldsymbol{r}}{\mathrm{d}s}\frac{\mathrm{d}\boldsymbol{r}}{\mathrm{d}s^*} = \frac{\mathrm{d}x_i}{\mathrm{d}s}\frac{\mathrm{d}x_i}{\mathrm{d}s^*}.$$

It changes to

$$\cos\vartheta = g_{\alpha\beta}\frac{\mathrm{d}q_\alpha}{\mathrm{d}s}\frac{\mathrm{d}q_\beta}{\mathrm{d}s^*} \tag{4.2.9}$$

because of (4.2.5) and (4.2.7). Let us calculate an area element $\mathrm{d}F$ (Figure 62). Then the basis vectors $\boldsymbol{e}_\alpha = \boldsymbol{r}_{,\alpha}$, $\alpha = 1, 2$, are introduced, for which because of (4.2.7)

$$\boldsymbol{e}_\alpha\boldsymbol{e}_\beta \equiv \boldsymbol{r}_{,\alpha}\boldsymbol{r}_{,\beta} = x_{i,\alpha}x_{i,\beta} = g_{\alpha\beta} \tag{4.2.10}$$

is valid. Let

$$dF = |\boldsymbol{e}_1 \times \boldsymbol{e}_2| dq_1 dq_2 = |\boldsymbol{e}_1| \cdot |\boldsymbol{e}_2| \sin \vartheta \, dq_1 dq_2 \tag{4.2.11}$$

be true. Since for the chosen special curves in Figure 62, $dq_1/ds^* = dq_2/ds = 0$ is true for the coordinate lines, we simply get from (4.2.9)

$$\cos \vartheta = g_{12} \frac{dq_1}{ds} \frac{dq_2}{ds^*}. \tag{4.2.12}$$

From the first fundamental form we also obtain for $\alpha = \beta = 1$, $\alpha = \beta = 2$, respectively,

$$\left(\frac{ds}{dq_1}\right)^2 = g_{11}, \left(\frac{ds^*}{dq_2}\right)^2 = g_{22}. \tag{4.2.13}$$

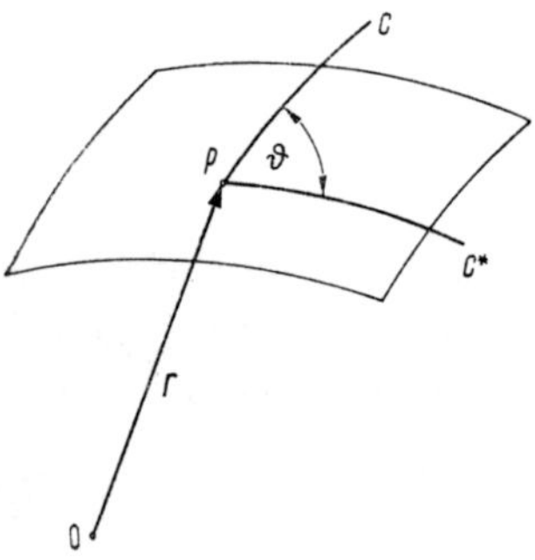

Figure 61

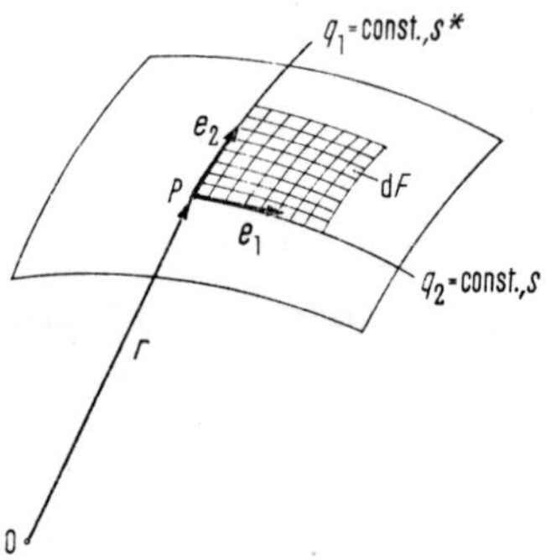

Figure 62

Inserting (4.2.13) in (4.2.12) it yields

$$\cos \vartheta = \frac{g_{12}}{\sqrt{g_{11} g_{22}}}$$

and from this, because of $\sin \vartheta = (1 - \cos^2 \vartheta)^{\frac{1}{2}}$,

$$\sin \vartheta = \left(\frac{g_{11} g_{22} - g_{12}^2}{g_{11} g_{22}}\right)^{\frac{1}{2}} \tag{4.2.14}$$

is obtained. We deduce

$$|\boldsymbol{e}_1| = \sqrt{g_{11}}, |\boldsymbol{e}_2| = \sqrt{g_{22}}, \boldsymbol{e}_1 \boldsymbol{e}_2 = g_{12} \tag{4.2.15}$$

from (4.2.10). Using (4.2.14) and (4.2.15), then (4.2.11) is changed to

$$
\begin{aligned}
dF &= (g_{11} g_{22} - g_{12}^2)^{\frac{1}{2}} dq_1 \, dq_2 = \sqrt{g} \, dq_1 \, dq_2 , \\
g &= \det (g_{\alpha\beta}).
\end{aligned}
\tag{4.2.16}
$$

We are now able to measure the lengths, angles and areas of the given surface by using (4.2.8), (4.2.9) and (4.2.16).

Let us now determine choosing the most suitable coordinate curves q_α: in point P of the area, a *normal cut* is made (Figure 63). This is done through a plane E, containing the *area normal* $\boldsymbol{n}$ of point P. On the surface, E cuts out the planar curve C, whose principal normal in P coincides in this case with the surface-normal $\boldsymbol{n}$ which, in general, is not-self-evident. Rotating plane E about $\boldsymbol{n}$, a number of intersecting curves C are obtained, whose curvatures in P are of various sizes. For two of these curves of intersection, vertical to each other, extremal curvatures are obtained, which are the *principal curvatures* in P. The two directions yielding the principal curvatures, are termed *principal directions*. When finding the curves on the surface whose tangents coincide at all points with the principal directions, then the *lines of curvature* are obtained. It is advisable to choose these as orthogonal coordinate lines.

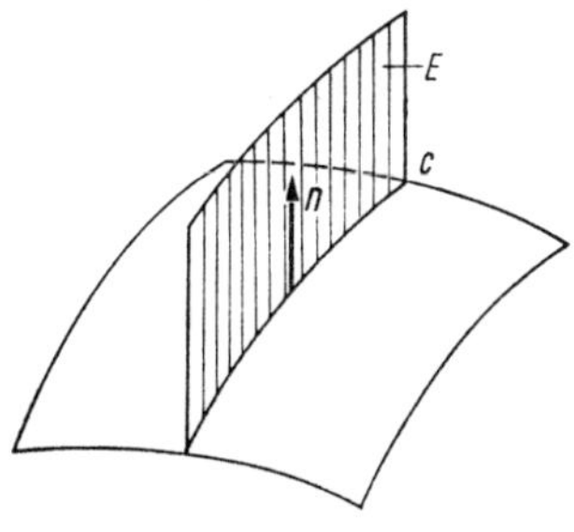

Figure 63

The principal directions in P are distinguished by the fact that the normals of points, adjacent to P in the principal directions, intersect with the normal of P at a point. In the limiting case, these points of intersection yield the *centres of curvature*. The distance of point P from the respective centre of curvature represents the respective *principal radius of curvature* of the surface in P.

Let

$$\boldsymbol{n} = \frac{\boldsymbol{e}_1 \times \boldsymbol{e}_2}{|\boldsymbol{e}_1 \times \boldsymbol{e}_2|}$$

be valid for the surface normal in P. Comparison of (4.2.11) with (4.2.16) indicates that

$$|\boldsymbol{e}_1 \times \boldsymbol{e}_2| = \sqrt{g} \tag{4.2.17}$$

is true. Hence, we can notate

$$\boldsymbol{n} = \frac{\boldsymbol{e}_1 \times \boldsymbol{e}_2}{\sqrt{g}}. \tag{4.2.18}$$

Let $\boldsymbol{r}$ be the radius vector and $\boldsymbol{n}$ be the normal of point P. Correspondingly, $\boldsymbol{r}+\mathrm{d}\boldsymbol{r}$ and $\boldsymbol{n}+\mathrm{d}\boldsymbol{n}$ belong to point Q, which is adjacent to P in the principal direction. It is required that adjacent normals intersect. This pre-supposes vector $\boldsymbol{n}$, $\mathrm{d}\boldsymbol{n}$ and $\mathrm{d}\boldsymbol{r}$ to be coplanar. Thus,

$$\boldsymbol{n}(\mathrm{d}\boldsymbol{n} \times \mathrm{d}\boldsymbol{r}) = 0 \tag{4.2.19}$$

must be valid. Let

$$\mathrm{d}\boldsymbol{n} = \boldsymbol{n}_{,\alpha}\mathrm{d}q_\alpha, \; \mathrm{d}\boldsymbol{r} = \boldsymbol{r}_{,\alpha}\mathrm{d}q_\alpha \tag{4.2.20}$$

be true.

Thus, (4.2.19) changes to

$$\boldsymbol{n}(\boldsymbol{n}_{,\alpha} \times \boldsymbol{r}_{,\beta})\mathrm{d}q_\alpha \mathrm{d}q_\beta = 0. \tag{4.2.21}$$

Since $\boldsymbol{r}_{,1}$ is tangential to $q_2 = \text{const}$, then $\boldsymbol{r}_{,1}$ is vertical to $\boldsymbol{n}$. Hence, $\boldsymbol{n}\boldsymbol{r}_{,1} = 0$, from which, after further derivation with respect to q_1,

$$\boldsymbol{n}_{,1}\boldsymbol{r}_{,1} + \boldsymbol{n}\boldsymbol{r}_{,11} = 0$$

results. Introducing

$$b_{\alpha\beta} = \boldsymbol{n}\boldsymbol{r}_{,\alpha\beta}, \tag{4.2.22}$$

the *fundamental quantities of the second order*, yield the relation

$$\boldsymbol{n}_{,1}\boldsymbol{r}_{,1} = -\boldsymbol{n}\boldsymbol{r}_{,11} = -b_{11}.$$

This result can be generalized to equation

$$\boldsymbol{n}_{,\alpha}\boldsymbol{r}_{,\beta} = -b_{\alpha\beta}. \tag{4.2.23}$$

From (4.2.22), the so-called Gauss' Formula

$$\boldsymbol{r}_{,\alpha\beta} = b_{\alpha\beta}\,\boldsymbol{n} \tag{4.2.24}$$

is easily obtained. Using coefficients $b_{\alpha\beta}$, the *second fundamental form* of the surface is set up,

$$B = b_{\alpha\beta}\,\mathrm{d}q_\alpha \mathrm{d}q_\beta, \tag{4.2.25}$$

which is important in connection with the curvature of the surface.

Vector $\boldsymbol{n}_{,\alpha}$ is vertical to $\boldsymbol{n}$, and is therefore in the tangential plane, i.e., in the plane of the surface which has been spanned by vectors $\boldsymbol{e}_1, \boldsymbol{e}_2$. Using unknown coefficients $c_{\alpha\beta}$, let us notate for now

$$\boldsymbol{n}_{,\alpha} = c_{\alpha\beta}\,\boldsymbol{e}_\beta \equiv c_{\alpha\beta}\,\boldsymbol{r}_{,\beta}. \tag{4.2.26}$$

Through (4.2.26) and (4.2.23), $\boldsymbol{n}_{,1}\boldsymbol{r}_{,1}$ and $\boldsymbol{n}_{,1}\boldsymbol{r}_{,2}$ changes to

$$\boldsymbol{n}_{,1}\boldsymbol{r}_{,1} = c_{11}\boldsymbol{r}_{,1}\boldsymbol{r}_{,1}+c_{12}\boldsymbol{r}_{,2}\boldsymbol{r}_{,1} = -b_{11},$$
$$\boldsymbol{n}_{,1}\boldsymbol{r}_{,2} = c_{11}\boldsymbol{r}_{,1}\boldsymbol{r}_{,2}+c_{12}\boldsymbol{r}_{,2}\boldsymbol{r}_{,2} = -b_{12}.$$

Because of $\boldsymbol{r}_{,\alpha}\boldsymbol{r}_{,\beta} = x_{i,\alpha}x_{i,\beta} \equiv g_{\alpha\beta}$, the system of equations

$$c_{11}g_{11}+c_{12}g_{12} = -b_{11},$$
$$c_{11}g_{12}+c_{12}g_{22} = -b_{12}$$

is obtained, from which

$$c_{11} = (b_{12}g_{12}-b_{11}g_{22}) : g, c_{12} = (b_{11}g_{12}-b_{12}g_{11}) : g \tag{4.2.27}$$

is yielded, by solving for c_{11}, c_{12}.

In the same way,

$$c_{21} = (b_{22}g_{12}-b_{12}g_{22}) : g, c_{22} = (b_{12}g_{12}-b_{22}g_{11}) : g \tag{4.2.28}$$

is determined.

Let $\boldsymbol{r}_{,\alpha} \equiv \boldsymbol{e}_\alpha$. Let us notate (4.2.18) for $\boldsymbol{n}$ and (4.2.26) for $\boldsymbol{n}_{,\alpha}$. Thus, (4.2.21) changes to

$$\left[\left(\frac{\boldsymbol{e}_1 \times \boldsymbol{e}_2}{\sqrt{g}}\right) c_{\alpha\beta}\,\boldsymbol{e}_\beta \times \boldsymbol{e}_\gamma\right] \mathrm{d}q_\alpha\,\mathrm{d}q_\gamma = 0. \tag{4.2.29}$$

Since $\sqrt{g} \neq 0$ is valid, then this quantity can be neglected in the following, and using $\boldsymbol{N} = \boldsymbol{e}_1 \times \boldsymbol{e}_2$, the equation (4.2.29) is expressed as

$$\begin{aligned}&[\boldsymbol{N}(c_{11}\,\boldsymbol{e}_1+c_{12}\,\boldsymbol{e}_2)\times \boldsymbol{e}_1]\,\mathrm{d}q_1^2+[\boldsymbol{N}(c_{11}\,\boldsymbol{e}_1+c_{12}\,\boldsymbol{e}_2)\times \boldsymbol{e}_2]\,\mathrm{d}q_1\,\mathrm{d}q_2\\&\quad+[\boldsymbol{N}(c_{21}\,\boldsymbol{e}_1+c_{22}\,\boldsymbol{e}_2)\times \boldsymbol{e}_1]\,\mathrm{d}q_2\,\mathrm{d}q_1+[\boldsymbol{N}(c_{21}\,\boldsymbol{e}_1+c_{22}\,\boldsymbol{e}_2)\times \boldsymbol{e}_2]\,\mathrm{d}q_2^2 = 0.\end{aligned}$$

Because of $\boldsymbol{N}\times \boldsymbol{e}_1 = g_{11}\boldsymbol{e}_2$, $\boldsymbol{N}\times \boldsymbol{e}_2 = -g_{22}\boldsymbol{e}_1$, and $\boldsymbol{e}_\alpha \boldsymbol{e}_\beta = g_{\alpha\beta}$ it changes to

$$c_{12}\,\mathrm{d}q_1^2+(c_{22}-c_{11})\,\mathrm{d}q_1\,\mathrm{d}q_2-c_{21}\,\mathrm{d}q_2^2 = 0. \tag{4.2.30}$$

Through (4.2.27) and (4.2.28) we obtain

$$\begin{aligned}&(g_{11}\,b_{12}-g_{12}\,b_{11})\,\mathrm{d}q_1^2+(g_{11}\,b_{22}-g_{22}\,b_{11})\,\mathrm{d}q_1\,\mathrm{d}q_2\\&\quad+(g_{12}\,b_{22}-g_{22}\,b_{12})\,\mathrm{d}q_2^2 = 0,\end{aligned} \tag{4.2.31}$$

following multiplication with -1. This represents an algebraic equation which is used to determine the principal directions $\mathrm{d}q_1/\mathrm{d}q_2$. Based on its definition, $\mathrm{d}q_1/\mathrm{d}q_2 = 0$ and $\mathrm{d}q_2/\mathrm{d}q_1 = 0$, respectively, must be true for the lines of curvature. From (4.2.30) follows that this is only possible if the conditions

$$c_{12} = c_{21} = 0, c_{22}-c_{11} \neq 0 \tag{4.2.32}$$

have been satisfied. But

$$c_{12}b_{22}-c_{21}b_{11} = 0$$

must then be true. This condition changes to

$$-(g_{11}b_{22}-g_{22}b_{11})b_{12} \equiv (c_{22}-c_{11})b_{12} = 0$$

through (4.2.27) and (4.2.28). Because of the third condition (4.2.32) it is realized that $b_{12} = 0$ applies to lines of curvature. This fact can be used to check whether arbitrarily given orthogonal Gaussian parametric curves are lines of curvature.

Because of the first two conditions of (4.2.32),

$$c_{12}g_{22}-c_{21}g_{11} = 0$$

must be true. Using (4.2.27), (4.2.28), it is changed to

$$(b_{11}g_{22}-b_{22}g_{11})g_{12} = (c_{22}-c_{11})g_{12} = 0.$$

Because of the third condition (4.2.32), $g_{12} = 0$ is obtained. It has been shown previously that $\cos\vartheta = g_{12}/\sqrt{g_{11}g_{22}}$ is valid for the angle of intersection pertaining to parametric curves. Therefore, we conclude from $g_{12} = 0$, that for lines of curvature, $\cos\vartheta = 0$ is valid, and that the *lines of curvature* are, therefore, in fact *orthogonal* to each other.

The centre of curvature of the surface in P has a radius vector $\boldsymbol{r}^* = \boldsymbol{r}+R\boldsymbol{n}$ (Figure 64), accordingly, $\boldsymbol{r}^*+\mathrm{d}\boldsymbol{r}^* = \boldsymbol{r}+\mathrm{d}\boldsymbol{r}+R\boldsymbol{n}+R\mathrm{d}\boldsymbol{n}+\boldsymbol{n}\mathrm{d}R$ is valid in Q. It becomes obvious that $\mathrm{d}\boldsymbol{r}^* = \mathrm{d}\boldsymbol{r}+R\mathrm{d}\boldsymbol{n}+\boldsymbol{n}\mathrm{d}R$ is true. Also, $\mathrm{d}\boldsymbol{r}^* \equiv \boldsymbol{n}\mathrm{d}R$ applies to points of the line of curvature. Hence, the formula by Rodrigues

$$\mathrm{d}\boldsymbol{r}+R\mathrm{d}\boldsymbol{n} = 0 \qquad (4.2.33)$$

is determined. Introducing the concept of curvature $\kappa = 1/R$, and using (4.2.20), (4.2.33) changes to

$$(\kappa\boldsymbol{r}_{,\alpha}+\boldsymbol{n}_{,\alpha})\mathrm{d}q_\alpha = 0. \qquad (4.2.34)$$

Multiplying by $\boldsymbol{r}_{,\beta}$ yields

$$(\kappa\boldsymbol{r}_{,\alpha}\boldsymbol{r}_{,\beta}+\boldsymbol{n}_{,\alpha}\boldsymbol{r}_{,\beta})\,\mathrm{d}q_\alpha = 0,$$

which because of (4.2.10) and (4.2.23) is equal to

$$(\kappa g_{\alpha\beta}-b_{\alpha\beta})\mathrm{d}q_\alpha = 0.$$

This equation system contains only then a non-trivial solution, if det $(\kappa g_{\alpha\beta}-b_{\alpha\beta}) = 0$ is true. Notating in full the condition for this determinant to be singular, then the algebraic equation

$$\kappa^2 g-\kappa(g_{11}b_{22}+g_{22}b_{11}-2g_{12}b_{12})+b = 0,$$
$$g = \det(g_{\alpha\beta}),\ b = \det(b_{\alpha\beta}) \tag{4.2.35}$$

is obtained. The contra-variant metric tensors $g^{\alpha\beta}$ are given by formula

$$g^{\alpha\beta} = \frac{\mathrm{co}\,(g_{\alpha\beta})}{g},$$

when co $(g_{\alpha\beta})$ represents the co-factor of element $g_{\alpha\beta}$ of matrix $(g_{\alpha\beta})$. Thus,

$$g^{11} = \frac{g_{22}}{g},\ g^{12} = -\frac{g_{21}}{g},\ g^{22} = \frac{g_{11}}{g},\ g^{21} = -\frac{g_{12}}{g} \tag{4.2.36}$$

is valid.

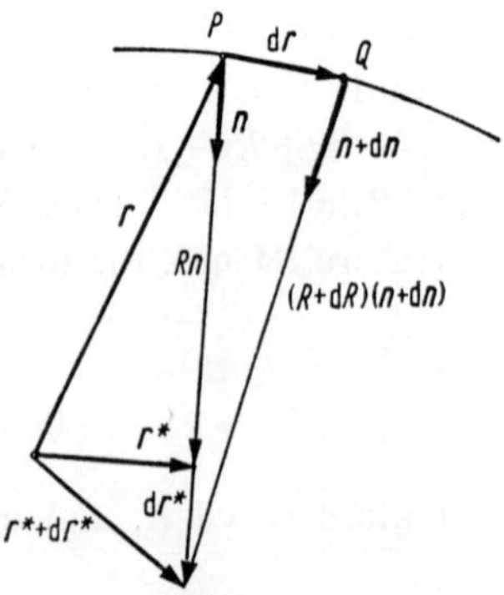

Figure 64

Using these relations for (4.2.35), then the formula is notated in short as

$$\kappa^2-g^{\alpha\beta}b_{\alpha\beta}\kappa+\frac{b}{g} = 0. \tag{4.2.37}$$

Introducing the following quantities:

$$g^{\alpha\beta} b_{\alpha\beta} = 2M, \tag{4.2.38}$$

where M is the *mean curvature*, and

$$\frac{b}{g} = G, \tag{4.2.39}$$

where G is *Gauss' curvature*, then

$$\kappa^2 - 2M\kappa + G = 0 \tag{4.2.40}$$

is true for (4.2.37). This yields for the *principal curvatures* κ_1, κ_2 the relations

$$\begin{aligned} &\kappa_1 = M + \sqrt{(M^2 - G)}, \kappa_2 = M - \sqrt{(M^2 - G)}, \\ &\kappa_1 + \kappa_2 = 2M, \kappa_1 \kappa_2 = G. \end{aligned} \tag{4.2.41}$$

From the definition equations (4.2.38) and (4.2.39),

$$2M = \frac{g_{11} b_{22} - 2g_{12} b_{12} + g_{22} b_{11}}{\det (g_{\alpha\beta})}, \quad G = \frac{\det (b_{\alpha\beta})}{\det (g_{\alpha\beta})}$$

is obtained because of (4.2.36). It becomes evident that the ratio of curvature of a given surface is known, when the coefficients of the first and the second fundamental forms are available.

Of special importance for practical applications are the *surfaces of revolution* (Figure 65). For these, the axis of rotation may be chosen as the q_2 axis. The surfaces are produced by rotating the *meridian line* $R_0 = R_0(q_2)$ about the q_2 axis. For the coordinate curves, the *meridians* $q_1 =$ const, and the *parallels of latitude* $q_2 =$ const may be chosen. Components x_i of radius vector $\boldsymbol{r}$ of the surface points are given by the equations (4.2.2) in the special form

$$x_1 = R_0(q_2) \cos q_1, x_2 = R_0(q_2) \sin q_1, x_3 = q_2. \tag{4.2.42}$$

At first, let us calculate the components $g_{\alpha\beta}$ of the metric tensor according to (4.2.7) and obtain

$$\begin{aligned} x_{i,1}x_{i,1} &= R_0^2 \sin^2 q_1 + R_0^2 \cos^2 q_1 = R_0^2 = g_{11}, \\ x_{i,1}x_{i,2} &= g_{12} = 0, \\ x_{i,2}x_{i,2} &= R_0'^2 \cos^2 q_1 + R_0'^2 \sin^2 q_1 + 1 = R_0'^2 + 1 = g_{22}. \end{aligned} \tag{4.2.43}$$

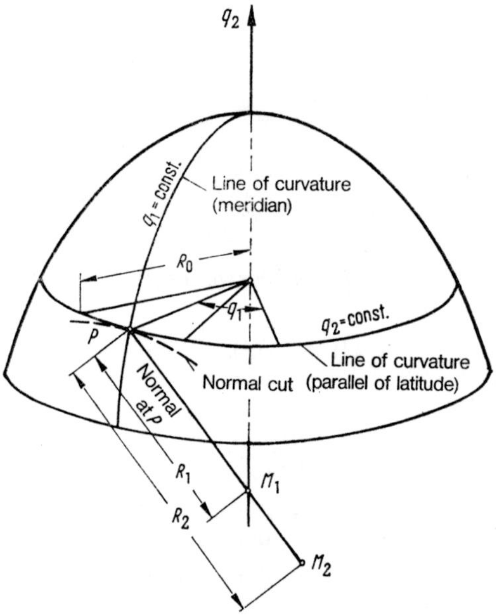

Figure 65a

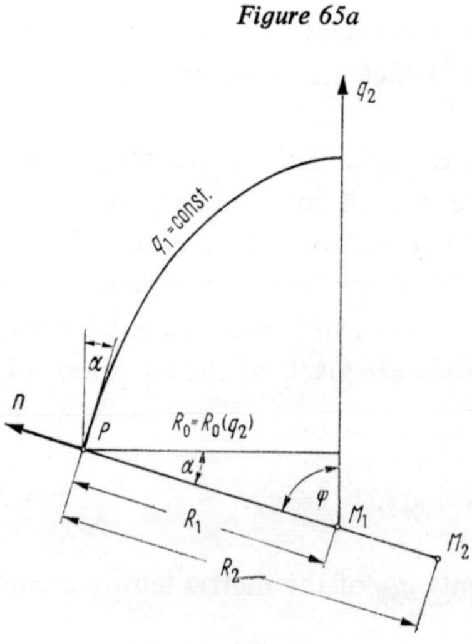

Figure 65b

Dashes (′) represent derivation with respect to q_2. The coordinate curves form an orthogonal net because of $q_{12} = 0$.

The surface normal vector reads, according to (4.2.18),

$$\boldsymbol{n} = \frac{1}{\sqrt{g}}(\boldsymbol{e}_1 \times \boldsymbol{e}_2) = \frac{1}{\sqrt{g}}(\boldsymbol{r}_{,1} \times \boldsymbol{r}_{,2})$$

$$= \frac{1}{\sqrt{g}} \begin{vmatrix} -R_0 \sin q_1 & R_0 \cos q_1 & 0 \\ R_0' \cos q_1 & R_0' \sin q_1 & 1 \end{vmatrix} = \frac{1}{\sqrt{g}} \begin{pmatrix} R_0 \cos q_1 \\ R_0 \sin q_1 \\ -R_0 R_0' \end{pmatrix}.$$

We can now calculate the fundamental quantities of the second order, $b_{\alpha\beta}$, following (4.2.23):

$$\begin{aligned} -\boldsymbol{n}_{,1}\boldsymbol{r}_{,1} &= -\frac{1}{\sqrt{g}}(R_0^2 \sin^2 q_1 + R_0^2 \cos^2 q_1) = -\frac{R_0^2}{\sqrt{g}} = b_{11} \\ -\boldsymbol{n}_{,1}\boldsymbol{r}_{,2} &= b_{12} = 0, \\ -\boldsymbol{n}_{,2}\boldsymbol{r}_{,2} &= -\frac{1}{\sqrt{g}}(R_0'^2 \cos^2 q_1 + R_0'^2 \sin^2 q_1 - R_0'^2 - R_0 R_0'') \\ &= \frac{R_0 R_0''}{\sqrt{g}} = b_{22}. \end{aligned} \tag{4.2.44}$$

We realize through $b_{12} = 0$ that the coordinate lines, i.e., the meridians and the parallels of latitude, as we have stated previously, are indeed lines of curvature.

To calculate the principal curvatures from (4.2.35), the values g and b must be determined from (4.2.43) and (4.2.44). Let us find

$$\begin{aligned} g &= \det(g_{\alpha\beta}) = R_0^2(1+R_0'^2), \\ b &= \det(b_{\alpha\beta}) = -\frac{R_0^3 R_0''}{g}. \end{aligned} \tag{4.2.45}$$

Using (4.2.43), (4.2.44), and (4.2.45), then (4.2.35) is changed to

$$R_0^2(1+R_0'^2)\kappa^2 - \kappa\left[\frac{R_0^2 R_0''}{(1+R_0'^2)^{\frac{1}{2}}} - R_0(1+R_0'^2)^{\frac{1}{2}}\right] - \frac{R_0 R_0''}{(1+R_0'^2)} = 0.$$

The solutions of this quadratic equation present the principal curvatures

$$\kappa_1 = -\frac{1}{R_0(1+R_0'^2)^{\frac{1}{2}}}, \quad \kappa_2 = \frac{R_0''}{(1+R_0'^2)^{\frac{3}{2}}}. \tag{4.2.46}$$

Principal curvature κ_2 is easily interpreted. It is the curvature of the meridian line, and hence the radius of the meridian line is the principal radius of curvature R_2 of the surface. To interpret also κ_1, we read from Figure 65 that $\tan\alpha = R_0'$ is true. Therefore, the known formula $\cos\alpha = (1+\tan^2\alpha)^{-\frac{1}{2}}$ yields $\cos\alpha = (1+R_0'^2)^{\frac{1}{2}}$ in this case, and thus $\kappa_1 = -\cos\alpha/R_0$. But according to definition, $\kappa_1 = -1/R_1$ (the minus sign results from the sense of $\boldsymbol{n}$), so that correspondingly for the other principal radius of curvature

$$R_1 = R_0/\cos\alpha$$

is obtained, which is in agreement with *Meusnier's theorem*. R_1 is determined according to this formula by extending the normal through point P to its intersection with axis q_2. The intersection with the axis is the centre of curvature M_1. The segment $\overline{PM_1}$ on the normal is R_1. The above statement on the minus sign is now proven as being correct: The radius R_1 and normal $\boldsymbol{n}$ have opposite sense, which is read off Figure 65. Finally, let us mention that R_2 has also to be marked off on the backward elongation of the normal. Hence, R_1 and R_2, as well as the centres of curvature M_1 and M_2 are all on one and the same straight line.

4.2.3 *Basic equations of the technical theory of shells*

Let us assume several conditions for the theory of shells which have already been used in the plate theory: Let us assume the shells to be *thin* and that the points of a shell on a normal are on this normal of the deformed shell's middle surface, even after deformation has taken place. Hence, we justify omission of the strain components ε_{i3}, $i = 1, 2$. However, because of Hooke's law, also $\sigma_{13} = \sigma_{23} = 0$ holds true. Similarly to the plate theory, the strain component ε_{33} and stress component σ_{33} will be neglected, the latter as being small in comparison with the remaining stress components.

Let us now determine the *strain tensor components*. Analogous to the plate theory, let us always refer to the middle surface of the shell in the following considerations. Let us use the curvilinear coordinates q_1, q_2 on the middle surface of the shell. The remaining points of the shell outside the middle surface are determined through q_1, q_2 and a third coordinate q_3 which is marked off on the respective normal of the middle surface belonging to the point specified by q_1, q_2 (Figure 66). Since curvilinear coordinates are being used, the strain components are set up by using (3.1.64). Confusion is impossible, since we shall always refer to curvilinear coordinates, then the asterisks from the formulae in Section 3.1.3 may be omitted. We obtain for the component ε_{11}:

$$\varepsilon_{11} = \frac{1}{h_1} \frac{\partial u_1}{\partial q_1} + \frac{u_2}{h_1 h_2} \frac{\partial h_1}{\partial q_2} + \frac{u_3}{h_1 h_3} \frac{\partial h_1}{\partial q_3}. \tag{4.2.47}$$

Based on the conditions specified,

$$u_1 = u_1^{(m)} + q_3 u_{1,3}^{(m)}, \; u_2 = u_2^{(m)} + q_3 u_{2,3}^{(m)}, \; u_3 = u_3^{(m)} \tag{4.2.48}$$

is notated for the displacements,

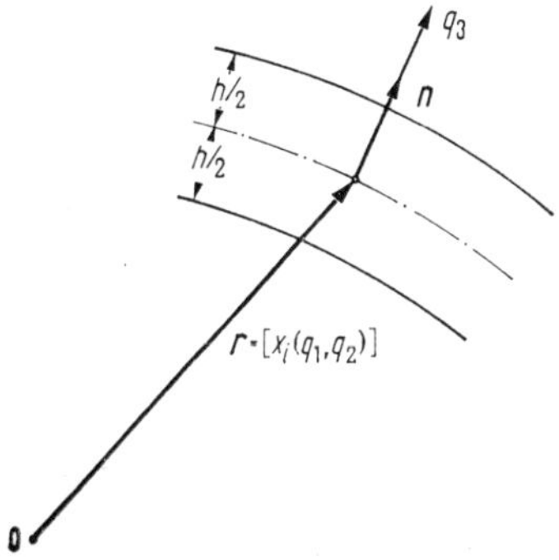

Figure 66

whereby the upper index m indicates in the quantities concerned that they are to be taken on the middle surface of the shell. Equation (3.1.64) yields

$$\begin{aligned} 2\varepsilon_{13} &= \frac{1}{h_3} \frac{\partial u_1}{\partial q_3} - \frac{u_1}{h_1 h_3} \frac{\partial h_1}{\partial q_3} + \frac{h_3}{h_1} \frac{\partial}{\partial q_1} \left(\frac{u_3}{h_3}\right) = 0, \\ 2\varepsilon_{23} &= \frac{1}{h_3} \frac{\partial u_2}{\partial q_3} - \frac{u_2}{h_2 h_3} \frac{\partial h_2}{\partial q_3} + \frac{h_3}{h_2} \frac{\partial}{\partial q_2} \left(\frac{u_3}{h_3}\right) = 0. \end{aligned} \tag{4.2.49}$$

These relations have been equated to nought because of the initial approximation $\varepsilon_{13} = \varepsilon_{23} = 0$. Let us consider

$$h_\alpha = h_\alpha^{(m)}\left(1 - \frac{q_3}{R_\alpha}\right), \quad \alpha = 1, 2, \; h_3 = 1. \tag{4.2.50}$$

Then,

$$\frac{\partial h_1}{\partial q_3} = -\frac{h_1^{(m)}}{R_1}$$

is valid. It yields for $q_3 = 0$ from the first line of (4.2.49)

$$\left(\frac{\partial u_1}{\partial q_3}\right)^{(m)} + \frac{u_1^{(m)}}{h_1^{(m)}}\frac{h_1^{(m)}}{R_1} + \frac{1}{h_1^{(m)}}\frac{\partial u_3^{(m)}}{\partial q_1} = 0,$$

and

$$\left(\frac{\partial u_1}{\partial q_3}\right)^{(m)} \equiv u_{1,3}^{(m)} = -\frac{u_1^{(m)}}{R_1} - \frac{1}{h_1^{(m)}} u_{3,1}^{(m)}, \tag{4.2.51}$$

respectively. In the same manner,

$$u_{2,3}^{(m)} = -\frac{u_2^{(m)}}{R_2} - \frac{1}{h_2^{(m)}} u_{3,2}^{(m)} \tag{4.2.52}$$

is obtained.

Inserting (4.2.51) and (4.2.52) in (4.2.48), then for the displacements

$$\begin{aligned} u_1 &= u_1^{(m)} - q_3\left(\frac{u_1^{(m)}}{R_1} + \frac{1}{h_1^{(m)}} u_{3,1}^{(m)}\right), \\ u_2 &= u_2^{(m)} - q_3\left(\frac{u_2^{(m)}}{R_2} + \frac{1}{h_2^{(m)}} u_{3,2}^{(m)}\right), \\ u_3 &= u_3^{(m)} \end{aligned} \tag{4.2.53}$$

is valid. These formulae are obviously analogous to those in (4.1.130), and which are valid for the von Kármán plate theory. This analogy is due to the fact that as in plates having large deflections, shells too will have the middle surface stretched.

Inserting (4.2.53), and from (4.2.50) the relations

$$\frac{\partial h_\alpha}{\partial q_3} = -\frac{h_\alpha^{(m)}}{R_\alpha}, \ \alpha = 1, 2,$$

in (4.2.47), as well as $h_1 = h_1^{(m)}$, $h_2 = h_2^{(m)}$ as an approximation, then

$$\begin{aligned} \varepsilon_{11} = {} & \left(\frac{1}{h_1^{(m)}}\frac{\partial u_1^{(m)}}{\partial q_1} + \frac{u_2^{(m)}}{h_1^{(m)}h_2^{(m)}}\frac{\partial h_1^{(m)}}{\partial q_2} - \frac{u_3^{(m)}}{R_1}\right) \\ & - q_3\left[\frac{1}{h_1^{(m)}}\frac{\partial}{\partial q_1}\left(\frac{u_1^{(m)}}{R_1} + \frac{1}{h_1^{(m)}}\frac{\partial u_3^{(m)}}{\partial q_1}\right)\right. \\ & \left. + \frac{1}{h_1^{(m)}h_2^{(m)}}\left(\frac{u_2^{(m)}}{R_2} + \frac{1}{h_2^{(m)}}\frac{\partial u_3^{(m)}}{\partial q_2}\right)\frac{\partial h_1^{(m)}}{\partial q_2}\right] \end{aligned} \tag{4.2.54}$$

is obtained. Let us introduce the *strain components of the middle surface*

$$\begin{aligned} \varepsilon_{11}^{(m)} &= \frac{1}{h_1^{(m)}}\frac{\partial u_1^{(m)}}{\partial q_1} + \frac{u_2^{(m)}}{h_1^{(m)}h_2^{(m)}}\frac{\partial h_1^{(m)}}{\partial q_2} - \frac{u_3^{(m)}}{R_1} \\ \varepsilon_{22}^{(m)} &= \frac{1}{h_2^{(m)}}\frac{\partial u_2^{(m)}}{\partial q_2} + \frac{u_1^{(m)}}{h_1^{(m)}h_2^{(m)}}\frac{\partial h_2^{(m)}}{\partial q_1} - \frac{u_3^{(m)}}{R_2}, \\ 2\varepsilon_{12}^{(m)} &= \frac{h_2^{(m)}}{h_1^{(m)}}\frac{\partial}{\partial q_1}\left(\frac{u_2^{(m)}}{h_2^{(m)}}\right) + \frac{h_1^{(m)}}{h_2^{(m)}}\frac{\partial}{\partial q_2}\left(\frac{u_1^{(m)}}{h_1^{(m)}}\right) \end{aligned} \tag{4.2.55}$$

as well as the *curvature changes of the middle surface*

$$\begin{aligned} \kappa_{11} = {} & \frac{1}{h_1^{(m)}}\frac{\partial}{\partial q_1}\left(\frac{u_1^{(m)}}{R_1} + \frac{1}{h_1^{(m)}}\frac{\partial u_3^{(m)}}{\partial q_1}\right) \\ & + \frac{1}{h_1^{(m)}h_2^{(m)}}\left(\frac{u_2^{(m)}}{R_2} + \frac{1}{h_2^{(m)}}\frac{\partial u_3^{(m)}}{\partial q_2}\right)\frac{\partial h_1^{(m)}}{\partial q_2}, \\ \kappa_{22} = {} & \frac{1}{h_2^{(m)}}\frac{\partial}{\partial q_2}\left(\frac{u_2^{(m)}}{R_2} + \frac{1}{h_2^{(m)}}\frac{\partial u_3^{(m)}}{\partial q_2}\right) \\ & + \frac{1}{h_1^{(m)}h_2^{(m)}}\left(\frac{u_1^{(m)}}{R_1} + \frac{1}{h_1^{(m)}}\frac{\partial u_3^{(m)}}{\partial q_1}\right)\frac{\partial h_2^{(m)}}{\partial q_1}, \end{aligned}$$

$$2\kappa_{12} = \frac{h_2^{(m)}}{h_1^{(m)}} \frac{\partial}{\partial q_1} \left(\frac{u_2^{(m)}}{h_2^{(m)} R_2} + \frac{1}{h_2^{(m)2}} \frac{\partial u_3^{(m)}}{\partial q_2} \right) + \frac{h_1^{(m)}}{h_2^{(m)}} \frac{\partial}{\partial q_2} \left(\frac{u_1^{(m)}}{h_1^{(m)} R_1} + \frac{1}{h_1^{(m)2}} \frac{\partial u_3^{(m)}}{\partial q_1} \right). \tag{4.2.56}$$

Then instead of (4.2.54), $\varepsilon_{11} = \varepsilon_{11}^{(m)} - q_3 \kappa_{11}$ is notated. The remaining strain components are similarly found so that altogether

$$\begin{aligned} \varepsilon_{11} &= \varepsilon_{11}^{(m)} - q_3 \kappa_{11}, \\ \varepsilon_{22} &= \varepsilon_{22}^{(m)} - q_3 \kappa_{22}, \\ \varepsilon_{12} &= \varepsilon_{12}^{(m)} - q_3 \kappa_{12} \end{aligned} \tag{4.2.57}$$

is obtained. Thus, all those *components of the strain tensor* have been found which are not to be neglected but must be included in the calculation. The quantities $h_1^{(m)}$ and $h_2^{(m)}$ of formulae (4.2.55) and (4.2.56) are obtained when $\sqrt{g_{11}}$ and $\sqrt{g_{22}}$ are calculated for the middle surface of the shell.

Let us change over to *Hooke's law*. Since the same approximations for the technical theory of shells as for the plate theory will be used, for example, $\sigma_{i3} = 0$, $i = 1, 2, 3$, then

$$\begin{aligned} \sigma_{11} &= \frac{E}{1-\nu^2} (\varepsilon_{11} + \nu \varepsilon_{22}), \quad \sigma_{22} = \frac{E}{1-\nu^2} (\varepsilon_{22} + \nu \varepsilon_{11}), \\ \sigma_{12} &= \frac{E}{1+\nu} \varepsilon_{12} \end{aligned}$$

will be obtained in the same way. Introducing (4.2.57) to these equations yields

$$\begin{aligned} \sigma_{11} &= \frac{E}{1-\nu^2} [\varepsilon_{11}^{(m)} + \nu \varepsilon_{22}^{(m)} - q_3(\kappa_{11} + \nu \kappa_{22})], \\ \sigma_{22} &= \frac{E}{1-\nu^2} [\varepsilon_{22}^{(m)} + \nu \varepsilon_{11}^{(m)} - q_3(\kappa_{22} + \nu \kappa_{11})], \\ \sigma_{12} &= \frac{E}{1+\nu} [\varepsilon_{12}^{(m)} - q_3 \kappa_{12}]. \end{aligned} \tag{4.2.58}$$

For the practical treatment of shells the use of *internal forces* is recommended. They are defined as follows (Figure 67):

$$N_{11} = \int_{-h/2}^{+h/2} \sigma_{11} \left(1 - \frac{q_3}{R_2}\right) dq_3, \; N_{22} = \int_{-h/2}^{+h/2} \sigma_{22} \left(1 - \frac{q_3}{R_1}\right) dq_3,$$
$$N_{12} = \int_{-h/2}^{+h/2} \sigma_{12} \left(1 - \frac{q_3}{R_2}\right) dq_3, \; N_{21} = \int_{-h/2}^{+h/2} \sigma_{21} \left(1 - \frac{q_3}{R_1}\right) dq_3,$$
$$Q_{11} = \int_{-h/2}^{+h/2} \sigma_{13} \left(1 - \frac{q_3}{R_2}\right) dq_3, \; Q_{22} = \int_{-h/2}^{+h/2} \sigma_{23} \left(1 - \frac{q_3}{R_1}\right) dq_3, \quad (4.2.59)$$
$$M_{11} = \int_{-h/2}^{+h/2} \sigma_{11} q_3 \left(1 - \frac{q_3}{R_2}\right) dq_3, \; M_{22} = \int_{-h/2}^{+h/2} \sigma_{22} q_3 \left(1 - \frac{q_3}{R_1}\right) dq_3,$$
$$M_{12} = -\int_{-h/2}^{+h/2} \sigma_{12} q_3 \left(1 - \frac{q_3}{R_2}\right) dq_3,$$
$$M_{21} = -\int_{-h/2}^{+h/2} \sigma_{21} q_3 \left(1 - \frac{q_3}{R_1}\right) dq_3.$$

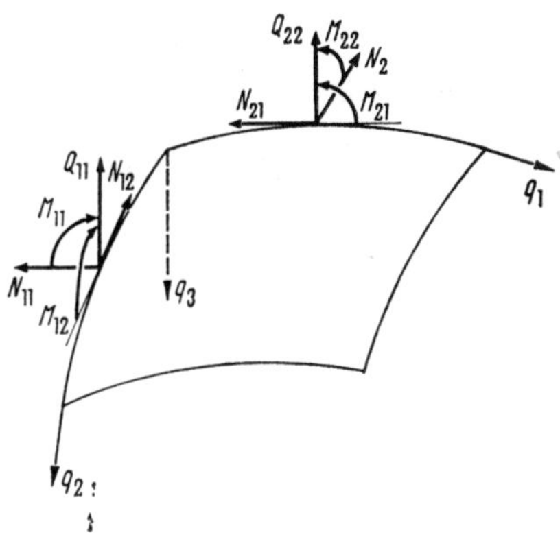

Figure 67

It is worth noting, as far as formula (4.2.59) are concerned, that contrary to our assumption $\sigma_{13} = \sigma_{23} = 0$, the shearing forces Q_{11}, Q_{22} are to be different from nought. In this way, the very contradiction contained already in the plate theory is tolerated a second time. In general, it becomes evident that the technical theory of shells is, in actual fact, a consequent development of the plate theory and in particular of the von Kármán version. But finally it is also realized, that for shells in general, $N_{12} \neq N_{21}$, $M_{12} \neq M_{21}$ is valid, whereas this is not the case for plates. It is only for *thin shells*, when q_3/R_1 and q_3/R_2 may be omitted as quantities of a magnitude much smaller than one, that this difference will disappear. Cancelling terms q_3/R_1 and q_3/R_2 and inserting the equations (4.2.58) in (4.2.59), we obtain for the internal forces:

$$\begin{aligned}
N_{11} &= \frac{Eh}{1-\nu^2}(\varepsilon_{11}^{(m)}+\nu\varepsilon_{22}^{(m)}),\ N_{22} = \frac{Eh}{1-\nu^2}(\varepsilon_{22}^{(m)}+\nu\varepsilon_{11}^{(m)}),\\
N_{12} &= N_{21} = \frac{Eh}{1+\nu}\varepsilon_{12}^{(m)},\\
M_{11} &= -D(\kappa_{11}+\nu\kappa_{22}),\ M_{22} = -D(\kappa_{22}+\nu\kappa_{11}),\\
M_{12} &= M_{21} = D(1-\nu)\kappa_{12}, D = \frac{Eh^3}{12(1-\nu^2)}.
\end{aligned} \tag{4.2.60}$$

Let us set up the *conditions of equilibrium* pertaining to the theory of shells. They yield the differential equations for the internal forces. The differential equations are simply obtained by going back to the formula in Section 3.1.3 for conditions of equilibrium relative to curvilinear coordinates. The first condition of equilibrium is given by (3.1.75). Integrating it over the shell's thickness, the first differential equation for $N_{11}, N_{12}, N_{21}, N_{22}$ and Q_{11} is obtained.

To begin with let us change (3.1.75) to

$$\begin{aligned}
&\frac{\partial}{\partial q_1}h_2 h_3 \sigma_{11}+\frac{\partial}{\partial q_2}h_1 h_3 \sigma_{12}+\frac{\partial}{\partial q_3}h_1 h_2 \sigma_{13}+h_3\frac{\partial h_1}{\partial q_2}\sigma_{12}+\\
&h_2\frac{\partial h_1}{\partial q_3}\sigma_{13}-h_3\frac{\partial h_2}{\partial q_1}\sigma_{22}-h_2\frac{\partial h_3}{\partial q_1}\sigma_{33}=0
\end{aligned} \tag{4.2.61}$$

by omitting the mass forces and the asterisks. Let us in addition consider (4.2.50) and carry out the integration in (4.2.61) over the shell's thickness h. Thus, the following is yielded: The first term of (4.2.61) because of (4.2.59) yields

$$\int_{-h/2}^{+h/2}\frac{\partial}{\partial q_1}\left[h_2^{(m)}\sigma_{11}\left(1-\frac{q_3}{R_2}\right)\right]\mathrm{d}q_3 = \frac{\partial}{\partial q_1}(h_2^{(m)}N_{11}),$$

and the second term, correspondingly, yields

$$\int_{-h/2}^{+h/2}\frac{\partial}{\partial q_2}\left[h_1^{(m)}\sigma_{12}\left(1-\frac{q_3}{R_1}\right)\right]\mathrm{d}q_3 = \frac{\partial}{\partial q_2}(h_1^{(m)}N_{21}).$$

The third term yields quite obviously

$$[h_1 h_2 \sigma_{13}]_{-h/2}^{+h/2} = h_1^{(m)} h_2^{(m)} p_{11},$$

whereby p_{11} represents the components of the shell's load p per unit area in the q_1 direction. Because of (4.2.50), the fourth term leads to

$$\begin{aligned} &\int_{-h/2}^{+h/2} \frac{\partial}{\partial q_2}\left[h_1^{(m)}\left(1-\frac{q_3}{R_1}\right)\right] \cdot \sigma_{12}\, dq_3 \\ &\quad = \int_{-h/2}^{+h/2}\left[\frac{\partial h_1^{(m)}}{\partial q_2} - q_3 \frac{\partial}{\partial q_2}\left(\frac{h_1^{(m)}}{R_1}\right)\right] \cdot \sigma_{12}\, dq_3 . \end{aligned} \tag{4.2.62}$$

To continue calculation of (4.2.62), let us calculate the following. Considering that we are using lines of curvature as coordinate lines, then the integrability conditions of the surface theory by Mainardi-Codazzi are

$$\frac{\partial}{\partial q_2}\left(\frac{b_{11}}{\sqrt{g_{11}}}\right) = \frac{b_{22}}{g_{22}} \frac{\partial \sqrt{g_{11}}}{\partial q_2}, \quad \frac{\partial}{\partial q_1}\left(\frac{b_{22}}{\sqrt{g_{22}}}\right) = \frac{b_{11}}{g_{11}} \frac{\partial \sqrt{g_{22}}}{\partial q_1}. \tag{4.2.63}$$

From (4.2.33), which is the equation by Rodrigues,

$$\boldsymbol{n}_{,\alpha} = -\frac{1}{R_\alpha} \boldsymbol{r}_{,\alpha}$$

is obtained. We insert this in (4.2.23), and it yields

$$\frac{1}{R_\alpha} \boldsymbol{r}_{,\alpha} \boldsymbol{r}_{,\beta} = b_{\alpha\beta} .$$

Because of (4.2.7) we obtain

$$R_1 = \frac{g_{11}}{b_{11}}, \quad R_2 = \frac{g_{22}}{b_{22}} . \tag{4.2.64}$$

Through (4.2.64), (4.2.63) changes to

$$\frac{\partial}{\partial q_2}\left(\frac{\sqrt{g_{11}}}{R_1}\right) = \frac{1}{R_2} \frac{\partial \sqrt{g_{11}}}{\partial q_2}, \quad \frac{\partial}{\partial q_1}\left(\frac{\sqrt{g_{22}}}{R_2}\right) = \frac{1}{R_1} \frac{\partial \sqrt{g_{22}}}{\partial q_1}, \tag{4.2.65}$$

thus, we obtain

$$\frac{\partial}{\partial q_2}\left(\frac{h_1^{(m)}}{R_1}\right)=\frac{1}{R_2}\frac{\partial h_1^{(m)}}{\partial q_2},\quad \frac{\partial}{\partial q_1}\left(\frac{h_2^{(m)}}{R_2}\right)=\frac{1}{R_1}\frac{\partial h_2^{(m)}}{\partial q_1} \tag{4.2.66}$$

with respect to the middle surface of the shell. Let us use this for (4.2.62). Through the first equation (4.2.66), and because of (4.2.59), (4.2.62) finally changes to

$$\int_{-h/2}^{+h/2}\frac{\partial}{\partial q_2}\left[h_1^{(m)}\left(1-\frac{q_3}{R_1}\right)\right]\sigma_{12}\,\mathrm{d}q_3=\frac{\partial h_1^{(m)}}{\partial q_2}\int_{-h/2}^{+h/2}\sigma_{12}\left(1-\frac{q_3}{R_2}\right)\mathrm{d}q_3=\frac{\partial h_1^{(m)}}{\partial q_2}N_{12}.$$

The fifth term, because of $\partial h_1/\partial q_3 = -h_1^{(m)}/R_1$ yields

$$\int_{-h/2}^{+h/2}h_2\frac{\partial h_1}{\partial q_3}\sigma_{13}\,\mathrm{d}q_3=-\int_{-h/2}^{+h/2}h_2h_1^{(m)}\frac{\sigma_{13}}{R_1}\mathrm{d}q_3=-\frac{h_1^{(m)}h_2^{(m)}}{R_1}\cdot\int_{-h/2}^{+h/2}\sigma_{13}\left(1-\frac{q_3}{R_2}\right)\mathrm{d}q_3=-\frac{h_1^{(m)}h_2^{(m)}}{R_1}Q_{11},$$

and the sixth term changes to

$$\int_{-h/2}^{+h/2}\frac{\partial}{\partial q_1}\left[h_2^{(m)}\left(1-\frac{q_3}{R_2}\right)\right]\sigma_{22}\,\mathrm{d}q_3=\frac{\partial h_2^{(m)}}{\partial q_1}N_{22},$$

by using the second equation (4.2.66). The seventh term changes to nought because of $\partial h_3/\partial q_1 \equiv 0$.

Summarizing these seven intermediate results, the first condition of equilibrium of the shell theory is obtained. Because of $h_\alpha^{(m)} = g_{(\alpha\alpha)}^{\frac{1}{2}}$, it reads as follows:

$$\frac{\partial}{\partial q_1}(g_{22}^{\frac{1}{2}}N_{11})+\frac{\partial}{\partial q_2}(g_{11}^{\frac{1}{2}}N_{21})+N_{12}\frac{\partial g_{11}^{\frac{1}{2}}}{\partial q_2}-\frac{(g_{11}g_{22})^{\frac{1}{2}}}{R_1}Q_{11}-N_{22}\frac{\partial g_{12}^{\frac{1}{2}}}{\partial q_1}+(g_{11}g_{22})^{\frac{1}{2}}p_{11}=0. \tag{4.2.67}$$

In the same manner, two more of the conditions of equilibrium are obtained:

$$\begin{aligned}
&\frac{\partial}{\partial q_1}(g_{22}^{\frac{1}{2}} N_{12})+\frac{\partial}{\partial q_2}(g_{11}^{\frac{1}{2}} N_{22})+N_{21}\frac{\partial g_{22}^{\frac{1}{2}}}{\partial q_1}-\frac{(g_{11}g_{22})^{\frac{1}{2}}}{R_2}Q_{22}\\
&\quad -N_{11}\frac{\partial g_{11}^{\frac{1}{2}}}{\partial q_2}+(g_{11}g_{22})^{\frac{1}{2}}p_{22}=0,\\
&\frac{\partial}{\partial q_1}(g_{12}^{\frac{1}{2}}Q_{11})+\frac{\partial}{\partial q_2}(g_{11}^{\frac{1}{2}}Q_{22})+N_{11}\frac{(g_{11}g_{22})^{\frac{1}{2}}}{R_1}\\
&\quad +N_{22}\frac{(g_{11}g_{22})^{\frac{1}{2}}}{R_2}+(g_{11}g_{22})^{\frac{1}{2}}p_{33}=0.
\end{aligned} \tag{4.2.68}$$

Then in all three equations, g_{11} and g_{22} are the components of the metric tensor pertaining to the middle surface of the shell, and p_{22}, p_{33} are the components of the surface load p in the q_2 and q_3 direction. The significance of p_{11} has already been explained.

Two conditions of equilibrium for the moments are obtained when the first condition of equilibrium (3.1.75) and the corresponding second for the stresses are multiplied by q_3 and integrated over the shell's thickness h. The mass forces are omitted thereby, and (4.2.50) and (4.2.59) are considered.

Let us exemplify the calculation by using (4.2.61). The first term yields

$$\int_{-h/2}^{+h/2} q_3\frac{\partial}{\partial q_1}(h_2\sigma_{11})\mathrm{d}q_3=\frac{\partial}{\partial q_1}\int_{-h/2}^{+h/2} q_3\left[h_2^{(\mathrm{m})}\left(1-\frac{q_3}{R_2}\right)\right]\sigma_{11}\,\mathrm{d}q_3=\frac{\partial}{\partial q_1}(h_2^{(\mathrm{m})}M_{11}).$$

The second term changes to

$$\int^{+h/2} q_3\frac{\partial}{\partial q_2}(h_1\sigma_{12})\mathrm{d}q_3=\frac{\partial}{\partial q_2}\int_{-h/2}^{+h/2} q_3\left[h_1^{(\mathrm{m})}\left(1-\frac{q_3}{R_1}\right)\right]\sigma_{12}\,\mathrm{d}q_3=-\frac{\partial}{\partial q_2}(h_1^{(\mathrm{m})}M_{21}).$$

The third and fifth term are summarized. Through integration by parts, the third term yields

$$\int_{-h/2}^{+h/2} q_3\frac{\partial}{\partial q_3}(h_1h_2\sigma_{13})\mathrm{d}q_3=[q_3h_1h_2\sigma_{13}]_{-h/2}^{+h/2}-\int_{-h/2}^{+h/2} h_1h_2\sigma_{13}\,\mathrm{d}q_3.$$

Let

$$[q_3 h_1 h_2 \sigma_{13}]_{-h/2}^{+h/2} = h_1^{(m)} h_2^{(m)} m_{22}$$

be true, when m_{22} is the component of the shell's moment load m in the q_2 direction. The remainder from the third term is summed up together with the integral from the fifth term, and because of $\partial h_1/\partial q_3 = -h_1^{(m)}/R_1$,

$$-\int_{-h/2}^{+h/2} h_1 h_2 \sigma_{13} \, dq_3 + \int_{-h/2}^{+h/2} q_3 h_2 \frac{\partial h_1}{\partial q_3} \sigma_{13} \, dq_3 =$$
$$\int_{-h/2}^{+h/2} h_2 \left(-h_1 - \frac{h_1^{(m)} q_3}{R_1}\right) \sigma_{13} \, dq_3$$

is obtained. Using (4.2.50), this is changed to

$$-\int_{-h/2}^{+h/2} h_1 h_2 \sigma_{13} \, dq_3 + \int_{-h/2}^{+h/2} q_3 h_2 \frac{\partial h_1}{\partial q_3} \sigma_{13} \, dq_3 =$$
$$-h_1^{(m)} h_2^{(m)} \int_{-h/2}^{+h/2} \left(1 - \frac{q_3}{R_2}\right) \sigma_{13} \, dq_3 = -h_1^{(m)} h_2^{(m)} Q_{11} \, .$$

The fourth term leads to

$$\int_{-h/2}^{+h/2} q_3 \frac{\partial h_1}{\partial q_2} \sigma_{12} \, dq_3 = \int_{-h/2}^{+h/2} q_3 \left[\frac{\partial h_1^{(m)}}{\partial q_2} - q_3 \frac{\partial}{\partial q_2}\left(\frac{h_1^{(m)}}{R_1}\right)\right] \sigma_{12} \, dq_3 \, ,$$

which changes to

$$\int_{-h/2}^{+h/2} q_3 \left(\frac{\partial g_1^{(m)}}{\partial q_2} - \frac{q_3}{R_2} \frac{\partial h_1^{(m)}}{\partial q_2}\right) \sigma_{12} \, dq_3 = \frac{\partial h_1^{(m)}}{\partial q_2} \int_{-h/2}^{+h/2} q_3 \left(1 - \frac{q_3}{R_2}\right) \sigma_{12} \, dq_3$$
$$= -\frac{\partial h_1^{(m)}}{\partial q_2} M_{12}$$

because of (4.2.66). The sixth term yields

$$-\int_{-h/2}^{+h/2} q_3 \frac{\partial h_2}{\partial q_1} \sigma_{22} \, dq_3 = -\int_{-h/2}^{+h/2} q_3 \frac{\partial}{\partial q_1} \left(h_2^{(m)} - \frac{h_2^{(m)} q_3}{R_2}\right) \sigma_{22} \, dq_3$$

$$= -\int_{-h/2}^{+h/2} q_3 \left(\frac{\partial h_2^{(m)}}{\partial q_1} - \frac{1}{R_1}\frac{\partial h_2^{(m)}}{\partial q_1} q_3\right) \sigma_{22}\, dq_3$$

$$= -\frac{\partial h_2^{(m)}}{\partial q_1}\int_{-h/2}^{+h/2} q_3 \left(1 - \frac{q_3}{R_1}\right) \sigma_{22}\, dq_3 = -\frac{\partial h_2^{(m)}}{\partial q_1} M_{22},$$

when (4.2.66) is used again. The seventh term equals to nought because of $\partial h_3/\partial q_1 = 0$. Summarizing the results of the calculation, the first equation of moments pertaining to the theory of shells is obtained,

$$\begin{aligned} &\frac{\partial}{\partial q_1}(g_{22}^{\frac{1}{2}} M_{11}) - \frac{\partial}{\partial q_2}(g_{11}^{\frac{1}{2}} M_{21}) - M_{12}\frac{\partial g_{11}}{\partial q_2} - M_{22}\frac{\partial(g_{22}^{\frac{1}{2}})}{\partial q_1} \\ &-(g_{11}g_{22})^{\frac{1}{2}} Q_{11} + (g_{11}g_{22})^{\frac{1}{2}} m_{22} = 0. \end{aligned} \tag{4.2.69}$$

We have also used relation $h_\alpha^{(m)} = g_{(\alpha\alpha)}^{\frac{1}{2}}$. From the second equilibrium condition for the stress components, which corresponds to (3.1.75), the second equation of moments of the theory of shells is obtained in a similar way:

$$\begin{aligned} &\frac{\partial}{\partial q_2}(g_{11}^{\frac{1}{2}} M_{22}) - \frac{\partial}{\partial q_1}(g_{22}^{\frac{1}{2}} M_{12}) - M_{21}\frac{\partial g_{22}^{\frac{1}{2}}}{\partial q_1} - M_{11}\frac{\partial g_{11}^{\frac{1}{2}}}{\partial q_2} \\ &-(g_{11}g_{22})^{\frac{1}{2}} Q_{22} - (g_{11}g_{22})^{\frac{1}{2}} m_{11} = 0. \end{aligned} \tag{4.2.70}$$

Then m_{11} is defined by

$$(g_{11}g_{22})^{\frac{1}{2}} m_{11} = -[q_3 h_1 h_2 \sigma_{23}]_{-h/2}^{+h/2},$$

and represents the component in the q_1 direction of the moment load m of the shell. It is left to the reader to find the calculation leading to (4.2.70). Thus, all fundamental equations of the theory of shells have been found. From the five equations (4.2.67), (4.2.68), (4.2.69) and (4.2.70), the internal forces may be calculated when the boundary conditions are considered. Through (4.2.60), and when the internal forces are known, the strain components, and the changes of curvature of the middle surface are obtained. Finally, the stress components are determined through (4.2.58).

It is also possible to obtain the equations for the shell's internal forces in another way, i.e., as Euler equations of a variational problem through the principle of the stationary value of the total potential energy. This method as applied to shells of revolution has been discussed by W. Müller [35], to cylindrical shells by H. L. Langhaar [25, p. 193], and to shells of general shape by E. Klingbeil [36].

4.2.4 *Membrane theory, examples*

In plate problems it was assumed that the amount of stretching of the middle surface was of a higher order small as compared to the deflections of the plate. Hence, the strain in the middle surface was neglected in Kirchhoff's theory and only considered in von Kármán's theory when large deflections occurred. In shell problems, this is entirely different. Only in very specific cases, for example, in the deformation of an open cylindrical shell when the lines generating the surface of the cylinder remain parallel after bending, is it possible for shells to be bent without deformation. In all other cases, bending occurs together with stretching of the shell's middle surface, and as compared with bending, this stretching is actually prevalent. Thus, it might be possible to neglect bending stresses in comparison with stretching forces, and hence to develop the so-called *membrane theory*.

Membrane equations are obtained from the general equilibrium conditions of the shell theory when $M_{11} = M_{22} = M_{21} = M_{12} = 0$ and $Q_{11} = Q_{22} = 0$ is notated. In addition, also $\boldsymbol{m} \equiv 0$ has naturally to be assumed. Thus, the entire problem becomes statically determinate, and when the outer load $\boldsymbol{p}$ is given, the *membrane forces* N_{11}, N_{22} and N_{12} can be calculated from the equilibrium conditions, without consideration of the stress-strain relations.

Let us restrict ourselves to shells of revolution with rotational symmetric load. In that case, additional, $N_{12} = N_{21} = 0$ is valid. According to these assumptions, the second equation (4.2.68) changes into

$$\frac{N_{11}}{R_1} + \frac{N_{22}}{R_2} + p_3 = 0. \tag{4.2.71}$$

This represents already the *first fundamental equation of the membrane theory*

Let us change over to new coordinates, i.e., coordinates $q_1 = \vartheta$, $q_2 = \varphi$ (Figure 68) for discussion of the shells of revolution. Thus, in particular, the significance of the coordinate q_2 in comparison with the previous statements has been changed (compare Figure 68 with Figure 65). We read

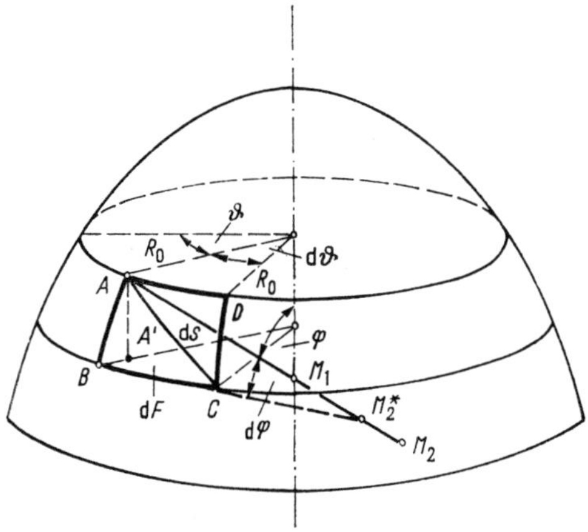

Figure 68

from Figure 68 that the principal radius of curvature R_2 for corner point A of the surface element dF is being given by segment $\overline{AM_2}$ which is on the surface-normal belonging to A. At the same time, R_2 is the radius of curvature of the meridian $q_1 = \vartheta = \text{const}$, passing through A. The principal radius R_1 of corner point A is on the same surface-normal and is represented by segment $\overline{AM_1}$, when the centre of curvature M_1 is on the axis of rotation. The radius of the parallel of latitude passing through A is R_0. The curved segments $\overset{\frown}{AB}, \overset{\frown}{AD}$, respectively, are equal to $R_2 \mathrm{d}\varphi, R_0 \mathrm{d}\vartheta$, respectively, which is read from Figure 68. But $\mathrm{d}s^2 = R_0^2 \mathrm{d}\vartheta^2 + R_2^2 \mathrm{d}\varphi^2$, and comparison with (4.2.8) yields that if ϑ is identified with q_1, φ with q_2, then for the present choice of coordinates

$$g_{11} = R_0^2, g_{22} = R_2^2 \tag{4.2.72}$$

is valid. Hence, from the first equation we obtain

$$\frac{\partial}{\partial q_2}(R_0 N_{22}) - N_{11} \frac{\partial R_0}{\partial q_2} + R_0 R_2 p_{22} = 0, \tag{4.2.73}$$

when all other assumptions are considered. But because of (4.2.71), and with $R_1 = R_0/\cos\alpha$,

$$N_{11} = -\frac{N_{22} R_0}{R_2 \cos\alpha} - \frac{p_{33} R_0}{\cos\alpha}$$

is true, which is inserted in (4.2.73). Thus,

$$\frac{\partial}{\partial q_2}(R_0 N_{22}) + \frac{N_{22} R_0}{R_2 \cos\alpha} \cdot \frac{\partial R_0}{\partial q_2} + \frac{p_{33} R_0}{\cos\alpha} \cdot \frac{\partial R_0}{\partial q_2} + R_0 R_2 p_{22} = 0 \quad (4.2.74)$$

is obtained. We read off Figure 68

$$\frac{\partial R_0}{\partial q_2} \mathrm{d}q_2 \equiv \frac{\partial R_0}{\partial \varphi} \mathrm{d}\varphi \equiv \overline{A'B} \approx \overline{AB} \cdot \cos\varphi = R_2 \,\mathrm{d}\varphi \cos\varphi.$$

But then $\partial R_0/\partial\varphi \equiv \partial R_0/\partial q_2 = R_2 \cos\varphi$ is also required. Comparison of Figure 68 with Figure 65 indicates that $\sin\varphi = \cos\alpha$ is true. Based on this fact, (4.2.74) is calculated first into

$$\sin\varphi \frac{\partial}{\partial q_2}(R_0 N_{22}) + R_0 N_{22} \cos\varphi + R_0 R_2 (p_{22} \sin\varphi + p_{33} \cos\varphi) = 0,$$

and then into

$$\frac{\partial}{\partial q_2}(R_0 N_{22} \sin\varphi) + R_0 R_2 p = 0. \quad (4.2.75)$$

We have also used $p_{22} \sin\varphi + p_{33} \cos\varphi = p$ for this calculation. The significance of p is evident from Figure 69. It is the vertical component of the shell's load.

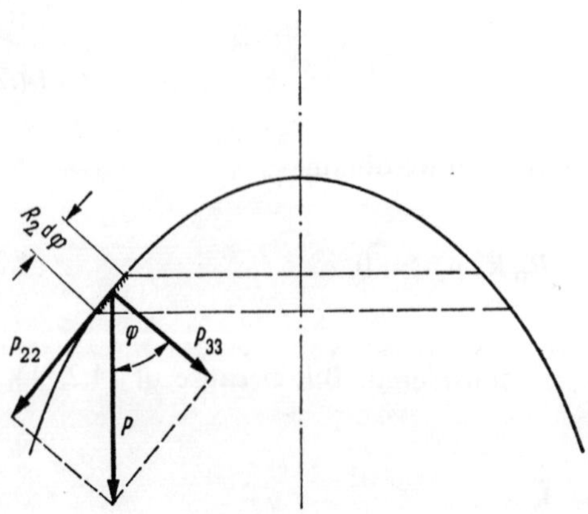

Figure 69

Integrating (4.2.75) over $q_1 = \vartheta$ from 0 to 2π and over $q_2 = \varphi$ from 0 to φ, then we obtain as the *second fundamental equation of the membrane theory*

$$2\pi R_0 N_{22} \sin \pi + P = 0. \tag{4.2.76}$$

In this formula, P is the resultant of that portion of the total load which is located above the parallel of latitude φ on the shell.

With respect to the special choice of coordinates $q_1 = \vartheta$, $q_2 = \varphi$, the two equations (4.2.71) and (4.2.76) are changed to

$$\begin{aligned} &\frac{N_{\vartheta\vartheta}}{R_1} + \frac{N_{\varphi\varphi}}{R_2} + p_3 = 0, \\ &2\pi R_0 N_{\varphi\varphi} \sin \varphi + P = 0 \end{aligned} \tag{4.2.77}$$

by renaming the indices of the internal forces. Then, $N_{\vartheta\vartheta}$, $N_{\varphi\varphi}$ are the radial, tangential internal force, p_3 is the component of $\boldsymbol{p}$ in the direction of the shell's normal, P is the resultant of the load above the parallel of latitude $\varphi = \text{const}$, R_0 is the radius of the parallel of latitude, and R_1, R_2 represent the principal radii of curvature of the cut through the shell. Through (4.2.77), the shells of revolution loaded symetrically with respect to the axis of rotation are completely determined.

As a first example for the use of (4.2.77), let us discuss the simple problem of a spherical shell with radius R and internal pressure $p_3 = -p_i$. Because of the point symmetry of the problem, $N_{\vartheta\vartheta} = N_{\varphi\varphi} = N$, and $R_1 = R_2 = R$ are true, so that from the first equation of (4.2.77), already the complete solution of the problem is obtained, i.e.,

$$\frac{2N}{R} = -p_3\,,\; N = \frac{p_i R}{2}\,, \text{ respectively,} \tag{4.2.78}$$

since, if the internal force N is known, the spherical shell, assumed to be a closed membrane, can be designed. We should add that R is in actual fact the radius of the spherical shell's middle surface.

The second equation of (4.2.77) does not yield anything new. Let $R_0 = R \sin \varphi$, $P = -p_i \pi R_0^2$ be true, and hence, following the said equation:

$$2\pi R \sin \varphi N \sin \varphi = p_i \pi R^2 \sin^2 \varphi$$

is also true. As before, (4.2.78) is obtained from this.

For a second example, let us assume a conical shell having an opening angle α, and filled with fluid to level H (Figure 70). It is

$$p_3 = -\gamma_F(H-x_3),\ R_1 = R_0/\sin\varphi,\ R_2 = \infty,$$

when γ_F represents the specific weight of the fluid. Thus, the first equation (4.2.77) changes to

$$N_{\vartheta\vartheta} = -p_3 R_1 = \frac{\gamma_F(H-x_3)}{\sin\varphi} R_0 .$$

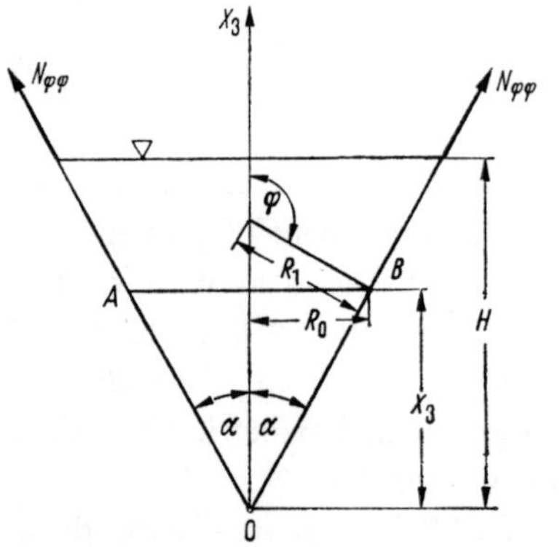

Figure 70

But, as shown in Figure 70, $R_0 = x_3 \tan\alpha$, $\sin\varphi = \cos\alpha$ is true, so that finally for the radial internal force

$$N_{\vartheta\vartheta} = \frac{\gamma_F(H-x_3)x_3 \tan\alpha}{\cos\alpha} \tag{4.2.79}$$

is obtained. The tangential internal force $N_{\varphi\varphi}$ is yielded from the second equation (4.2.77) to

$$N_{\varphi\varphi} = -\frac{P}{2\pi R_0 \sin\varphi} = -\frac{P}{2\pi R_0 \cos\alpha} . \tag{4.2.80}$$

In this case, P is the weight of the head of liquid on shell portion AOB. Thus, for P

$$P = -\pi\gamma_F x_3^2(H - \tfrac{2}{3}x_3)\tan^2\alpha$$

is valid.

Inserting this and $R_0 = x_3 \tan\alpha$ in (4.2.80), then we obtain for the tangential internal force

$$N_{\varphi\varphi} = \frac{\gamma_F x_3(H - \frac{2}{3}x_3)\tan\alpha}{2\cos\alpha}. \tag{4.2.81}$$

The formulae (4.2.79) and (4.2.81) enable us to calculate the internal forces for the conically shaped shell when the values given are γ_F, H and α, and to design the shell. Thus, the problem has been solved.

Further references to the theory of membranes, for cylindrical shells which may be of special practical significance, as well as for shells with any middle surface are given in [29] and in [33].

4.2.5 *Bending of shells, examples*

When it becomes necessary to consider the influence of bending in shells, then leaving aside the membrane theory, the general fundamental equations in Section 4.2.3 have to be used. The membrane theory cannot be used, for example, when either a concentrated load is acting vertically to the middle surface of the shell, or when the boundary conditions prevent it. However, it becomes evident that the solutions of the membrane theory can always be used as particular solutions of the bending theory, which offers an advantage in the calculation, and that the 'disturbances' in the solutions of the membrane theory which result from bending, are usually limited to a small part of the shell, because these disturbances do in general disappear at a distance from the place of its origin.

To limit our calculations (explanations of a more thorough nature are given in literature dealing with shell problems), let us discuss a cylindrical shell with load symmetrical with respect to the axis of rotation (Figure 71). For a shell of this kind, $R_0 = \text{const} = a$ is valid, and hence $R_0' = \mathrm{d}R_0/\mathrm{d}q_2 = 0$, so that from (4.2.43), $g_{11} = a^2$, $g_{22} = 1$ is obtained. Without calculating further, it becomes evident that $R_2 = \infty$, $R_1 \equiv R_0 = a$ is valid for the principal radii of curvature. If, according to the particular coordinates $q_1 = \vartheta$, $q_2 = z$, $q_3 = x$, we rename the indices in the internal forces as well as in the components of the external load, i.e., if

$$\begin{aligned}
&N_{11} = N_{\vartheta\vartheta}, N_{22} = N_{zz}, N_{12} = N_{21} = N_{\vartheta z},\\
&M_{11} = M_{\vartheta\vartheta}, M_{22} = M_{zz}, M_{12} = M_{21} = M_{\vartheta z},\\
&Q_{11} = Q_{\vartheta\vartheta}, Q_{22} = Q_{zz},\\
&p_{11} = p_{\vartheta\vartheta}, p_{22} = p_{zz}, p_{33} = p_{xx},\\
&m_{11} = m_{\vartheta\vartheta}, m_{22} = m_{zz}
\end{aligned} \tag{4.2.82}$$

is valid, then we obtain from (4.2.67) and (4.2.68)

$$\begin{aligned}
&\frac{\partial N_{\vartheta\vartheta}}{\partial\vartheta} + a\frac{\partial N_{\vartheta z}}{\partial z} - Q_{\vartheta\vartheta} + a p_{\vartheta\vartheta} = 0,\\
&\frac{\partial N_{\vartheta z}}{\partial\vartheta} + a\frac{\partial N_{zz}}{\partial z} + a p_{zz} = 0,\\
&\frac{\partial Q_{\vartheta\vartheta}}{\partial\vartheta} + a\frac{\partial Q_{zz}}{\partial z} + N_{\vartheta\vartheta} + a p_{xx} = 0,
\end{aligned} \tag{4.2.83}$$

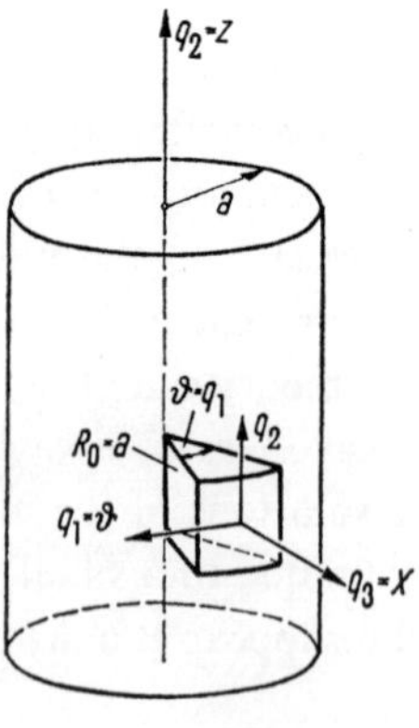

Figure 71a

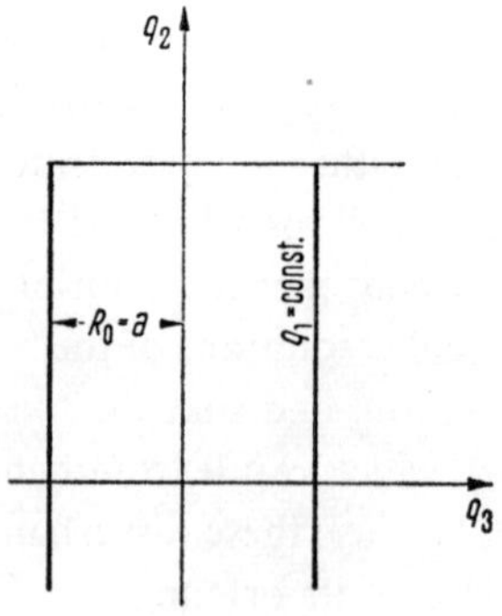

Figure 71b

and from (4.2.69) and (4.2.70)

$$\begin{aligned}
&\frac{\partial M_{\vartheta\vartheta}}{\partial\vartheta} - a\frac{\partial M_{\vartheta z}}{\partial z} - aQ_{\vartheta\vartheta} + am_{zz} = 0,\\
&a\frac{\partial M_{zz}}{\partial z} - \frac{\partial M_{\vartheta z}}{\partial\vartheta} - aQ_{zz} - am_{\vartheta\vartheta} = 0.
\end{aligned} \tag{4.2.84}$$

Through solving for the shearing forces, we obtain from (4.2.84)

$$Q_{\vartheta\vartheta} = \frac{1}{a}\frac{\partial M_{\vartheta\vartheta}}{\partial \vartheta} - \frac{\partial M_{\vartheta z}}{\partial z} + m_{zz}, \qquad Q_{zz} = \frac{\partial M_{zz}}{\partial z} - \frac{1}{a}\frac{\partial M_{\vartheta z}}{\partial \vartheta} - m_{\vartheta\vartheta}. \tag{4.2.85}$$

Inserting (4.2.85) in (4.2.83), then it yields equations

$$\begin{aligned}
&a\frac{\partial N_{zz}}{\partial z} + \frac{\partial N_{\vartheta z}}{\partial \vartheta} + ap_{zz} = 0,\\
&a\frac{\partial N_{\vartheta z}}{\partial z} + \frac{\partial N_{\vartheta\vartheta}}{\partial \vartheta} + \frac{\partial M_{\vartheta z}}{\partial z} - \frac{1}{a}\frac{\partial M_{\vartheta\vartheta}}{\partial \vartheta} - m_{zz} + ap_{\vartheta\vartheta} = 0,\\
&a\frac{\partial M_{zz}}{\partial z^2} - 2\frac{\partial^2 M_{\vartheta z}}{\partial \vartheta\,\partial z} + \frac{1}{a}\frac{\partial^2 M_{\vartheta\vartheta}}{\partial \vartheta^2} - a\frac{\partial m_{\vartheta\vartheta}}{\partial z} + \frac{\partial m_{zz}}{\partial \vartheta} + N_{\vartheta\vartheta} + ap_{xx} = 0.
\end{aligned} \tag{4.2.86}$$

Within the framework of the bending theory, the relations between internal forces, strain components and displacements must be considered, since the problems are not statically determined anymore as they were in the membrane theory. Let us rewrite the equations (4.2.55) relating to the present situations and terms, i.e.,

$$q_1 = \vartheta,\ q_2 = z,\ q_3 = x,\ h_1^{(\mathrm{m})} = (g_{11})^{\frac{1}{2}} = a = \text{const},$$
$$h_2^{(\mathrm{m})} = (g_{22})^{\frac{1}{2}} = 1,\ R_1 = a,\ R_2 = \infty.$$

Thus, we obtain for the strain components of the middle surface the formulae

$$\varepsilon_{\vartheta\vartheta}^{(\mathrm{m})} = \frac{1}{a}\frac{\partial u_\vartheta^{(\mathrm{m})}}{\partial \vartheta} - \frac{u_x^{(\mathrm{m})}}{a},\quad \varepsilon_{zz}^{(\mathrm{mm})} = \frac{\partial u_z^{(\mathrm{m})}}{\partial z},\quad 2\varepsilon_{\vartheta z} = \frac{1}{a}\frac{\partial u_z^{(\mathrm{m})}}{\partial \vartheta} + \frac{\partial u_\vartheta^{(\mathrm{m})}}{\partial z}. \tag{4.2.87}$$

Following this, let us rewrite (4.2.56) in the same manner, and find for the changes of curvature of the middle surface

$$\begin{aligned}
\kappa_{\vartheta\vartheta} &= \frac{1}{a}\frac{\partial}{\partial \vartheta}\left(\frac{u_\vartheta^{(\mathrm{m})}}{a} + \frac{1}{a}\frac{\partial u_x^{(\mathrm{m})}}{\partial \vartheta}\right),\\
\kappa_{zz} &= \frac{\partial^2 u_x^{(\mathrm{m})}}{\partial z^2},\\
2\kappa_{\vartheta z} &= \frac{2}{a}\frac{\partial^2 u_x^{(\mathrm{m})}}{\partial \vartheta\,\partial z} + \frac{\partial u_\vartheta^{(\mathrm{m})}}{\partial z}.
\end{aligned} \tag{4.2.88}$$

Insert (4.2.87) in (4.2.60) which yields for the internal forces

$$
\begin{aligned}
N_{\vartheta\vartheta} &= \frac{Eh}{1-\nu^2}\left(\frac{1}{a}\frac{\partial u_\vartheta^{(m)}}{\partial\vartheta} - \frac{u_x^{(m)}}{a} + \nu\frac{\partial u_z^{(m)}}{\partial z}\right),\\
N_{zz} &= \frac{Eh}{1-\nu^2}\left[\frac{\partial u_z^{(m)}}{\partial z} + \nu\left(\frac{1}{a}\frac{\partial u_\vartheta^{(m)}}{\partial\vartheta} - \frac{u_x^{(m)}}{a}\right)\right],\\
N_{\vartheta z} &= \frac{Eh}{2(1+\nu)}\left(\frac{1}{a}\frac{\partial u_z^{(m)}}{\partial\vartheta} + \frac{\partial u_\vartheta^{(m)}}{\partial z}\right),\\
M_{\vartheta\vartheta} &= -D\left[\frac{1}{a}\frac{\partial}{\partial\vartheta}\left(\frac{u_\vartheta^{(m)}}{a} + \frac{1}{a}\frac{\partial u_x^{(m)}}{\partial\vartheta}\right) + \nu\frac{\partial^2 u_x^{(m)}}{\partial z^2}\right],\\
M_{zz} &= -D\left[\frac{\partial^2 u_x^{(m)}}{\partial z^2} + \frac{\nu}{a}\frac{\partial}{\partial\vartheta}\left(\frac{u_\vartheta^{(m)}}{a} + \frac{1}{a}\frac{\partial u_x^{(m)}}{\partial\vartheta}\right)\right],\\
M_{\vartheta z} &= D(1-\nu)\cdot\frac{1}{a}\left(\frac{\partial^2 u_x^{(m)}}{\partial\vartheta\,\partial z} + \frac{1}{2}\frac{\partial u_\vartheta^{(m)}}{\partial z}\right),
\end{aligned}
\tag{4.2.89}
$$

when (4.2.82) is considered.

Finally we arrive at the differential equations for the displacements of the shell's middle surface, when (4.2.89) is inserted in (4.2.86). For $m_{zz} = m_{\vartheta\vartheta} \equiv 0$, they are

$$
\begin{aligned}
&\frac{\partial^2 u_z^{(m)}}{\partial z^2} + \frac{1-\nu}{2a^2}\frac{\partial^2 u_z^{(m)}}{\partial\vartheta^2} + \frac{1+\nu}{2a}\frac{\partial^2 u_\vartheta^{(m)}}{\partial\vartheta\,\partial z} - \frac{\nu}{a}\frac{\partial u_x^{(m)}}{\partial z} + p_{zz}\frac{(1-\nu^2)}{Eh} = 0,\\
&\frac{1+\nu}{2a}\frac{\partial^2 u_z^{(m)}}{\partial\vartheta\,\partial z} + \frac{1-\nu}{2}\frac{\partial^2 u_\vartheta^{(m)}}{\partial z^2} + \frac{1}{a^2}\frac{\partial^2 u_\vartheta^{(m)}}{\partial\vartheta^2} - \frac{1}{a^2}\frac{\partial u_x^{(m)}}{\partial\vartheta}\\
&\quad + \frac{h^2}{12a^2}\left(\frac{\partial^3 u_x^{(m)}}{\partial z^2\partial\vartheta} + \frac{\partial^3 u_x^{(m)}}{a^2\partial\vartheta^3}\right) + \frac{h^2}{12a^2}\left(\frac{1-\nu}{2}\frac{\partial^2 u_\vartheta^{(m)}}{\partial z^2} + \frac{\partial^2 u_\vartheta^{(m)}}{a^2\partial\vartheta^2}\right)\\
&\quad + \frac{p_{\vartheta\vartheta}(1-\nu^2)}{Eh} = 0,\\
&\nu\frac{\partial u_z^{(m)}}{\partial z} + \frac{1}{a}\frac{\partial u_\vartheta^{(m)}}{\partial\vartheta} - \frac{u_x^{(m)}}{a} - \frac{h^2}{12}\left(a\frac{\partial^4 u_x^{(m)}}{\partial z^4} + \frac{2}{a}\frac{\partial^4 u_x^{(m)}}{\partial z^2\partial\vartheta^2} + \frac{1}{a^3}\frac{\partial^4 u_x^{(m)}}{\partial\vartheta^4}\right)\\
&\quad - \frac{h^2}{12}\left(\frac{1}{a}\frac{\partial^3 u_\vartheta^{(m)}}{\partial z^2\partial\vartheta} + \frac{1}{a^3}\frac{\partial^3 u_\vartheta^{(m)}}{\partial\vartheta^3}\right) + \frac{a p_{xx}(1-\nu^2)}{Eh} = 0,
\end{aligned}
\tag{4.2.90}
$$

when no external moments are present.

Let us simplify the system of differential equations (4.2.90), which applies to any load $\boldsymbol{p}$, in the special case of the load being symmetrical with respect to the axis of rotation. All derivatives with respect to ϑ, the displacements $u_{\vartheta}^{(m)}$, and the load components $p_{\vartheta\vartheta}$ are then equal to nought. Therefore, (4.2.90) changes into the two differential equations

$$\frac{\mathrm{d}^2 u_z^{(m)}}{\mathrm{d}z^2} - \frac{\nu}{a}\frac{\mathrm{d}u_x^{(m)}}{\mathrm{d}z} + \frac{p_{zz}(1-\nu^2)}{Eh} = 0,$$
$$\nu\frac{\mathrm{d}u_z^{(m)}}{\mathrm{d}z} - \frac{u_x^{(m)}}{a} - \frac{h^2 a}{12}\frac{\mathrm{d}^4 u_x^{(m)}}{\mathrm{d}z^4} + \frac{a p_{xx}(1-\nu^2)}{Eh} = 0. \tag{4.2.91}$$

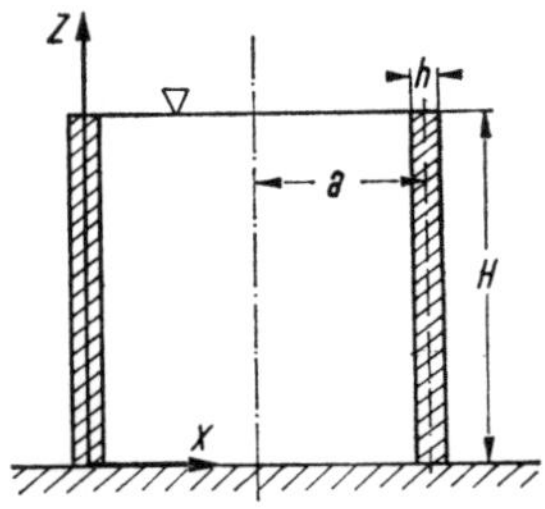

Figure 72

Let us look at a cylindrical container for example, which has been filled with fluid to the level $z = H$. The fluid has the specific weight γ_F (Figure 72). Let the cylindrical container have average radius a and thickness h. According to this load, $p_{zz} = 0$ is valid. Thus, we obtain from the first equation (4.2.91) through integration,

$$\frac{\mathrm{d}u_z^{(m)}}{\mathrm{d}z} - \frac{\nu}{a}u_x^{(m)} = c, \tag{4.2.92}$$

containing the constant of integration c. The second equation (4.2.89) yields

$$\frac{\mathrm{d}u_z^{(m)}}{\mathrm{d}z} - \frac{\nu}{a}u_x^{(m)} = \frac{N_{zz}(1-\nu^2)}{Eh} \tag{4.2.93}$$

under the present assumptions. Comparing (4.2.92) with (4.2.93) yields that internal force N_{zz} is constant. From (4.2.93)

$$\frac{\mathrm{d}u_z^{(m)}}{\mathrm{d}z} = \frac{\nu}{a}u_x^{(m)} + \frac{N_{zz}(1-\nu^2)}{Eh} \tag{4.2.94}$$

is derived. Inserting this into the second equation (4.2.91), we obtain after simple changes the differential equation

$$\frac{\mathrm{d}^4 u_x^{(\mathrm{m})}}{\mathrm{d}z^4} + 4\beta^4 u_x^{(\mathrm{m})} = \frac{p_{xx}}{D} + \frac{\nu N_{zz}}{D}, \; \beta^4 = \frac{3(1-\nu^2)}{a^2 h^2}, \; D = \frac{Eh^3}{12(1-\nu^2)}. \tag{4.2.95}$$

The solution of this equation governs the entire problem of the cylinder filled with fluid.

The weight of the container has been omitted because the mass forces were omitted, and therefore no external load in the z direction is applied to the wall of the container. Hence, $N_{zz} = 0$ is true. The fluid pressure yields

$$p_{xx} = -\gamma_F(H-z),$$

which changes (4.2.95) to

$$\frac{\mathrm{d}^4 u_x^{(\mathrm{m})}}{\mathrm{d}z^4} + 4\beta^4 u_x^{(\mathrm{m})} = -\frac{\gamma_F(H-z)}{D}. \tag{4.2.96}$$

This represents an ordinary inhomogeneous differential equation for $u_x^{(\mathrm{m})}$, having the well-known general solution

$$u_x^{(\mathrm{m})} = \mathrm{e}^{-\beta z}(C_1 \cos \beta z + C_2 \sin \beta z) + \mathrm{e}^{\beta z}(C_3 \cos \beta z + C_4 \sin \beta z) - \\ - \frac{\gamma_F(H-z)a^2}{Eh}. \tag{4.2.97}$$

It has to be adapted to the boundary conditions so that the constants of integration C_1 to C_4 can be determined.

If H is large in comparison with $\sqrt{ah}$, then the container can be assumed as being infinite in length. Therefore, the constants C_3 and C_4 must equal nought.

Let us assume the walls of the container to be clamped at their bases. Hence, the conditions

$$u_x^{(\mathrm{m})} = 0, \; \frac{\mathrm{d}u_x^{(\mathrm{m})}}{\mathrm{d}z} = 0 \text{ for } z = 0 \tag{4.2.98}$$

are true. Using the conditions (4.2.98) in (4.2.97), and considering $C_3 = C_4 = 0$, then for the two remaining constants

$$C_1 = \frac{\gamma_F a^2 H}{Eh}, \quad C_2 = \frac{\gamma_F a^2}{Eh}\left(H - \frac{1}{\beta}\right)$$

is obtained, through which (4.2.97) changes to

$$u_x^{(m)} = -\frac{\gamma_F a^2}{Eh}\left\{H - z - e^{-\beta z}\left[H \cos \beta z + \left(H - \frac{1}{\beta}\right) \sin \beta z\right]\right\}. \quad (4.2.99)$$

Finally, let us discuss the internal forces. In the case of the container filled with fluid, the equations (4.2.89) are considerably simplified because of $\partial/\partial\vartheta = 0$, $u_\vartheta^{(m)} = 0$, and they read

$$\begin{aligned} N_{\vartheta\vartheta} &= \frac{Eh}{1-\nu^2}\left(\nu \frac{du_z^{(m)}}{dz} - \frac{u_x^{(m)}}{a}\right), \\ N_{\vartheta z} &= M_{\vartheta z} = 0, \\ M_{\vartheta\vartheta} &= -D\nu \frac{d^2u_x^{(m)}}{dz^2}, \quad M_{zz} = -D\frac{d^2u_x^{(m)}}{dz^2}. \end{aligned} \quad (4.2.100)$$

The internal force N_{zz} is already known to be nought. Therefore, from (4.2.94),

$$\frac{du_z^{(m)}}{dz} = \frac{\nu}{a} u_x^{(m)}$$

is true, whereby the first equation of (4.2.100) yields

$$N_{\vartheta\vartheta} = -\frac{Eh}{a} u_x^{(m)}. \quad (4.2.101)$$

Inserting solution (4.2.99) into the last two equations of (4.2.100) and (4.2.101), then the internal forces

$$\begin{aligned} N_{\vartheta\vartheta} &= \gamma_F a \left\{H - z - e^{-\beta z}\left[H \cos \beta z + \left(H - \frac{1}{\beta}\right) \sin \beta z\right]\right), \\ M_{\vartheta\vartheta} &= \nu M_{zz} = \frac{\nu\gamma_F aHh}{\sqrt{12(1-\nu^2)}} e^{-\beta z}\left[-\sin \beta z + \left(1 - \frac{1}{\beta H}\right) \cos \beta z\right] \end{aligned} \quad (4.2.102)$$

are obtained. The maximum bending moment occurs at the lower edge of the cylinder and has the value

$$M_{zz,\,\max} = \left(1 - \frac{1}{\beta H}\right) \frac{\gamma_F \, aHh}{\sqrt{12(1-\nu^2)}} \,.$$

Since the exponential function in the formulae (4.2.102) contains the negative exponent $-\beta z$, then the bending moments from this maximum value quickly fade to nearly nought, since it can be deduced from (4.2.95) that β represents a very large value for thin shells with a small thickness h. Therefore, it becomes evident that the influence of bending is limited to a small portion at the lower edge of the container, as has been predicted previously.

References

[1] S. F. Borg: *Matrix-Tensor Methods in Continuum Mechanics*, D. van Nostrand Comp., Inc., Princeton/New Jersey/Toronto/London/New York 1963.

[2] I. S. Sokolnikoff: *Tensor Analysis*, John Wiley & Sons, Inc., New York/London 1960.

[3] F. Seitz: *Théorie moderne des solides*, Masson et Cie, Paris 1949.

[4] A. Seeger: *Moderne Probleme der Metallphysik*, Springer-Verlag, Berlin/Heidelberg/New York 1965.

[5] E. Kröner: *Kontinuumstheorie der Versetzungen und Eigenspannungen*, Springer-Verlag Berlin/Göttingen/Heidelberg 1958.

[6] M. Reiner: *Lectures on Theoretical Rheology*, North-Holland Pub. Co., Amsterdam 1960.

[7] M. Lagally: *Vorlesungen über Vektor-Rechnung*, Akad. Verlagsges. Becker und Erler K. G., Leipzig 1944, pp. 197–204 and pp. 248–249.

[8] I. S. Sokolnikoff: *Mathematical Theory of Elasticity*, McGraw-Hill Book Comp., Inc., New York/London 1956, pp. 14–16.

[9] A. Sommerfeld: *Mechanik der deformierbaren Medien*, Akad. Verlagsges, Geest und Portig K.G., Leipzig 1964, p. 351.

[10] I. N. Sneddon und D. S. Berry: *The Classical Theory of Elasticity*. Handbuch der Physik VI, Springer-Verlag, Berlin/Göttingen/Heidelberg 1958, pp. 15–17.

[11] W. Prager: *Einführung in die Kontinuumsmechanik*, Birkhäuser Verlag, Basel/Stuttgart 1961, p. 171, 175.

[12] H. Kauderer: *Nichtlineare Mechanik*, Springer-Verlag, Berlin/Göttingen/Heidelberg 1958.

[13] D. Morgenstern and J. Szabó: *Vorlesungen über Theoretische Mechanik*, Springer-Verlag, Berlin/Göttingen/Heidelberg 1961, p. 90, 91.

[14] L. Collatz: *The Numerical Treatment of Differential Equations*, Springer-Verlag, Berlin/Göttingen/Heidelberg 1960.

[15] L. W. Kantorowitsch und W. Krylow: *Näherungsmethoden der Höheren Analysis*, Deutscher Verlag der Wissenschaften, Berlin 1956.

[16] S. G. Michlin: *Variationsmethoden der Mathematischen Physik*, Akademie-Verlag, Berlin 1962.

[17] N. J. MUSKHELISHVILI: *Some Basic Problems of the Mathematical Theory of Elasticity*, P. Noordhoff Ltd., Groningen 1953.

[18] C. E. PEARSON: *Theoretical Elasticity*, Harvard Univ. Press., Cambridge, Mass., 1959, p. 107.

[19] C. B. BIEZENO und R. GRAMMEL: *Technische Dynamik*, Springer-Verlag, Berlin/Göttingen/Heidelberg vol. 2, 1953, Bd. I, p. 235.

[20] S. TIMOSHENKO und J. N. GOODIER: *Theory of Elasticity*, McGraw-Hill Book Comp., Inc., New York/Toronto/London 1951, p. 292.

[21] A. and L. FOPPL: *Drang und Zwang*, R. Oldenbourg Verlag, München/Berlin vol. 4, 1944, Bd. 2, p. 57.

[22] A. J. DURELLI, E. A. PHILLIPS, C. H. TSAO: *Introduction to the Theoretical and Experimental Analysis of Stress and Strain*, McGraw-Hill Book Comp., Inc., New York/Toronto/London, 1958, pp. 126–131.

[23] R. A. EUBANKS and E. STERNBERG: On the Completeness of the Boussinesq-Papkovich Stress Functions, *J. Rational Mech. and Analysis 5* (1956) p. 735.

[24] J. L. SYNGE: *The Hypercircle in Mathematical Physics*, Cambridge Univ. Press, Cambridge 1957.

[25] H. L. LANGHAAR: *Energy Methods in Applied Mechanics*, John Wiley & Sons, Inc., New York/London 1962, p. 128.

[26] A. KNESCHKE: *Differentialgleichungen und Randwertprobleme*, B. G. Teubner Verlagsges. Leipzig 1962, pp. 354–472.

[27] U. WEGNER: Ein Beitrag zur Lösung von Balken- und Plattenproblemen, *Beton- und Stahlbetonbau 9* (1961) pp. 210–218.

[28] G. N. SAWIN: *Spannungserhöhung am Rande von Löchern*, VEB Verlag Technik, Berlin 1956, pp. 324–329.

[29] K. GIRKMANN: *Flächentragwerke*, Springer-Verlag, Wien 1954, p. 292.

[30] E. GIENCKE: *Über die Berechnung von Flugkonstruktionen*, Jahrbuch der WGLR (1964) p. 94.

[31] A. S. WOLMIR: *Biegsame Platten und Schalen*, VEB Verlag für Bauwesen, Berlin 1962.

[32] A. KACNER: Bending of Plates with Variable Thickness, *Arch. Mech. stos. 3*, 13 (1961) p. 393.

[33] S. TIMOSHENKO and S. WOINOWSKY-KRIEGER: *Theory of Plates and Shells*, McGraw-Hill Book Comp., Inc., New York/Toronto/London 1959, pp. 396–428.

[34] CHI-TEH WANG: *Non-linear Large Deflection Boundary Value Problems of Rectangular Plates*, NACA, TN 1425, Mar. 1948.

[35] W. MÜLLER, *Theorie der elastischen Verformung*, Akad. Verlagsges. Geest und Portig K.G., Leipzig 1959, p. 125.

[36] E. KLINGBEIL: *Tensorrechnung für Ingenieure*, B.I.-Hochschultaschenbuch 197/197a, Bibliogr. Inst., Mannheim 1966, p. 165.

[37] H. WILKEN, Über eine funktionentheoretische Behandlung von Randwertaufgaben der Kirchhoffschen Plattentheorie, *Forsch. Ing.-Wes. 37* (1971) pp. 12–20.

Subject index

LIBRARY COPY
RETURN TO
LEWIS LIBRARY
CLEVELAND, OHIO